Granville Bradshaw

a flawed genius?

2nd edition

Barry M Jones

Panther Publishing

Second Edition Published by Panther Publishing Ltd in 2012
First published 2008

Panther Publishing Ltd
10 Lime Avenue
High Wycombe
Buckinghamshire HP11 1DP, UK
www.panther-publishing.com
info@panther-publishing.com

© Barry M Jones

The rights of the author have been asserted in accordance with the Copyright, Designs and Patents Act 1988.

Also by Barry M Jones:
The Story of Panther Motorcycles (Patrick Stephens Ltd)
The Panther Story (Panther Publishing Ltd)
Biographer to The Rugged Road (Panther Publishing Ltd)
The Village Bus Remembered (Barry M Jones)
Brenell - True to Life Performance (Barry M Jones)
A Truvox Product (Barry M Jones)
A Guide to British Tape-recorders (Barry M Jones)

Acknowledgements

I am indebted to the late Geoffrey Bradshaw and his family, Antony, Angela and Josanne for their help and more recently to Millicent Bradshaw (Peter's wife), Adrian Bradshaw (Ewart's son), Doug Brew, The Brooklands Museum, David Hales, Nick Kelly, Arthur Ord-Hume, The Royal Air Force Museum, Kim Siddron, Michael Smith, Sid Wilkinson, Don Williams and many others including my faithful friend, Percy Verance. Special thanks in this second edition go to Derek Jones for recent research into W H Dorman & Co, to Ken Gasmier on the Dorman KNO and other engine designs, and to Peter J Grey, all of whom have inspired a more thorough account of Walter Lawson Adams, now included as Appendix 5.

Where possible, photographic credits have been given to their original source. Where an abbreviation is given as the source of the illustration these are as below. Where no credit is given, these are from miscellaneous uncredited loose cuttings and scrap-books.

GB	Granville Bradshaw
BMJ	The author
CMC	Classic Motor Cycle, Mortons Media Group Archives

All rights reserved, no part of this publication may be reproduced, stored in a retrieval system or transmitted, in any form or by any means, electronic or mechanical including photocopying, and recording without the prior permission of the publisher and/or the copyright owner.

ISBN 978-09564975-7-4

Contents

Foreword	iv
Preface	v
Introduction	1
Aviator	7
Aero-engines	33
ABC Radial Engines	43
ABC Powered Aeroplanes	61
Post-war ABC Aero-engines	79
ABC Motorcycles	99
The Sopwith-ABC Motorcycle	107
Skootamota	121
Oil Cooled Motorcycles	128
Bradshaw's Prophecies	141
Phelon & Moore	155
ABC Cycle-car	165
Belsize-Bradshaw	177
Inventor Extraordinaire	183
Bradshaw at War	198
Peoples' Car	209
Bradshaw Pulsation Motor	226
Toroidal Engines - Project Omega	230
Wankel Opposition Engine	247
Twin-toroidal Gas Generator	250
Conclusion: Patentee Extraordinaire - A Personal View	259
Appendix 1. Walton Motors	266
Appendix 2. The Family	272
Appendic 3. Addresses	276
Appendix 4. Patents	277
Appendix 5. Walter Lawson Adams	285
Bibliography	303
Index	305

Foreword

I didn't know my grandfather, Granville, very well. My father, Geoffrey, used to talk about him when we were children and tell us stories from his childhood. Kids love that! It's hard to imagine a time in our parents' lives before we existed. He told us about visiting Granville and his second family in the country mansion where they lived, Lowfield Park, on the site where Gatwick Airport now is.

Granville loved to invent things. He also loved hot baths. In the mansion he would keep the boilers running through the night so there was plenty of hot water. He'd sit in the bath creating his designs (sometimes through the night when everyone else was asleep). He kept the water temperature constant by drilling a small hole in the bath plug and running the hot tap at the same rate as the water was running out through the hole in the plug. That story fascinated me so much that I used to do my maths homework in the bath, in the hope that it would make me clever like him!

After the many negative things that have been said and written about Granville over the years, it's good to finally read a book that shows him as he really was - I wonder what he would have made of the 21st century - he would have been in his element I've no doubt.

Josanne Bradshaw

Preface

Born in the age of steam, Granville Eastwood Bradshaw was in a unique position to see in the birth of the motorcycle, motor-car, aircraft, space flight and the transformation of sail into high speed motor launches. He also witnessed the birth of cinema, radio, television and coin operated amusement machines and became intimately involved in all their development, he even pioneered three-dimensional (3-D) colour television in the 1950s which is now enjoying a revival

In fact there was not much left in the world of engineering for him to discover, but it was as a pioneer aviator, aero-engineer, motor engineer and patentee that he was to become most famous, even if his visionary dreams were, at that time, often not supported by contemporary technology which led him to being ridiculed when his promised ideas 'failed', for what ever reason. Yet for critics to castigate Bradshaw for every one of his 'failures' without also castigating his contemporaries for their far more numerous, often fatal but less well publicised, failures is akin to a witch hunt! If anything, Bradshaw's failing was his public charisma of over enthusiastic self-belief and self-publicity.

A millionaire in the 1930s from having made a fortune in coin operated amusement machines, he was declared bankrupt in 1936 in the aftermath of the famous Hatry swindle and the Wall Street crash, from which he never fully recovered, yet it was his post-war years which proved his most difficult with the continuation of post-war rationing, an 'export or die' philosophy, the confusion of ever changing advances in technology and the fickle consumer demands which made life difficult to predict for any inventor and investor.

Bradshaw saw his salvation in a toroidal engine - a brilliantly simple design which proved impossibly difficult to perfect, not only by Bradshaw, but many other engineers around the world. One person at least had huge faith in his potential, Francesca King, his housekeeper and companion, whom he met in the war when she was head driver in the Admiralty's driving pool - she routinely drove for Lord Louis Mountbatten. Francesca invested heavily in him and became joint rights-holder to his many post-war patents.

He never stayed long in any one place for, like the nomad, he moved close to wherever his next project took him, renting a house as needed. Indeed, he rarely had the funds in the post war years to buy a home of his own and relied heavily on Francesca to provide a roof over his head especially when in ill-health, until 1964 when they parted. Granville died on 13th April 1969.

I first became intrigued by the legendary Granville Bradshaw when writing *The Story of Panther Motor Cycles*, the history of Phelon & Moore, for whom he did much work. Researching his many patents, I got to meet his son, the late Geoffrey Bradshaw who, in the mid-1980s, had written a short biography of his father for the *Brooklands Gazette*, the journal of the Brooklands Society, which honours Britain's pioneering aviators, racing drivers and motor-cyclists at the famous Brooklands airfield and motor circuit.

My biography is based on my own research and interviews with Geoffrey who then held his father's personal archives and which, through the kindness of his children, I have now been able to fully read for myself. These archives are far from complete and, sadly, many documents and magazine cuttings have, over the passage of time, decayed, been cut about, or been lost. There are many superbly crafted engineering drawings as well as negatives, of varying quality, taken from a simple box or vest-pocket camera, which Granville processed and developed himself. There is also evidence of Granville's own incomplete 60 page draft autobiography, *The life of an inventor - who made a million*, which he intended offering to Temple Press in 1964. Alas, the manuscript no longer survives, so there is little supporting evidence of the inter-war years which were covered by such tantalising subject headings as: "The first tarred road: Second invention a failure: Nearly electrocuted: Henry Ford and line assembly: Saving Kipling's *If*: Killing a man and helping a murderer: Meeting Rudolph Diesel" .... and many others which will probably forever remain a mystery.

Those who have never been involved with pioneering work, or in developing patents, can never possibly appreciate the painstaking and heartbreaking work involved. Many of his most vociferous critics are blessed with the benefit of hindsight gained from the failures of others; yet they still remain ignorant of the background and toil in Granville's pioneering work. I have tried then to write this biography with a proper perspective, but only the reader can determine, in his or her own mind, whether Granville Bradshaw was, as many have persistently tried to prove, 'a plausible rogue, a charlatan, a fraud' or was he, I suggest, more truthfully, simply "a flawed genius"?

At the end of this book, I have presented my own personal view of these charges. Whether the reader wishes to read this first, or last, is their choice, but whatever one's conclusion, the chapter on amusement machines and his photographic booth, proves that he certainly did had a sense of humour!

Barry M Jones, May 2008

Notes on the second edition

It is always the case that, having gone to print, new information comes to light. While little new has been unearthed on Granville Bradshaw himself or, more frustrating, his missing archives (which will I am sure dutifully turn up after this second edition is published), there is much new allied information especially on Walter Lawson Adams, which adds to and reinforces that recorded in the first edition. Some critics, aviation in particular, quite rightly pulled me up on too lax or vague generalisation in some background information - but then this book was not written as a treatise on early manned flight readily available from the bibliography! Nevertheless, such criticism is acknowledged and corrections made, including my unforgivable muddling between the two 'George Bulmans':

George Purvis Bulman, a graduate engineer, joined the Royal Flying Corps in 1916 and entered the Engine Branch of the Aeronautical Inspection Department of the War Office where he was appointed Deputy Chief Inspector (Engines) in late 1918.

Paul 'George' W S Bulman, a reservist in the Honourable Artillery Company, transferred to the Royal Flying Corps at the age of 22, first flying in 1917. Commissioned at the end of the war he became an Experimental Pilot for the RAF 'ironing out the bugs of the air cooled radial'. He resigned his commission in Spring 1925 to become Chief Test Pilot at Hawkers, working with Fred Raynham and became a Director in 1936. His nick-name 'George' came through his inability to remember names; to him, all servicemen were 'Colonel' or 'General' and civilians 'George' - by which he too became known.

It must also never be forgotten that Bradshaw and his contemporaries relied heavily on news reports (many still contentious) and scientific knowledge available to them. Today's readers and 'experts' benefit greatly from hindsight, often gleaned from histories rewritten by others to reflect modern science and opinion, which are often much divorced from that contemporaneous to Bradshaw and others.

Barry M Jones, May 2012

1
Introduction

Ancestors

Every one of us has interesting ancestors, many of whom we know nothing about and, just occasionally some of whom we wish we didn't! Granville Bradshaw's immediate ancestry is traced to his grandfather, William Bradshaw of Chorley, Lancashire; he was an 'overlocker' in a textile mill. He married Alice Jackson in 1839 and they settled at 6, Commercial Road, Chorley, Lancashire.

With this link to Chorley, Granville 'calculated mathematically' that they could, with some certainty trace their ancestors right back to the reign of Edward III when Bradshaws served in the 14th century Parliament; and the 15th century Anthony Bradshaw, Deputy Steward of the village of

Granville Bradshaw 1886-1969 (Photo Millicent Bradshaw)

Duffield Frith near Derby, within the Royal Manor associated with Sherwood Forest; and also his blood relative, Judge John Bradshaw from the Civil War, (appointed by Cromwell to the Presidency of a special High Court of Justice to try King Charles I for tyranny, treachery and murder). There is also a claimed association with George Bradshaw author of the famous *Bradshaw Railway Guides*, the forerunner of today's timetables - but how accurate Granville's assesment is remains to be tested. His brother Ewart remained very sceptical and no link has been found.

As a place name, Bradshaw in Derbyshire is recorded as Bradschag in 1345; in Lancashire as Bradeshaghe (1246). Its literal root in Old English are Brad (broad, wide) and scaga (a thicket, shaw, wood). Hence 'Bradshaw' was descriptive of a settlement in a broad-shaw.

By the time Granville Bradshaw was born in 1886, the railway network had already spread across Britain. (The last *Bradshaw Guide* was published in June 1961 having been ousted by British Railways' own version which itself ceased publication in the spring of 2007 due to modern internet technology!) Granville's son, Geoffrey, recalled one briefing as an Officer in the Royal Artillery when it was suggested, tongue in cheek by a superior, that with the name Bradshaw, he should know the train times to the new posting. To the astonishment of all, Geoffrey rattled off times and connections! Unbeknown to them, and fearing such a question would arise, he had taken the precaution of earlier checking and memorising the times.

Granville Eastwood Bradshaw

William and Alice Bradshaw's sixth surviving child was William Septimus (Granville's father), born on 9th April 1847; he became a commercial traveller and moved to Woodhead Road, Eccleshall, Sheffield where he met and married on 21st November 1872, the 22 year old Annie Virginia Mathews of Washington Road, Sharrow, Sheffield. Her father, John, was a veterinary surgeon of Lord Street, Sheffield. They returned to Lancashire, where William set himself up as a clockmaker/repairer and jeweller at Fishergate, Preston. They had no children. Sadly, Annie Virginia died of acute peritonitis on 29th June 1878, aged 27 years.

William Septimus married again in 1879 to Annie Eastwood, who bore him five children in Preston: Ruby (b.1870), Violet V (b.1881), Ewart Gladstone (b. 16/6/1884), Granville Eastwood (b. 8/12/1886) and Henrietta W, ('Hettie' or 'Ethel' b.1888 d. 12/10/59). While Ewart was tall and slim, Granville was quite short, yet his own son, Geoffrey, was tall and thin. The family home was at 148, Friargate in Preston's town centre where they employed a Scottish servant, Jessie Moffat (aged 16), but by 1889, they had moved up the road to 59, Tulketh Crescent, Ashton-on-Ribble and by 1891 to 138, Friargate. William Septimus then became an 'Optologist' (with a London Diploma) making and fitting spectacles. With Ewart's help, he operated as Bradshaw & Son, opening further shops in St.Annes and Fleetwood. It appears he later made motoring goggles.

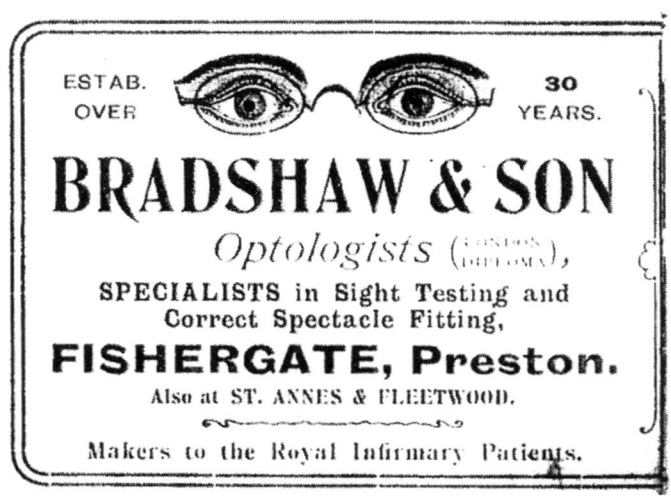

Bradshaw & Son label (Adrian Bradshaw)

Annie died of double pneumonia in 1904, aged 55 years. William Septimus remarried and had a son, John, who it appears lived mainly in London. By 1910 the family had moved to 7, Park View, Tulketh Road, Preston. Hettie eventually married a Mr Granfell of Preston, who had a large motor factors, while Ruby by now a dressmaker, became a fairground clairvoyant, although Ewart did his very best to dissuade her from public appearance by offering her an attractive allowance!

It was TOM Sopwith who first nick-named Granville, 'Braddles' during their time together at Brooklands, and it was by this name that he was affectionately known by family, friends and business acquaintances. Braddles died peacefully in his sleep on April 13th, 1969, aged 82 years, at his younger son's home in Hitchin*. His obituaries were few and brief, but that in *The Motor* perhaps best sums up the man as: 'one of the most advanced designers of his time. Always an independent, who sold his ideas at the prototype stage'.

His wealth was indeed derived from selling his ideas or patents** outright , mostly only as provisional patents, for cash or company shares. Though he had a successful engineering company, ABC, after his removal from that company in 1920, when it was bought by Harper Bean, it was rare indeed that he personally put into production a patented idea except through a company for whom he then worked as a consulting engineer. Though he often over-estimated the financial worth of his inventions, there is clear evidence he derived more pleasure in seeing his designs come to fruition than in money alone. This may have been through his personal experience of financial collapse in the 1930s following the Clarence Hatry fraud case and being declared bankrupt. After the war, when he was financially embarrassed following a run in with the taxman who did not always allow him to write off development costs against tax, Granville conceded to his brother Ewart, that 'it is a pity I was not born with less technical vision and with more commercial acumen. You were the lucky one!'

The apprentice

Preston's major employer, originally a railway and tramcar company, was a subsidiary of Dick, Kerr & Company of the Britannia works, Kilmarnock. Formed by W B Dick and John Kerr in 1880, they established a Preston base in 1897 as the Electric Railway & Tramcar Carriage Company but soon changed its name to the United Electric Car Company and in 1900 built a new works in West Strand Road, Preston for electric traction motors and equipment, operating under a new subsidiary company, the English Electric Manufacturing Company (EEMC). In 1903 these works came under direct control of Dick, Kerr & Co and remained so until 1919 when Dick, Kerr & Company, Siemens Bros Dynamo Works Ltd, Willans & Robinson Ltd, the Phoenix Dynamo Company and the Coventry Ordnance Works Ltd merged to form The English Electric Company.

Ewart undertook his apprenticeship at EEMC before leaving to help his father in his opticians shop in Friargate, possibly inspiring the making of motoring goggles. On leaving school at 13 years of age, Granville however, was sent to work as a boy clerk in a solicitor's office at 5/- (five shillings) a week, where he learned to type. Meanwhile, he studied engineering at night school and at the age of 14, built a model dynamo which incorporated an AC/DC converter and weighed only 4lbs, winning him a prize and much coverage in *Model Engineering Magazine*. He started his apprenticeship at the EEMC works at the age of $14^1/_2$ in the summer of 1901.

Ewart had a profitable sideline buying and selling second-hand bicycles and motorcycles, fostering in Granville an interest in motorcycles and motor cars. Granville began riding motorcycles

---

See Appendix 2 for some notes on his family and Appendix 3 for the addresses where Granville made his home

**For a list of patents see Appendix 4*

in 1900, at the age of 14, well before the minimum age was set under the 1903 Motor Car Act. While good in the matters of business, Ewart left the repairs and servicing for Granville to tackle during the evenings and weekends. One weekend, the brothers decided to go to London on Ewart's Belgian made, belt driven, single speed Antoine 750cc motorcycle, with Granville riding pillion. Unimpressed with the dense London traffic, chock-a-block with horse-drawn wagons and no sense of any rules of the road, they found solace in some digs carrying the 120lbs motorcycle up two flights of stairs to their small room.

The early years

Progressing through the various shop floor departments at EEMC he moved into the drawing office to train as a draughtsman and was soon brought into the design of their new 3,000KW alternator; their largest ever built. But, by the time he was 17, he was tiring of his work at EEMC though was quite undecided on his next move. As the family was deeply religious, they and family friends were keen to see him enter the Methodist ministry, so he dutifully studied towards it, but the deeper he read, the less convinced he became of the religious arguments and the more he turned his thoughts to engineering. Then one day in 1905, standing idly at a street corner awaiting a friend, a passing French motorist in a large, raucous, racing car stopped to seek directions to Blackpool for the 1905 motorcycle and motor car race. Mesmerised by the car, and unable to understand a word the foreigner was saying, Granville unhesitatingly indicated that he would show him the way, at which the delighted driver pointed to the dashboard mounted fuel pressure pump indicating that Granville should operate it! They were soon on their way and once pointing in the right direction Granville got out and walked home and, determined to go to the races, borrowed Ewart's motorcycle.

Captivated by the day's racing, Granville was awe struck by Stanley Edge's 6-cylinder Napier and a Peugeot 1,700cc V-twin racing motorcycle (without brakes) ridden by a Monsieur Henri Cissac. He then persuaded Ewart to sell his $1^{1}/_{4}$hp motorcycle and buy a $2^{1}/_{4}$hp single-cylinder, overhead inlet valve, cast-iron head Peugeot; but in Granville's enthusiasm to boost its power, the piston seized forcing the barrel to fracture and it was now that Granville's inventiveness came to light. Casting a new barrel was considered out of the question, so Granville decided to have one turned, complete with cooling fins, from a billet of steel. All the experienced engineers at EEMC laughed at this saying it would seize, but he responded that in his considered wisdom, as steel had a higher coefficient of expansion than the cast iron piston, all would be well. In common with most engineering works, apprentices and staff persuaded the machine operators to do the odd 'foreigner' or 'Government job' in their spare time. Granville managed to persuade his works foreman to turn one up on his treadle lathe in his garden shed but quaking at the though, the foreman had a better idea and armed with Bradshaw's drawing (his first to prove fruitful) got the cylinder, resplendent with its turned fins, turned up on the work's huge lathes during the quieter night-shifts and meal breaks. Duly completed, it was 'despatched' over the wall at the back of the factory one prearranged evening.

It was fitted straight away and, as he had expected, worked perfectly. He then modified the timing and cam lift and then proceeded to build his own screw cutting lathe. Many years later, when a manufacturing problem was found, Granville was often seen in the workshop, coat off, overseeing the work with an engineer's critical eye, for Bradshaw was indeed as skilled with a hand tool as any craftsman and made most of his prototypes by his own hand - usually to perfection.

Granville was by now determined to follow a career in engineering and study the subject at university. By chance he spotted an advertisement in a technical journal for a Junior Draughtsman in Edinburgh and promptly applied. To his amazement, he received an offer by return from Messers Bruce Peebles Ltd, a heavy electrical engineering company of high standing who had recently won a

Granville on, presumably, Ewart's Peugeot V-twin in 1907. CK196 is a Preston number (Photo Adrian Bradshaw)

contract to construct both the Portmadoc & South Snowdon railway and a hydroelectric power station at Cwym Dyli, nearby. They were more than happy for him to study at the 'university in Edinburgh'*.

These were for him exciting days and, with a friend, he bought a motor-cycle and sidecar by which to tour Scotland. He then bought a JAP 8hp V-twin engine and duly fitted it into a motorcycle 'bitza' frame of his own design.

In 1906, Granville left Edinburgh to join Vickers & Company as a draughtsman at their massive River Don Works, built in 1863 alongside the River Don in Brightside, Sheffield. Badly damaged by a major flood from a breached dam in 1864, it was completed in 1866. Originally millers and steel rollers, Vickers soon grew into one of Britain's largest industrial engineering conglomerates.

He lodged in digs at 317, Eccleshall Road, Sheffield (possibly his relatives) and in October 1906, he and a John King, hairdresser, of 72 High Street, Sheffield, co-patented a quick-acting adjustable spanner under Patent No. 1906-23043. This was a rather clever device. A cammed lever moved a serrated bar carrying the adjustable head towards the fixed head, against a spring. The adjustable head was locked in place on the serrated bar thus catering for any size of nut. It does not appear to have entered production for no references to this design are known, besides which, the established screw wrench and slip/wedge wrench gave far greater purchase. (It should be noted that John King and Abingdon Engineering's 'King Dick' adjustable screw-wrenches are not related; that brand derives its name from the owner's prize winning bulldog, 'King Dick'.)

At Vickers, he was put onto helping design a 2,000bhp motor for their rolling mills which constantly had to stop and reverse its direction. Having studied the mathematics of stress analysis at Edinburgh, his experience and education instilled a depth of knowledge in metallurgy and stress

---

* *It is not clear if this was the then University of Edinburgh or the Heriot-Watt College, for while neither establishment has any record of Granville attending, this does not mean he didn't attend. I suspect he attended Heriot-Watt technical college as the then University of Edinburgh only offered engineering degrees .*

which in due course made him an expert in those fields. He soon had two senior draughtsmen working under him. But his stay at Brightside was short lived, for in late 1907/early 1908, at the age of 21, he joined The Electric Construction Corporation at Bushbury, Wolverhampton as Assistant Chief Draughtsman on £3 a week. Formed by Thomas Parker, they began life as horse-shoe makers, but in 1883 began making electrical lifting equipment for mines and were soon manufacturing powerful electric motors and dynamos, as well as omnibuses. They also became the main contractors for Liverpool's new overhead railway, opened in 1893. By the late 1890s, Parker had produced several novel electric and petrol driven car designs which incorporated four wheel steering and hydraulic brakes.

Granville was soon involved in the development of powerful electric motors of up 12,000bhp for driving massive 250 ton flywheels used in steel rolling mills. One had recently been installed at Bilston, Staffordshire under a contract which had a hefty hourly penalty clause against mechanical breakdown. Granville quickly noticed a potential design flaw in the way in which the power was transmitted between the motor and the rollers by a four jaw coupling and universal joint which allowed the roller to rise and fall as the smouldering steel billets were reduced in thickness. He predicted a stress fracture would occur at a precise point - and it did, the very next day! He promptly designed a lighter, more tolerant three jaw coupling, saving the day and ECC a small fortune.

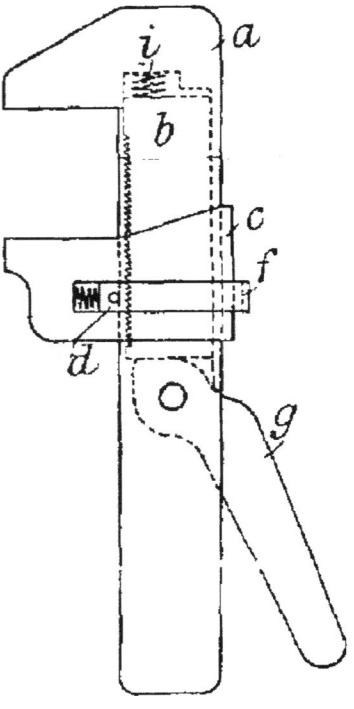

Bradshaw's adjustable wrench
Pat No 23043

But his time at ECC was, once again, short lived for he started to take an interest in the new fangled aeroplanes and increasingly became disillusioned with heavy engineering. In 1909, he handed in his notice to join Joe Lisle at the Star Motor Company.

2
Aviator

Several attempts had been made to fly before the Wright brothers. Clément Ader of France achieved a 150ft uncontrolled 'hop' on 9th October, 1890 with his steam engine powered Eole monoplane. The New Zealander, Richard Pearse, first rose into the air in March 1902 with a powered aircraft which, like the engine, he had built himself; but though 'powered and had sustained flight', Pearse had no control over it once airborne; his claims were later dismissed as fabrication. Accreditation to the first 'powered, sustained and controlled' flight went to two cycle shop owners from Dayton, Ohio: Orville and Wilbur Wright. Their first 'hop' at Kitty Hawk, North Carolina on 14th December 1903, with Wilbur at the controls by toss of a coin, used a 9° declined steel-capped wooden rail. This was followed with Orville taking his turn, lying prone in the Wright biplane, on 17th December 1903 taking off into a strong, 20mph headwind achieving a successful 120ft straight 'flight'. Their 'Flyer' biplane was powered by a single 12hp Wright engine driving two propellers by cycle-chain. Many argue that this 'flight' was more correctly a 'powered glide' for without a sufficiently strong headwind, they were unable to take off under engine power alone to develop their flying controls and techniques. Their subsequent test flights, now from Hubbard Prairie in late 1904, used a weight assisted launch off the track.

Their first, witnessed, flight over a 1,000 feet circuit was not until 6th May 1908; this was followed by a conclusive demonstration at Le Mans, France on 9th August 1908 for the French military trials, in which Wilbur completed a circuit, dispelling myths that the Wright brothers could not have flown earlier. This was again catapult assisted as the Wright Flyer was not fitted with wheeled undercarriage, allowing unassisted take off, until 1910.

Many pioneering aviators developed their 'heavier than air flying machines' around boxed kites developed for the army's aerial surveillance work, such as those by the American showman, 'Colonel' Cody (born Samuel Frankin Cowdrey - he modelled himself on 'Buffalo' Bill Cody) who achieved his first 'hop' in September 1908 with his British 'Army Aeroplane No.1'. Cody was at the time employed at the Army Balloon factory, Farnborough as their Chief Designer of man-lifting observation kites. In these pioneering days, aviators learnt to fly by trial and error. There were no instruction manuals and, until 1911, very few flying schools. As a consequence, there were far more crashes than successful flights - these were par for the course and accepted as an occupational hazard which did little to dampen enthusiasm by others to fly. The slow airspeed and flimsy construction allowed most pilots to walk away from a crash and happily relate their experiences to their fellow pilots!

Daily Mail prizes

The British army saw little need for aircraft, as their balloons and static kites served well enough for airborne artillery observation posts and as a result, Britain was a late starter in flying. However, the visionary Alfred Harmsworth (later Lord Northcliffe) of *The Daily Mail*, was so inspired by the new fangled aeroplane that he promptly offered several £1,000 prizes for crossing the Straits of Dover, for endurance and distance records, and for completing a 1 mile circuit by a British pilot in a British built aeroplane. These were followed by a £5,000 prize for an around Britain course and, later, an ambitious £10,000 trans-Atlantic prize. Driven by these prizes, British aviation developed rapidly and on 19th March 1909 the first British Aeroplane Show was held at the Olympia Exhibition Hall in west London.

The early contenders for crossing the Straits were two Frenchmen, Henri Farman and Louis Blériot and a solitary Englishman, Hubert Latham, who made the first attempt in 19th July 1909 only to land in the sea! On 25th July 1909, Louis Blériot successfully crossed the Channel.

Aviators

Not all aviators bothered to take the new Royal Aero-Club's Aviation Certificate as this only became a prerequisite for joining the fledgling Royal Flying Corps in 1911 and for that reason Granville Bradshaw, like many others, is not on the RAeC register. The first two certificates were issued on 8th March 1910 at Shellbeach (near Leysdown, Isle of Sheppey) to John Theodore Charles Moore-Brabazon, who first flew 2nd May 1909, and No.2 going to Charles Stewart Rolls of Rolls-Royce fame who, sadly, was killed a matter of weeks later. Moore-Brabazon later became Lord Brabazon and chaired the Brabazon Committee of 1942 to identify requirements for Britain's post-war civil aviation needs resulting in the gigantic Bristol Brabazon trans-Atlantic airliner.

Bradshaw's flying activities began in late 1909 and over the following years, his many crashes caused him to become affectionately known as the 'mad aviator'. He was often reproached for flying at Brooklands when wind speeds were deemed too high for safety, but he was determined to prove a point about the aeroplane's inherent safety and stability, as well as test his own modifications in such adverse conditions. He suffered many a flying mishap and on one occasion, when taking off, the aeroplane was blown over by a strong gust, breaking a wing spar. On another, very early one morning, the engine stalled and he crashed through the roof of a farmhouse, landing in the farmer's bed - luckily, both he and his wife had risen early for their daily chores! On that occasion, Bradshaw suffered no more than sprains and bruises - but others were not so lucky and the first reported fatality was the French flyer, Leon Delagrange on 4th January, 1910 during a sharp turn when his Blériot's wings collapsed.

The first British fatality was one of Bradshaw's earliest flying friends, the Hon. Charles Stewart Rolls. A skillful motorist, graduate engineer and the marketing man behind Rolls-Royce motor cars. Rolls had flown balloons and was the first Englishman to make a two-way crossing of the English Channel on 2nd June 1910 in a French built Wright biplane, for which he was awarded the Royal Aero-Club Gold Medal. However, anxious to get to Bournemouth for a race on 12th July 1910, he sought Bradshaw's help in preparing his aeroplane. Granville willingly gave the help but it had unforeseen and fatal consequences, making him wish he had not got involved, for as Charles Rolls swept across the spectators' stand at Bournemouth, attempting to dive steeply to land at the 'bulls-eye' target, the rear tail elevator and rudder became detached and, unable to recover control, Rolls plunged to his death. It appears that against Bradshaw's advice, Rolls had replaced the fixed outrigger tail plane with a French made moving tail plane which moved with the forward elevator; the two hinges employed were too close together and created torsional instability, placing considerable strain on the rear frame, twisting and fracturing it in flight. An expert on stress from his days at ECC, Bradshaw was consulted by other aviators such as Cody.

Samuel Cody, a true pioneer of aviation, was now flying his new 'Hydroplane' biplane, powered by an Austro-Daimler engine. Built for the 1913 *Daily Mail* Coastal Circuit of Britain course, it had a massive wing-span, a huge, suspended central float and outrigger pontoons. Flotation and taxiing tests were done at RFC Calshot on the Solent, and on the Basingstoke Canal, near Laffan's Plain, west of Farnborough Common, close to the Army Balloon factory where Cody was employed. He hoped to win the £5,000 prize and put it to a new aeroplane to compete in the trans-Atlantic race later that year. Now fitted with landing wheels, the Hydroplane successfully flew from Calshot to Brooklands to display this wonderful machine where he too consulted with Bradshaw who, though impressed, forewarned Cody of a fatal flaw in his design should he attempt any aerobatics.

Being the showman that he was, Cody took little heed of Bradshaw's advice. Scheduled to fly to Calshot on 7th August, 1913 for the Circuit of Britain race, where it would be fitted with improved floats for sea trials, he gave a demonstration flight that morning to his friend, Lt. Charles Keyser and offered another to Keyser's friend, W H B Evans, the famous Oxford University and Hampshire county cricketer, who gleefully accepted the joy-ride. But while flying at 500ft over Ball Hill, on the legendary Laffan's Plain, the biplane's rear elevator was seen by witnesses to rise straight into the air and the wings fold into each other. Being thrown out in the process, the aviators and the hydroplane plummeted to the ground killing Cody and Evans instantly. It was reported that 100,000 mourners lined the 2½ mile route from Ash Vale, North Camp to the Military cemetery at Thornhill to pay their respects to 'Colonel' Cody, the most famous of showmen and aviators.

With Cody out of the Circuit of Britain race, the Radley-England No.2, with its three Gnôme engines driving one propeller, now replaced by a single Sunbeam engine, also withdrew from the Daily Mail Seaplane Trial of August 1913, following a crash, and the Short brothers withdrew their entry through mechanical problems. This left Harry Hawker as the only entrant, but though he completed only two-thirds of the course, crashing at sea, he had flown some 1,043 miles in 20 hours and was duly awarded a special £1,000 prize for his 'splendid airmanship'!

Though the Royal Aero-Club's Accident Investigations Committee was later unable to confirm the cause of Cody's crash, it concluded that "the failure was due to inherent structural weakness" and that, had they been wearing seat belts, they might have survived. The Cody family preferred to believe a propeller blade had sheared off, severing the tail plane, but Bradshaw remained satisfied of his forewarning.

Others learnt valuable lessons too. Lt Wilfred Parke RN, discovered the secret of recovering from a potentially fatal spin was to do the exact opposite of what seemed natural. After an official investigation, this became standard practice from 1916. A similar, unnatural, tactic in the late 1940s allowed pilots to 'break through the sound barrier'.

Attracted by the £1,000 prizes, Bradshaw's first attempt at designing and building a light monoplane was while working at ECC. Based on the successful Antoinette, he spent his spare time in a rented stable close to his lodgings. He used the old JAP 8hp V-twin engine which he had fitted to a bicycle frame when in Edinburgh, but it proved greatly under powered, so he made a form of variable pitch propeller to eke out as much power as possible. The engine, with propeller, was duly bolted to a stout wooden bench and fired up - and up it went, pulling the bench up into the air, turning over and shattering the propeller, missing his head by inches! Undaunted, in September 1909, Bradshaw lodged a patent for his 'flying machines' (Prov Pat: 21903), but it was declared void.

Word of his experiments soon got around Wolverhampton and Joe Lisle, the owner of the Star Engineering Company, made contact. Impressed, he bought the aircraft, the proceeds from the sale of which enabled Bradshaw to purchase a new 500cc side-valve Triumph with belt drive and 3-speed Sturmey-Archer hub on which he would spend many a happy day riding and tested himself out on the Land's End to John O'Groats course. Lisle also offered him an aircraft design engineer's job at his Star works which, being fully-equipped, would allow him full use of all tools and labour necessary to design and build the aircraft properly. It was an offer too good to refuse and Granville promptly handed in his notice to ECC.

The Star Engineering Company

Teenage brothers Edward and Alfred Lisle, built their first bicycle at their father's tin-smith's shop in the early 1860s. By 1868 they had a catalogue of designs and by 1876, Edward had gone into partnership with Edwin John Sharratt as 'Sharratt & Lisle' at Stewart Street, Wolverhampton; that partnership was however soon dissolved and in 1896 Edward continued in business as the Star Cycle Co Ltd with his two sons, Edward (junior) and Joseph. It was not long before the enterprising Edward (senior) had established a string of local engineering companies primarily to supply his own cycle business; these included the famous Presto Gear Case Company. In 1899, Edward produced his first motor car and in 1902 formed the Star Motor Company to produce several models using deDion or Daimler engines such as the 'Star', the quite basic 'Starling' (produced by the cycle division under Edward junior) and the middle market 'Briton' model. In 1909, the business was reorganised as The Star Engineering Company with motor car production now centred on the Briton Motor Co under control of Edward (junior) for which many kit-parts were supplied to F Hopper & Co of Barton-on-Humber whose core business was the famous 'Elswick' and 'Falcon' bicycles. By late 1909, young Joseph Lisle was left to run the Star Engineering Company on his own while his father, Edward established the Star Aeroplane Company with a capital of £1,000 to build a Farman biplane.

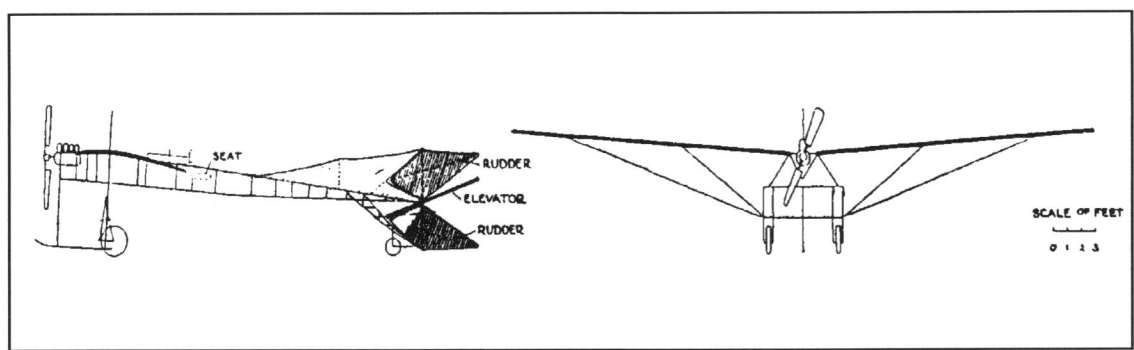

Top: The Star Monoplane as drawn by 'Flight' in April 1910, and below, as finally built and flown

Star monoplane

As can be seen, the Lisles were an enterprising family who were not slow to realise the possibilities offered by Bradshaw's work and, with him on board in late 1909, were able to build their second aeroplane developed from Bradshaw's Antoinette inspired monoplane. This was now fitted with a proven 15hp Star racing-car engine, with side mounted cylindrical radiator, which developed a peak output of 30hp. The fuselage remained of triangular cross-section using a lightweight open box girder design of exceptionally tough English ash timber. It was built in two sections, joined behind the pilot's seat, for ease of transport by road. The tail plane and rudder each comprised two members which could be worked together as a rudder and elevator, or in opposition as a tail-aileron. The main wing, of 38ft wingspan, was set at a 1:13 dihedral and 6° incidence and, like the tail plane, was covered in a close-woven, 'Dunlop' rubberised cotton canvas fabric. It was fitted with ski-undercarriage and 12" sprung wheels to act as shock-absorbers in a heavy landing.

Fully assembled by early January 1910, it was exhibited at the Midlands Aero Club's winter meeting, of which he was a member, and where Frederick Lanchester and Sir Hiram Maxim were giving lectures. Bradshaw confidently predicted that with its 6'-0" propeller it would have an airspeed of 36mph, but despite preliminary trials at Dunstall Park in February 1910, it was unable to lift off the ground. By March the wings had been extended to 42ft and the fuselage increased from 30ft to 32ft, in the hope of giving it extra lift at an estimated take off speed of 30mph. The date of its first flight is not recorded, but the Star monoplane was exhibited at the Olympia Aero Show in April 1910 with a price tag of £450. It attracted the attention of the new *Flight* magazine, describing it in the April 23rd 1910 edition as: "Altogether this machine gave evidence of being one of the most carefully thought out designs in the show". In hindsight, it must be appreciated that being gentlemen, they rarely vehemently criticised any design preferring to suggest opinions as to how a design might, in their estimation, be improved.

Granville's biographical notes record a flight at Dunstall Park on 10th June 1910, supported by a local newspaper photograph of what looks like the Star, said to be flown by Bradshaw at a height of about 30ft, accredited to 10th June. It was later published as a picture postcard carrying the dubious claim: "The first ever of a plane in flight"! Unfortunately the archive copy of this is now missing. (Jim Boulton's *Powered vehicles of the Black Country*, suggests that the Star's first flight was on 10th November, 1910, though this was most probably the first flight in its modified form).

The Star's public flying debut was scheduled for the first All-British meeting held between 27th June and 2nd July 1910 at the Midland Aero Club at Dunstall Park race-course, just outside Wolverhampton. Sponsored by Edward Lisle, the event included a 21 mile cross-country course. There were some dozen or so pioneer flyers entered, including Graham Gilmour, Claude Grahame-White, J Radley and Charles Rolls. Unfortunately for Lisle, bad weather, unpredictable strong gusts and heavy rain so hampered the event that the judges decreed that only those who held a RAeC Pilot's Certificate, or who could satisfy them of their ability to fly by 24th June, 'without being a danger to themselves or the public', would be permitted to take part. Understandably, due to the limited available flying time, this ruling meant several pilots and aircraft designers, including Granville Bradshaw, were unable to compete.

However, within weeks, a second attempt to fly was made, but though the 15hp engine, with its reduction gear, developed 180lbs thrust at the propeller at 1,200rpm, the heavy cast aluminium, steel reinforced 'Star' propeller sapped too much energy and the aeroplane failed to reach sufficient take off speed. In the last week of August, Granville achieved a 30 yard flight at 20 or 30 feet, followed by a 100 yard flight at 20ft on 3rd September. Then on 5th September he achieved a half-mile flight at 40ft but, avoiding a cow grazing on his landing path, crashed into the railings, damaging his wing and

The improved Star 40hp engine

propeller. Bradshaw then set about designing a much lighter 40hp 4 cylinder engine, replacing the usual cast iron, water-jacketed cylinders with lighter, machined steel cylinders wrapped with thin copper water-jackets to increase thermal efficiency and reduce weight. It is reported in Jim Boulton's book that these thin copper jackets were made by electroplating over a beeswax coated cylinder; the wax was then melted to leave a cavity and the thin copper jackets were then soldered to the barrel. The cylindrical radiator had pumped circulation, but was later replaced by a conventional, vertical, hydro-thermal radiator behind the engine, as seen in the photograph.

Back at Dunstall Park in November 1910, the repaired and further modified monoplane took off again on the 5th but, hampered by fog, he resumed flights on the 10th now fitted with the new four-cylinder, 3.9ltr engine with its simpler cigar shaped, horizontally split crankcase. Bore and stroke were 100mm x 125mm and it developed 40hp at 1,450rpm. Weighing 182 lbs, complete with water pump and Simms magneto, it drove an 80" twin-bladed 'Clarke' wooden propeller. The airframe was also modified, now adopting a more conventional tail-plane/rudder assembly; the wingspan was shortened by 5 feet to reduce the weight. *Flight* magazine later reported that the Star always was a 'steady flyer' but the tail had too much weight on it; however the heavier engine, with its rear vertical radiator, helped balance and trim the aeroplane. Joe Lisle's first attempt at flight so alarmed his father Edward, that he forbade him from ever flying again!

There are two different accounts of what happened next. One is in the late Jim Boulton's book which says that Edward had the airframe broken up and the engine sold to Mr. Bartelt for his 'Ornithopter' (a flapping wing aeroplane) exhibited at the Olympia Air Show in March 1911, suggesting the Star may have been broken up at Brooklands that winter. That engine survives today in the RAF Museum, Hendon. With a background in soap and laundries, Frederick Ludwig Bartelt became an ambitious patentee of 'aerial propulsion' proposing moving, collapsible linen pockets by which to 'suck' the aeroplane into the air, with further 'Improvements to...' leading to a proposed ornithopter-propelled car, and a cyclogyro aeroplane.

The other story, told by Granville, is that he flew the Star monoplane daily and took it to the newly established Brooklands aerodrome in Surrey in the hope of demonstrating the Star, attracting wider publicity and orders for its manufacture. Both accounts are probably correct, if chronologically inaccurate, but it does seem more likely the Star monoplane was transported by road to Brooklands, for during autumn 1910 Star were renting Shed No. 11 in the heart of the new Brooklands Flying Village, opposite A V Roe in sheds Nos. 14 and 15. His debut flight at Brooklands is recorded as 2nd

December, 1910 followed by several for prospective buyers, including James Valentine (RAeC No.47 of January 1911) who came 7th in the Circuit of Europe in 1911 flying a Depurdussin. During the war Major James Valentine was made O/C RFC Villacoublay, Paris, gaining the Legion d'Honeur then, after a special mission to Italy, won the DSO. In July 1917 he served with the British Government in Kiev, as acting Lt.Col, assisting the Imperial Russian government; he received their Order of Stanislau but died on a mission in Kiev on 7th August 1917. His widow, Louisa Eileen (daughter of Maj-Gen G W Knox and Lady Sybil Knox, niece of the Earl of Londsdale) married Ronald Charteris in the summer of 1919 (his first marriage to Edith Tryon in 1907 had ended in divorce in 1917).

Without doubt, Granville Bradshaw was indeed an early aviation pioneer. His later claim "that he was one of the first six Britons ever to get into the air and this in a plane, a monoplane, with engine and variable pitch propeller all of his own design; it was said the only man ever to have achieved this trio", is most probably true, for most early flyers flew aeroplanes built or designed by pioneer aviators such as Wrights, Cody and Farman. Monoplanes were certainly a rarity, and most were of French origin, such as the Blériot or Antoinette.

Brooklands

Built in a natural depression on prime agricultural land at Hollick Farm and Wintersells Farm in motoring enthusiast Hugh Locke-King's large estate, Brooklands was the world's first purpose built motor racing circuit. Constructed at a cost of £150,000, work began in the spring of 1906, requiring the diversion of part of the River Wey and bridging a large sewage settlement lagoon into which many a hapless aviator would plummet through their engine faltering 'in the foul air', soon after take off. The site benefits from a unique, and quite local, freak weather condition - there is little wind in the morning which made it ideal for pioneering aviators who demanded the stillest of days for their flying.

The $2^3/_4$ mile, 100 foot wide concreted circuit was scientifically banked to allow motor-cars and motorcycles to achieve speeds of up to 125mph. Following its official opening in July 1907, Brooklands became a mecca for racing and prompted a 30-year old Mancunian railway engineer, Edwin Alliott Verdon-Roe, who had been sent to school near Brooklands, to persuade Locke-King to allow him to experiment with his 6hp JAP V-twin powered flying machine. Based in a large wooden shed by the Test Hill, he freely experimented with the aeroplane built by himself at his brother's home in Putney. Alas, the JAP engine with its large belt driven propeller proved grossly underpowered, so he borrowed a 18-24hp Antoinette engine and on 8th June 1908, became the first man to 'take off' from Brooklands, achieving a 2-3 foot high 'hop' over 80 yards. He achieved further hops by being towed off the banked circuit and later managed a 100 yard 'flight', again having been towed into the air by car. But, as these were tow-assisted, his claim to be the first Briton to fly in a British built aeroplane was unacceptable to the Royal Aero-Club and in July 1908, the increasingly anti-aeroplane Brooklands' Committee evicted Verdon-Roe from his shed.

Verdon-Roe duly returned the Antoinette engine to France and moved to some railway arches on the Lea Marshes, North London to try out a new design of his own in a short lived enterprise with John Alfred Prestwich, of engine fame. They formed the JAP-Avroplane Company in September 1908 with a triplane which had triple main and tail planes. Two were scheduled to be built, the first with a 9hp JAP V-twin and the other with an engine of Verdon-Roe's own design, but Prestwich dismissed the design and duly built a 20hp JAP V4. Only one airframe was completed; it was named the 'Roe I Triplane' and first flew properly on July 13th 1909.

Evicted from Lea Marshes, he moved temporarily to an unsuitable site on Old Deer Park, Richmond before moving to Wembley in November 1909, whereupon the partnership with Prestwich was dissolved following a major disagreement. Verdon-Roe now sought funding from family and friends in Manchester and formed A V Roe & Company on 1st January 1910, based at Brownsfield Mills, Ancoats, Manchester. Flying his own triplane, he secured his Royal Aero-Club Certificate (No.18) on 1st May, 1910.

With Brooklands rapidly becoming the mecca for motor-racing, aviation and engine technology, Maj. Lindsay Lloyd opened it as an aerodrome and had built the first of six wooden aeroplane sheds next to the Wintersells Farm buildings in late 1909; the first for Louis Paulhan's Farman biplane, but by early 1910 this had been taken over by H P Martin and G H Handasyde, who then moved into a new, larger shed, No.12. Their old shed became the famous 'Blue Bird' restaurant, run by Noel Pemberton-Billing's brother and sister-in-law. Noel's company, Pemberton-Billing Ltd, became Supermarine Aviation in 1915. Verdon-Roe returned to Brooklands in March 1910 but was now based in shed No.14. Several, much larger sheds were built in 1911, plus a row of new sheds (Nos. 29-42) erected under the Byfleet bank where Vickers opened their flying school; these became Hawker's after the war. Also in 1911 *The Daily Mail* launched their famous £10,000 Circuit of Britain air race. Life at Brooklands in these events is quite accurately portrayed in *Those Magnificent Men in their Flying Machines*!

Turning point

It was at Brooklands in 1910 that the 23 year old Granville Bradshaw became acquainted with Ronald Louis Charteris, a fellow pioneer aviator who was in partnership with Walter Lawson Adams at Adams' fledgling Aeroplane Engine Company, outside Southampton. Charteris was able to offer Bradshaw far greater scope in aeroplane design than Star, so offered him equal partnership in a new aero-business of his own making. This was indeed a turning point for Bradshaw, for, as it happened, Joe Lisle had already decided to concentrate resources on the successful Star motorcar and abandon aeroplane work altogether. They parted on friendly terms and Bradshaw took over Lisle's lease of Shed No. 11 at Brooklands. Now on his own, Lisle continued to offer the Star monoplane to prospective buyers, but none came forward and nothing more was heard of the Star after late summer 1911.

Lisle's new Star 15hp car was of exceptional build quality with a high standard of appointment, using expensive, pressed steel chassis members. Indeed, before the Great War of 1914-1918, Star was considered to be one of the top six British car makers. Ironically, they were contracted during the war to build components and sub-assemblies for the aircraft industry, such as wings for the new Avro 504. In 1918 they were contracted to build the French Renault V8 450hp aero-engine adopted by the Royal Aircraft Factory for a new version of the Airco deHavilland DH9 - but war ended after only 12 of these engines had been built. Jim Boulton's book suggests that Star built some ABC Dragonfly radial engines, but this claim is not supported by other evidence and it is probable that they were simply sub-contractors because, it was reported, they had a much sought after crankshaft balancing machine.

War-time production at their Briton motor-car subsidiary concentrated on a car chassis modified to become the 25cwt Star UE military truck or heavy ambulance of 1916. Though Star had produced their first van in 1907 (several were bought by the Lever Brothers for their Port Sunlight soap works) the military chassis was extended to a 3 ton chassis. At war's end, production of both the high quality, hand built Star car, commercials and their budget Briton models resumed, but times had changed and they were no match for the new breed of post-war light cars and lorry chassis such as those mass produced models from Ford, Austin, Morris and Bean. Though there was a slight increase in sales in the early 1920s, by 1922, their Briton business had gone into liquidation, being reformed as Tractor Spares Ltd making parts for caterpillar tracked vehicles. Following a dramatic drop in demand, Joe Lisle approached Sydney Guy, their near neighbour at Guy Motors, with the view of selling him the Star company.

Sydney Slater Guy was the Chief Designer at the Sunbeam Motor Car company before he formed Guy Motors in 1914, but no sooner had his new business begun than war was declared and he was contracted by the Government to manufacture various munitions and, in 1918, to build the new ABC Wasp and Dragonfly radial aero-engines designed by Granville Bradshaw, formerly of Star Engineering. In 1926 Guy took on John Harper-Bean, and following Joe Lisle's approach to sell Star, Guy promptly bought a majority share holding and, in 1928, became the Managing Director of the new Star Motor Company from which point the Star lorry chassis now carried the famous Guy 'redskin' Indian badge. However, in the aftermath of the Wall Street crash of 1929 and the ever deepening economic depression in Britain, production ceased in March 1932 and the Star company closed soon after.

Aeroplane Engine Company

Born in 1874, and apprenticed as an engineer with a firm of millwrights, Walter Lawson Adams became a cycle maker in Lowestoft but moved to Northampton where, as a motor engineer, he developed his 'Precision' motorcycle engine. In 1904 he moved to 'Yewtrees', Redbridge, Southampton. Redbridge had been a significant boat-building centre; Adams had experience of boat engines.

Now operating as W L Adams Ltd, Redbridge Motor Works, he designed and built a handful of small engines (presumably 'Precision' based single, V-, in-line and water cooled engines) one of which, a 10-12hp engine (most likely a twin), was fitted to the *Bluebottle* in 1904, securing it a Class win in an endurance trial on Southampton Waters in 1905. He now developed an interest in aviation and proposed a V-8 engine for the dirigibles being used by the Government. His biographical notes say he exhibited this in 1905, but while firm evidence is yet to emerge, there is an obscure reference to a dirgible engine in 1906 and it may have been one of these which was fitted to an unidentified 'hydroplane' (a specially designed, stepped motor-boat hull as opposed to a sea-plane with 'hydro-plane' floats) using the Fauber principle of injecting air behind the stepped hull to break surface tension and increase lift and speed, as further developed in Britain by Samuel Saunders at Cowes.

His engine was developed in the 90° V-8 (4" x 4¾") 'Redbridge' marine engine which, like his earlier water cooled engine and the Antoinette V-8 of 1906, used cast-iron cylinders surrounded by corrugated copper water jackets, secured by shrunk-on external steel bands. But more significantly, this engine employed inlet and exhaust valves activated by the same cam driving a push-rod coupled to a push-pull rocker arm (with auxiliary spring to counter the inlet valve spring). The valve timing allowed engine rotation in either direction by rotating the camshaft 90° relative to the crankshaft and, by inducing a back-fire, the engine was now able to run in reverse with the inlet valve operating atmospherically. Although it had reduced power, it was sufficient to reverse a boat or dirigible.

In December 1907, he lodged a patent on 'Improvements to the internal combustion engine' (granted Pat No. 27877 in December 1908). This was for a two-stroke type cylinder wall exhaust port at bottom dead centre to both improve engine efficiency by more fully scavenging exhaust contaminants on the induction cycle, and relieve the pressure on the exhaust valve as it opened. There was a very slight loss in power, but this was compensated by a cooler running engine. His cylinder porting design was reported in *The Automotor Journal* in spring 1908 and it is worthy of note that similar porting was used successfully on the 1910 Darracq and 1914 Alvaston horizontal aero-engines.

A further patent was lodged in December 1907 (granted Pat No: 28432 in November 1908) for a water-jacketed (or exhaust gas heated) carburettor which improved atomisation. By using a rotary throttle control exposing different jets, it could reduce and then cut-off fuel supply allowing ambient air alone to be drawn through, both cooling the still rotating engine and, being incombustible, acting as an engine brake (rotary engines often used ignition cut-off for that purpose).

Adams' 1907/8 Redbridge Marine V-8 Engine

In March 1908, Adams exhibited his Redbridge V-8 marine engine at Charles Cordingley's Motor Show at the Agricultural Hall, Islington and announced that several had been ordered for fast motor-boats and a 'hydroplane' built at Cowes - this was probably one of Sam Saunders' although his success came from a Wolseley engine. The reversible marine engine was found to be equally useful for dirigibles and racing cars (for which a simple high and low ratio sliding dog clutch gear-train dispensed with the need of a reverse gear) and indeed it was reported at the show that a racing car was under development at Brooklands, but whose is not known.

V-8 Aeroplane engine

As orders for the 'Redbridge' V-8 increased, and encouraged by its potential for aeroplanes, a new business was formed with Ronald Charteris (b. 24/9/1879) in summer 1908 as The Aeroplane Engine Company. When Charteris' interest in aviation began, bringing him into contact with Adams, is not known but it does appear he helped finance Adams' work supported by his late father's estate, Captain The Hon Frederick William Charteris RN (1833-1887; the son of 9th Earl of Wemyss. Ronald's mother, Lady Louisa Charteris (née Keppel), was the daughter of 6th Earl of Albemarle.

The Aeroplane Engine Company duly moved into 'a new factory' at Redbridge in late summer 1908; its location is not yet confirmed. Their new '1909 pattern' 100hp V-8 'Aeroplane' non-reversible aero-engine was launched that autumn, but while closely following the marine engine's design, its revised push-pull valve timing allowed it to develop 113bhp at 2,000rpm.

Bradshaw says that after meeting Charteris at Brooklands, he joined a company based in Bournemouth in 1910. Adams was certainly living in a 9-room property, The Nest, 46 Surrey Road, Bournemouth by that time while Charteris is recorded as lodging at the Meryick Estate, Bournemouth, but working at The Stanpit Garage, Christchurch - was that the unknown 'Bournemouth factory' referred to by Adams in his biographical notes? Charteris had only joined the RAeC that June but while he regularly flew a Hanriot monoplane at Brooklands, it was not until March 1912 that he gained his Certificate (No. 197), on a Depurdussin monoplane. Adams does not appear to have flown.

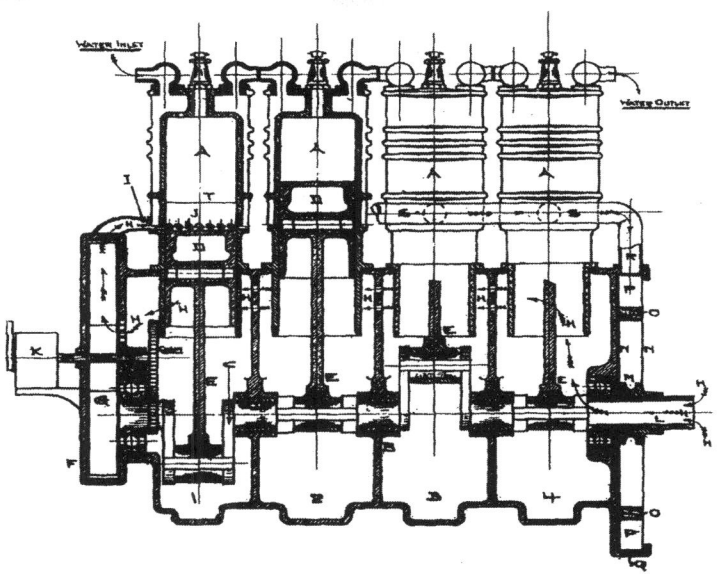

Adams' auxiliary exhaust scavenging system, hollow crankshaft breathing and fan assisted induction on a 2-stroke engine. Pat No. 27511

Adams' inventiveness led to a further patent (Pat No. 27511, lodged November 1909, granted December 1910) for a crankshaft driven, fan assisted induction and exhaust gas scavenging system on a two-stroke engine, once again designed to improve performance and efficiency. Air was now drawn through a hollow crankshaft and crankcase, helping cool the engine, which in turn heated the incoming air before passing under pressure from the exhaust gas driven fan through the carburettor into the cylinder: a primitive turbo-charger! Indeed, a reader had asked in *Flight*, of 2nd July 1910, if the expelled exhaust gases could be employed to drive a second propeller? The editor warned of the sap of energy through back-pressure from a directly coupled turbine, but Adams rose to the challenge in a letter that October, promoting his patented porting design (as used successfully in his Redbridge marine engine) by which he asserted the exhaust gases applied pressure of 96psi to the flywheel scavenging pump, which thereby improved the engine's efficiency. In August 1910, Adams improved his hollow crankshaft induction system with drilled oil-ways to further cool the bearings, however in this patent (Pat No. 20328, granted August 1911) the rate of air flow and pressure relied entirely on the forward motion of the aeroplane.

All British Company Ltd

Their earliest reported aero-engine was a 60hp V-8 (4" x 4³/₄") (probably the Redbridge marine for dirigibles); this was joined by the new 100hp V-8 Aeroplane engine exhibited at the Cordingley show in 1908 which developed 113bhp at 2,000rpm and is described variously as weighing between 452 and 488-lbs. This was followed in 1910 by a 4-cylinder in-line (4" x 4³/₄" - also reported in *Flight* as 4³/₄" x 4³/₄"). Exhibited at the Olympia Show it featured the air-cooled crankshaft and auxiliary exhaust ports (some reports suggest it lacked these ports). Adams and Charteris had intended submitting one of their new Aeroplane Engine Company's engines to the first Government trials of spring 1910, for which a prize was offered by Patrick Y Alexander for the best British aero-engine. But of which type, whether they failed to enter, or if it was not accepted is not known, for of the six other entrants only three competed. Only the Green completed the exhaustive tests but as it had not managed to maintain the required 35hp over 24hrs duration, it was announced in January 1911 that the Alexander Prize had not been awarded.

While demonstrating his Star monoplane and engine at Brooklands in late 1910, Charteris offered Granville Bradshaw an equal partnership in a new business of his own making to run in association with their Aeroplane Engine Company. The outcome was the All British Company, formed in February 1911 but renamed All British (Engine) Co in April 1911, for which Bradshaw designed the logo. Reputedly, the name reflected national pride following Lord Northcliffe's offer in the *Daily Mail* of handsome cash awards for flights made in aeroplanes of entirely British design, manufacture and materials while the initials, by chance or design, represented those of the three partners, Walter **A**dams, Granville **B**radshaw and Ronald **C**harteris. (There had been an earlier ABC motor-car company, established in Glasgow in 1906 by George Johnston, formerly of the Arrol-Johnston car company, but fortunately his ABC business had gone into liquidation in 1908).

The first ABC engines (still described by *Flight* as for dirigibles) were duly exhibited at the Olympia Show in March 1911: a 40hp in-line 4-cyl, an 80hp V8 and a 120hp V-12, all to Adams's $4^{3}/_{4}$" bore and stroke format.

Adams duly submitted another engine for the July 1911 Alexander Prize against seven competitors which included Green, Isaacson and ENV. This time the Green 4-cyl won the prize. ABC's hopes had been heightened earlier when a 100hp V-8 was ordered by Robert Macfie for his July 1911 Circuit of Britain race biplane, but not being ready in time, it was substituted by a 50hp Gnôme. But one of their first new ABC 40hp vertical in-line, water cooled aeroplane engines was used in Ellis Victor Sassoon's monoplane as a replacement for the 50hp Gnôme rotary which had been flown by H J Astley in the July 1911 race. Sassoon's Universal Aviation Company's monoplane was modelled on a Blériot Type XI and built at Brooklands. Now christened the 'Birdling', it was also entered for the autumn 1911 Michelin Cup No.2 contest, but in the event only Cody completed the course and, failing to win any prize money, Sassoon went bankrupt soon after. The Birdling was bought by Frank McClean of Short Brothers and, on being rebuilt, inspired the design of their new Short monoplane.

Meanwhile, now in partnership at ABC and having free reign to carry out his experiments in aeroplane and aero-engine designs, Bradshaw worked from his old Star shed at Brooklands developing a new ABC powered aeroplane (this may be that submitted in his second aeroplane patent in January 1911, Prov Pat: 2136, but the application was abandoned). The immediate problem of getting the engines from Adams' Redbridge works to Brooklands for flight testing was soon resolved by fitting the humble 4-cyl or V-8 (there is no record of the V-12 having flown) to the back of Charteris' now modified car, affectionately known as the 'Wind-waggon', propelling it along the road! It was first tested in early May 1911 with a V-8, which produced 330-lbs thrust at 900rpm; the road speed was not declared. The 'Wind-waggon' was later used at Brooklands for testing propellers, and is described in greater detail later.

Adams' departure

While Bradshaw was in his element, Walter Lawson Adams appears to have been turning away from aero-engines. He had already been approached by W H Dorman & Co of Stafford to buy his aero-engine designs and patents and was even offered the position of Manager, Chief Engineer and designer in their new Aero-engine Section of the Dorman Internal Combustion Engine Department (Dormans merely describe his position as Engineering and Sales Expert). It is quite possible that the War Office's dismissal of the value of aeroplanes in warfare had persuaded Adams to sell his W L Adams Ltd Redbridge Motor Works to Dormans in May 1912.

Then on 24th July 1912, the Aeroplane Engine Company went into voluntary liquidation with the appointed liquidators, Frank Shalders and Maj D W Sweitzer disposing of the assets: the Adams 90°, 60hp V-8, parts and production machinery went to W H Dorman & Co and duly became their short-lived 1912 Pattern Dorman Engine (4" x 4¾", also stated as 112 x 120mm), described later. Adams' other aero-engine designs appear to have gone to ABC. Ironically, in early July, Ronald Charteris had been elected to the RAeC's British Aero-engine Manufacturers Sub-Committee. With the loss of their Redbridge manufacturing base, Bradshaw now transferred their manufacture to Armstrong-Whitworth Ltd. How many Adams/ABC engines had been built since 1905, and of what type, is simply not known.

Adams did not stay long with Dormans for he is next reported as being a Consulting Engineer in London developing a cycle-car engine, yet his own account suggests he returned to ABC at Brooklands for two years "developing my own engines and aeroplanes... and while there, acted as consulting engineer" on the car engine design. Nevertheless, in July 1915 he was commissioned into the Royal Naval Volunteer Reserve and served in the Admiralty' Air Department, with distinction, in aero-engine production. (His biography appears in Appendix 5).

Lightweight air-cooled V-8

Bradshaw submitted a further patent, jointly with ABC, in May 1912 (Prov Pat: 10891), for a new aeroplane design. This too was abandoned and it appears he then abandoned all further ambitions in aircraft production preferring to concentrate on developing and selling aero- and motor-cycle engines. But he still promoted his ideas and theories on stress for safe aeroplane design, and of aero-engine technology in *Flight* in summer 1912. Whether these reflected Adams' influence is not known, but Bradshaw emphasised strength of materials, with lightness, encouraging short stroke multi-cylinder designs to minimise the size and thus weight of the already heavy crankcase. He duly redesigned the ABC V-engines with 5" bore x 4¼" stroke and now listed a full range of V-6, V-8, V-12 and V-16, 90° aero-engines with forged steel cylindrical crankcases, a pair of magnetos and multiple carburettors described as 'floatless', being capable of 'working at any angle', ensuring a rapid rate of climb.

That summer he introduced a new range of compact 3¾" x 3⅛" water cooled V-engine designs and duly listed a 35-45hp V-4, 45-65hp V-6 and a 60-85hp V-8 using a forged steel crankcase of a new modular, spiggoted and bolted, construction allowing any V-configuration (with or without reduction gear) to be assembled. This American inspired method of manufacture ensured low production costs, but the new engines appear to have remained only as paper designs or mock-ups; little else is known of them.

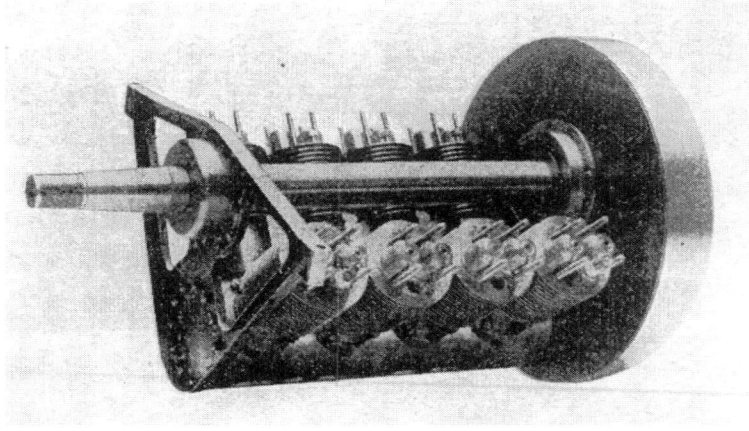

The simple V-8 design showing huge spur-gear reduction drive for the inboard propeller shaft-cum-camshaft for the ohv rocker system

Ronald Charteris holding the V-8 lightweight engine

His articles and a lecture to the Scottish Aeronautical Society in December 1913 were remarkably astute and were reflected in a new lightweight, 1.4ltr (2½" x 2¼") high speed 90° V-8, air cooled aero-engine, developed in 1912 possibly for his abandoned aeroplane. Despite its all-steel construction, it weighed only 112 lbs and developed 50hp at 4,200rpm, compared to 60hp at 1,450rpm for his new 3¾" x 3⅛" water cooled V-8, which weighed 175 lbs. The 4-bearing crankshaft was fitted with a massive spur gear which turned, at a ratio of 6:1, the propeller shaft laying within the 'V', making for a particularly compact design. Uniquely, the propeller shaft also served as a cam-shaft, driving the ohv rocker arms. It featured many advanced ideas on big-end bolts, splash lubrication (already proven on the ABC flat twin motorcycle engine at 6,500rpm) and pumped air-cooling for the machine-turned, finned steel cylinders.

However, unlike his later Gnat and radial aero-engines, the cylinders were not copper coated. Indeed it may have been work on this V-8 which prompted investigation into copper coated cylinders and that some of his abandoned patent applications on crankshafts and internal combustion engines may also relate to this but, unfortunately, provisional patents which were declared void, withdrawn or abandoned are not archived, so we have no record of their claims. It is tempting to wonder if Granville Bradshaw had also set his sights with this V-8 lightweight on the race track but, alas, nothing more was heard of it and all that survives is an *Autocar* article of January 1913 with a photograph of Ronald Charteris holding the part-assembled engine in his arms.

Armstrong-Whitworth

Following the closure of their Redbridge works, Bradshaw transferred manufacture to Armstrong-Whitworth Ltd at their Elswick, Newcastle-upon-Tyne works. Armstrong-Whitworth were a highly respected engineering and armaments company who first became involved in aviation in 1910 when they rebuilt a Farman biplane which was eventually sold to Verdon-Roe at Brooklands.

Production had begun by December 1912 of both the ABC 4-cyl 40hp and the V-8. One 4-cyl was fitted into the 'Wong Tong Mei' biplane in early 1913. Built and tested at Shoreham by a Chinese, Tseo K Wong, his intention was to produce it in China but whether that also applied to the ABC engine is not known; his later 2-seater biplane of 1914 used an Anzani radial. Nothing more was heard of either but by twist of fate, the 'Tong Mei' came to haunt ABC - it is Chinese for Dragonfly!

Armstrong-Whitworth's first 100hp ABC V-8 engine was fitted into a cradle with chain driven overhead propeller shaft, and delivered in May 1913 to Filey for fitting to an unidentified hydroplane for M. Henri Salmet, the famous French aviator who, in 1912, had become the chief pilot at the Blériot Flying School, Hendon. This engine was returned to Armstrong-Whitworth on 3rd October 1913 for overhaul and was test run for 12 hours non-stop before taking to the air again on 29th October, 1913. News of these new Armstrong-Whitworth built engines was not announced by *Flight* until June 1913 by which time Bradshaw had preparing revised designs for his improved V-8 for which detailed specifications were submitted to the Elswick works in August 1913. These survive in Granville's archives with the comment that pre-production design and prototyping was to be carried out by Armstrong-Whitworth, rather than by Bradshaw or ABC.

Early Armstrong-Whitworth built ABC V-8 as fitted to Henri Salmet's hydroplane of 1913. (GB)

Ricardo-Halford-Armstrong V-12 of 1917

Specification of 100hp ABC V8 engine

8 cylinders of 5" bore, 4¼" stroke, set four aside of a circular steel crankcase at 90°, developing 120bhp at 1,350rpm. Water cooled by gear driven circulation pump. Two magnetos with dual ignition to twin spark plugs per cylinder. Gear driven pump for pressurised oil lubrication. The engine, less radiator, to weigh 380-lbs.

The cylindrical crankcase of cast, nickel gun steel machined in one piece, closed at the propeller end by steel nose with central ball bearing and thrust washer. The back-end coupled to a cast aluminium timing gear case. The crankcase to have five white metal main bearings in phosphor bronze castings.

The crankshaft of 'special steel' drilled throughout its length for lightness. Choke tubes fitted inside the crank pins to force oil up the connecting rods to the gudgeon pins. The big ends of connecting rods of opposing cylinders to be bolted to the same crankpin in pairs. Keyways at the tapered ends of the crankshaft to allow sprockets or a flywheel to be fitted.

Cylinders of one piece machined steel with adaptors screwed into the head to allow free circulation of water around the valves chambers.

The water jacket of spun copper, corrugated *(to allow cylinder expansion)* and attached to the cylinders by silver soldering. A watertight joint between the jacket using rubber rings between the V-shaped ferrules terminating the upper and lower cross pipes.

Pistons to be of best close grained cast iron. The gudgeon pins being case hardened and locked with copper tubes and steel washers.

Connecting rods of steel stampings drilled for lightness. White metal lined big end bearing.

Valves enclosed in cast iron valve domes operated by overhead rocking levers by tappet rods off a case hardened camshaft running in phosphor bronze camshaft case, bolted to the top of the crankcase between the cylinders. The gear train for their operation and that of the magneto to be housed in the aluminium timing case.

Two carburettors to be used, feeding each side of the engine through thin copper induction pipes.

Ignition by two Bosch HT magnetos bolted to platforms on camshaft casing, synchronised for simultaneous ignition through dual coils with provision for a standby accumulator ignition.

Lubrication by powerful direct, gear driven pump drawing oil from a suspended thin sheet steel sump, force lubricating the five main bearings, entering the hollow crankshaft to the big-end bearings and then via the hollowed con-rods to the small end brass bushes. The system to have a pressure relief valve and over flow to the sump.

Cooling by geared pump situated in the front of the aluminium timing gear casing. The water is delivered to the bottom of the jackets of the first pair, passing out of the top of the last pair to the radiator.

Tests to be performed:

Weight of engine complete with all ancillaries not to exceed *(unspecified)* pounds,

Horsepower; engine shall develop 100bhp at 1,200rpm and make a non stop run for 6 hours at that rating; and shall develop 130bhp at 1,600rpm and make an hour run at that rating.

Slow running; capable of slow running at 250rpm

Fuel consumption shall not exceed 8½ gallons at 100bhp; 4½ gallons at 50bhp and 2½ gallons at 25bhp

Oil consumption shall not exceed 0.4 gallon at 100bhp; 0.3 gallon at 50bhp and 0.25 gallon at 25bhp.

That August, 1913, Armstrong-Whitworth announced a major expansion at Selby for the production of aero-engines and dirigibles, doubtless using the ABC V-8 which duly entered production; the first was installed into an Avro 504 in February 1914. It appears that Armstrong-Whitworth may have bought this aeroplane for flight testing. After extensive engine run-ups, it was flown by Avro's chief test pilot, Fred Raynham in April 1914, but it appears the engine was not acceptable and was never flown again. A second engine was fitted experimentally to an unidentified Sopwith in March 1914 and flown by Howard Pixton at Brooklands, but no more is known of that. Two further ABC V-8s were to be submitted to the 1914 Military trials of April 1914, but again there are no further details.

Why these early Armstrong-Whitworth built ABC V-8 engines failed to live up to expectation is not known (similar problems would arise during war-time sub-contracted production of Bradshaw's radials). In a curious twist to this story, Armstrong-Whitworth now demanded that Bradshaw move to Elswick to oversee their development and that of their new V-12 (of which little is known) which was undergoing prototype development at Elswick. Bradshaw declined, most likely as its pre-production development lay with Armstrong-Whitworth and as a result they abandoned all further ABC work, terminating their contract in June 1914. Their Elswick works then assembled BE.2a fighters for much of the war but in 1917, Armstrong-Whitworth constructed six Harry Ricardo and Frank Halford designed experimental R-H-A 45° V-12 260hp engines which, when fitted with an under-piston supercharger (not dissimilar to Bradshaw's later Pulsation Motor) developed 300-360hp, but only two had been completed and flown by the Armistice.

Dorman V8

Dormans were better known at that time for their footwear, printing and machine-shop machinery, but had started making motor car engines in 1903. Among Adams' assets they acquired was the water-cooled V8 aero-engine which Dormans then built as 'The 1912 Pattern Dorman Aero-engine'. This had a bore and stroke of 4" x 4.75" and developed 80hp at 1,300rpm. It had cast iron cylinders, flange jointed to the cylindrical crankcase with spun, seamless copper water-jackets. Air was induced through

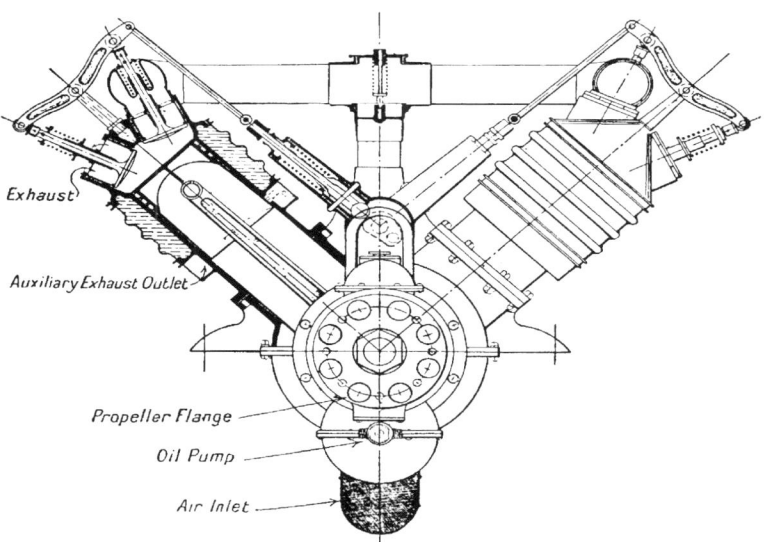

FIG. 41.—Diagram of Dorman Vee engine.

Dorman water cooled V-8

the crankcase and crankshaft as per Adams' patents. By how much the Dorman engine differed from the original Adams is not clear, but while it retained his hollow crankcase breathing and auxiliary exhaust ports, *Flight* describes that 'the cylinders are mounted exactly opposite each other so that that in order to avoid the use of offset or forked connecting rods, one piston of each pair has two rods attaching it to the crankshaft'. Adams' Redbridge and aeroplane engines used conventional offset spacing. Mention of the Dorman V-8 ceased after Spring 1913, so how many were built and whether it proved a successful design is not known, but no further aero-engines were produced by Dormans as their war effort concentrated on their 'JO' 4-cylinder War Office Subsidy Scheme lorry engine and the development of the Constantinescu 'Gun Synchronising Interrupter Gear' which synchronised the firing of machine guns through the sweep of the spinning propeller, greatly improving the accuracy and potency of British fighters.

Unlike the earlier mechanically operated Scarff-Dibovsky interrupter gear which tended to vibrate loose, causing mistiming and the shooting away of the propeller blade, the Constantinescu design used hydraulic action from an eccentric pump which was far more responsive and, thus, faster acting than a conventional mechanical cam and push rod design. Regarded as 'Top Secret', it was built at the closely guarded Sonic Works in Alperton, Middlesex. The hydraulic pulsing 'wave generator' system was covered under Gogu (George) Constantinescu's patent (No. 5152 of 1913); this principle remains in use even today. Gogu Constantinescu was a Romanian citizen who gained British citizenship in 1910. He was also granted several patents between 1908-1914 relating to carburettors and other automotive applications.

Bradshaw's own connections with Dorman came immediately after the war when he designed for them a chain driven twin cam overhead valve design which Dormans first exhibited at the 1919 Motor Show and was, he claimed, promptly copied by others, as was evident at the 1920 show. But what exactly he designed, and how deeply involved he was, remains a mystery, however recent research

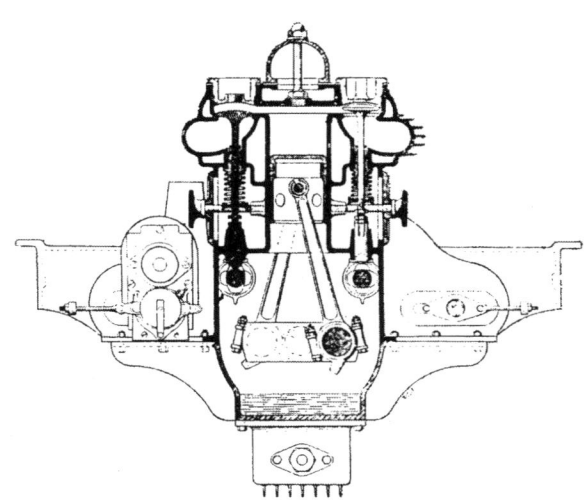

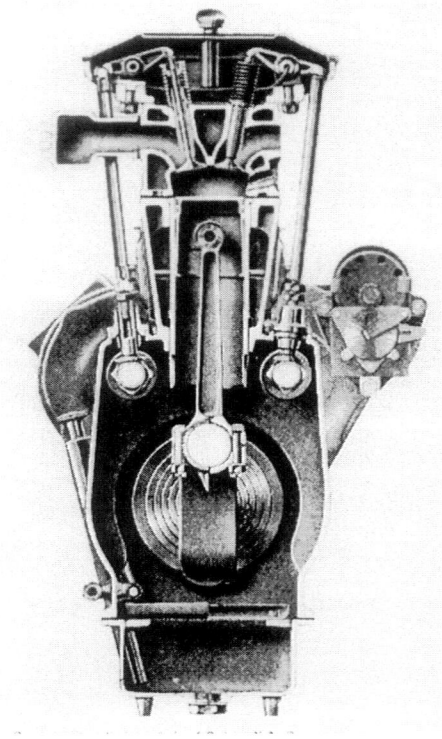

Left, Mendip twin-cam side valve engine of 1913.
Right, Dorman KNO twin-cam ohv engine of 1919

by others into Dorman suggests this may be related to their KNO engine but the lack of reference to Bradshaw in surviving Dorman Board Room minutes of that period suggests his involvement was possibly informal.

Twin-cam push-rod operated 'T-head' sv engines were not uncommon in early aero-engines, such as the Benz and Maybach; indeed Dorman's publicity material on the KNO was at pains to emphasise the influence of war-time aero-engine design and aluminium-alloy technology, on which Bradshaw was a known expert. It had also been found on early motorcycles and some cars, such as the Lanchester with its opposed valves, and Adams' pre-war twin-cam Mendip engine (see Appendix 5). Indeed there are some subtle Adams' design features in the Dorman KNO, adding to the mystery, so was this engine based on a fledgling Adams/Bradshaw/ABC 4-cyl twin-cam aero- or car engine which Adams had taken with him to Dormans in 1912? The truth is, we simply don't know!

ABC Brooklands

Leasing the only brick built workshop at Brooklands, Bradshaw established a well equipped machine shop in the centre of the new Flying Village, adjacent to the old Wintersells Farm and next door but one, to his (former Star) shed No.11. Behind this building was a generator room and in the back yard, a smithy's forge. By late 1911 or early 1912, Bradshaw had moved ABC's aero-engine development and team of six engineers from Redbridge to Brooklands, having changed the company's name to the All British (Engine) Company to better reflect their business activities. Although a small number of hand built aero and flat-twin motorcycle engines were built at Brooklands, aero-engine production was contracted to Armstrong-Whitworth.

Among ABC's pre-war staff were John Black (later Sir John of Standard Cars fame) and John Griffiths who, as a war veteran of the last war, lost both hands prompting him to design controls for disabled drivers at the British School of Motoring.

ABC Works, Brooklands, seen from Birdcage Walk. Note the Wind-Waggon in its original state (GB)

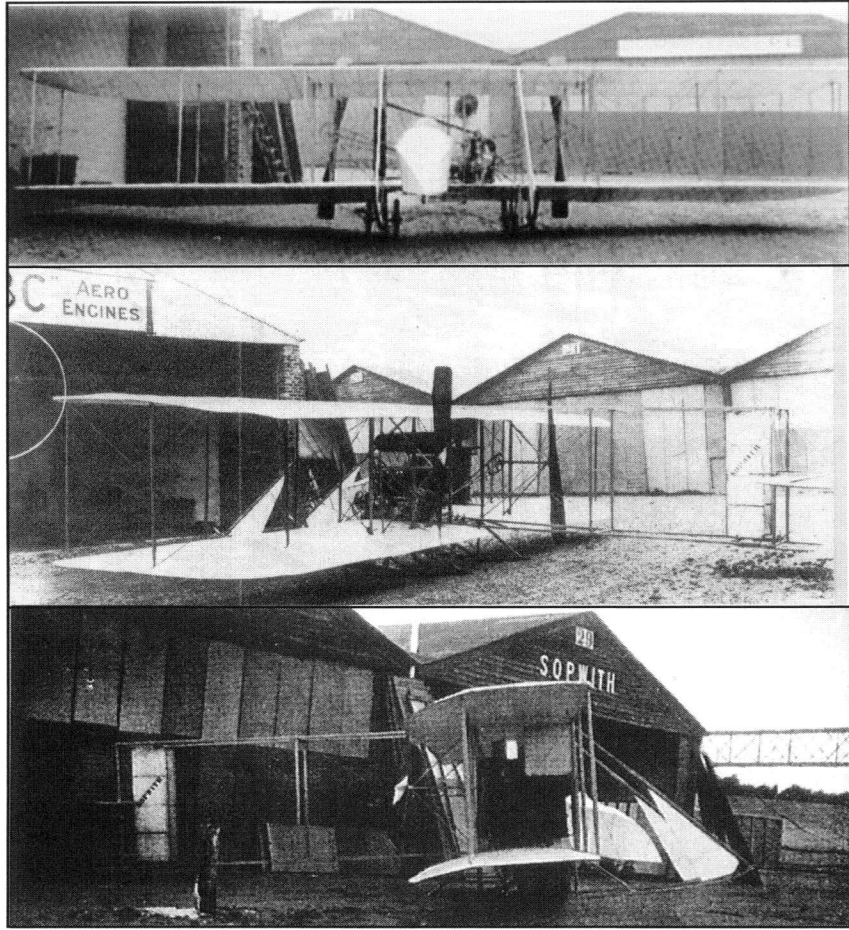

The streamlined ABC 40hp powered Sopwith-Burgess flown by Harry Hawker in the 1912 Michelin Trophy endurance race, seen outside the ABC Hanger in Swallow Walk (top and centre) and Shed 29 (bottom) with the footbridge visible in the background.

With his experience of the Star monoplane and the success of his in-line engines, Bradshaw had high hopes of developing a new observation aircraft for the Army, and in early 1912 sought permission from the (now) Royal Aircraft Factory at Farnborough to fit one of the modified Adams designed 100hp V8, engines in a Flanders B2 tractor biplane, replacing its standard 40hp in-line engine. However Bradshaw received a stiff response that the War Office saw "... no military value in the aeroplane whatsoever", as had Alliott Verdon-Roe and many other experienced aviators who had also offered their own inspirational designs for evaluation, or badgered the Government to form an air force. Determined efforts from other aeroplane makers and the RAeC forced the government to have a change of heart and announce that the first military aircraft trials of privately built, as opposed to 'approved' Army Aircraft Factory designs, was to be held at Larkhill, on Salisbury Plain. Contenders had to register by 21st June 1912, but the RAeC, Charteris (for ABC) and Howard T Wright got this postponed to 31st July, as they needed more time!

The August 1912 trials were intended to subject the aircraft to the most searching, rigorous tests and scrutiny so far devised, doubtless to prove the superiority of the Royal Aircraft Factory's designs.

Verdon-Roe had also begun work on a 'military biplane' in early 1912; this was renamed the Avro 500 Type E. Powered by a 60hp ENV engine, it first flew at Brooklands in March 1912 and then at Farnborough in June. It exhibited extraordinary flying agility and was returned to Brooklands in August 1912 to be fitted with the new ABC 60hp 4-cylinder in-line engine and then flight tested by Verdon-Roe's test pilot, Fred P Raynham on 31st August 1912. After several adjustments to both airframe and engine, it was successfully flown by ABC's Ronald Charteris on 18th October 1912 putting in an impressive performance; but having served its purpose as ABC's flying test-bed, the ENV engine was refitted. The 500 Type E was then developed into the new Avro 500 tandem military trainer using the Gnôme rotary. Many were ordered by the fledgling RFC's Central Flying School and the design was developed into the famous, and much loved, Avro 504 basic trainer which remained in RAF service at the CFS until 1930.

Bradshaw intended submitting two ABC 100hp V8 engines for the 1912 Military trials in two tractor biplanes; an Avro Type G, to be flown by Charteris, and a Brooklands built Howard Flanders B2 to be flown by Fred Raynham. Unfortunately, neither engine was ready in time and the Avro was hastily refitted with a spare Green engine. The Flanders was submitted to static tests, which it easily passed, and was later flown at 55mph with three adults aboard powered by a 40hp ABC 4-cyl, but too late to impress the authorities. Howard Flanders began his aviation career assisting Alliott Verdon-Roe before joining Vickers in 1915, designing both the Vickers FB 11 and the prototype Vimy before joining English Electric at Preston in 1925. He left the industry to design radio sets but returned to aviation in 1939, joining Bristol; sadly, he died later that year. Another pioneer aviator who joined the Vickers team was Alec Ogilvie who in 1908 famously used the flat Camber Sands, near Rye, from which to fly his Wright biplane. During the war, Ogilvie, (now Commander RN), joined the English Electric Phoenix works in land-locked Bradford on their flying boat programme and in 1918 became Controller, Technical Department (Design) at the Air Board.

In autumn 1912 an improved 40hp ABC in-line engine, with two massive twin-blade propellers, was fitted to Tom Sopwith's Burgess-Wright pusher biplane (some sources describe it as a Sopwith-Wright), replacing the Gnôme rotary. He then fitted a streamlined 'nacelle' ahead of the wings using, at first, a perambulator hood and later, a larger canvas nacelle built by Sopwith's engineer, Fred Sigrist. In that October, Harry Hawker entered for the British Empire Michelin Trophy No.1 Cup for endurance. Aviators were allowed several attempts at the Cup but on his first attempt in the ABC powered Burgess-Wright, Hawker landed early with a painfully raw finger and thumb, having

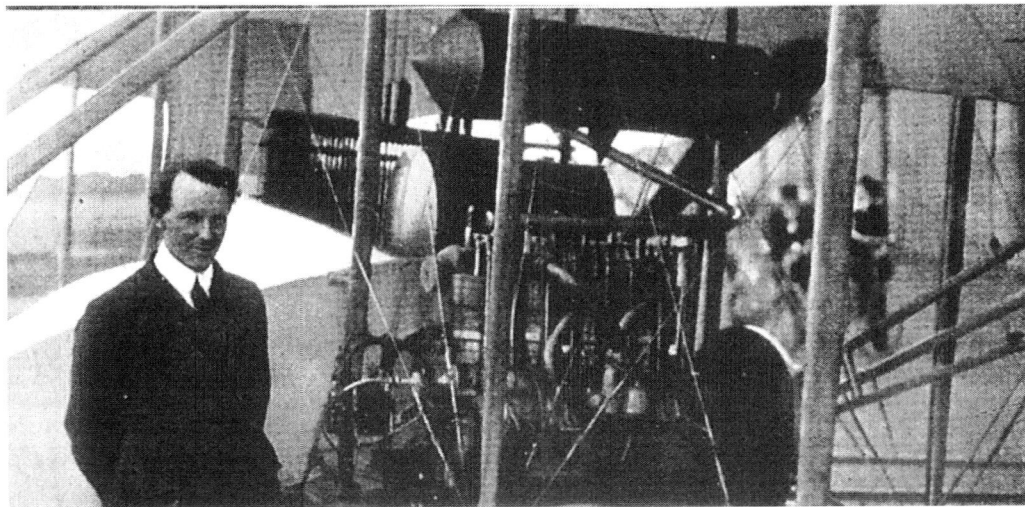

Harry Hawker with the record-breaking ABC powered Sopwith-Burgess biplane

held for 1½ hours between his gloved fingers, an exceedingly hot, broken exhaust valve spring - a not uncommon problem! Further attempts were thwarted by poor weather, but on 24th October he flew the Burgess-Wright in competition with Fred Raynham until his Green in-line engined Avro was forced down, after 7hrs 31mins, with engine trouble. Now flying alone at 400ft, quite content with his Thermos flask of hot cocoa, and sandwiches, flying at an incredibly slow speed to conserve fuel, he was soon joined by Tom Sopwith and Ronald Charteris, shouting encouragement from a Sopwith tractor biplane, as Hawker spluttered along into the record books with a flight duration of 8hrs 23mins, winning the coveted Michelin Trophy and cash prize of £500. Though the record breaking flight went without mishap and the ABC had proven to be a reliable, if underpowered engine, it was replaced by a more powerful 60hp British designed, but French built, ENV engine with which Sopwith had greater success.

ABC engine used in the Sopwith-Burgess

War

In May 1914, No.1 Squadron Royal Flying Corps moved from Farnborough to Brooklands to begin training on their new aeroplanes. On 14th August, at the outbreak of war, Hugh Locke-King offered exclusive use of the whole of Brooklands to the Government in the national interest and within 24 hours the RFC had taken charge, closing it to the public in September 1914 and transforming it into their Aircraft Acceptance Park and training school. The famous banked circuit was now used for trials of military motorcycles and lorries until May 1915 when, due to serious damage caused by the lorries, the Brooklands track authorities banned any further use of the track. The RFC were not popular either with the aviators when in 1917 The Blue Bird cafe, which had become the Officer's Mess, was destroyed by fire along with the adjacent aeroplane sheds.

Many experienced flyers, such as Ronald Charteris, (who had enlisted in the Royal Flying Corps (Special Reserve) on 17th July 1912) were immediately commissioned to train new pilots. Second Lieutenant Charteris was promoted Lieutenant RFC Military Wing (Special Reserve) on 30th August 1914, regraded Flying Officer on 12th October and became acting Captain in April 1915. Moore-Brabazon was commissioned as Lieutenant and made Officer in charge of the RFC's new Wing Photographic section, perfecting aerial photography. Granville Bradshaw volunteered for war service along with H M Cave-Brown-Cave (later Wing Commander), but his decaying archival notes report he was turned down and later suffered the disgrace of being offered a white feather one day when in Shaftsbury Avenue, London after attending a Ministry of Munitions meeting! Walter Lawson Adams was commissioned into the RNVR and served in the RNAS.

All useful aircraft and engineering facilities (including ABC's comprehensively equipped workshop) at Brooklands were immediately commandeered by the RFC and those companies not producing aircraft were given 24 hours notice to quit. Tom Sopwith moved to his main works in

Left, ABC 500cc auxilliary engine with kick start. Below, hand start balloon blower set. Bottom left, searchlight dynamo set. (GB)

Canbury Park Road, Kingston-upon-Thames where he employed seven men. Avro transferred to their main factory in Brownsfield Mills and Clifton Street, Manchester to meet military orders. In 1916 Avro built a new factory at Hamble, on Southampton Water, for naval aircraft production, including the aborted Avro Pike and the ABC Dragonfly powered Avro 533 Manchester bombers. ABC moved to nearby Hersham, Walton-on-Thames, two miles to the north of Brooklands.

ABC Hersham

Bradshaw was then living at Darby House, Sunbury-on-Thames. Built for Vice-Admiral George Darby, who had broken the French siege of Gibraltar in 1781, the magnificent early 18th century house had been badly damaged by fire in 1907; all that remained was the rebuilt, smaller, central section.

In his ambitious prewar plans, Bradshaw had already identified a vacant 1¾ acre (also described as a 2 acre) site at Hersham Road (the section below The Barley Mow was renamed Moseley Road after WW2), in Hersham close to Hersham Lodge. There already existed on the site a brick built smithy run by local engineer, William Faulkner (part of which survives). On being expelled from Brooklands, work began at Hersham on a relatively small, brick built single storey factory into which they installed what they were able to salvage from the Brooklands works, quickly re-establishing a well equipped machine shop. The old 'Wind-waggon', having served its purpose, was now parked up behind their

ABC 250cc portable wireless set (GB)

new factory where she fell into decay. Their telephone number was Walton-on-Thames 220 and their telegraphic address, 'Revs'. In later years they added a two storey test house at the rear.

Early in the war, ABC was contracted to produce auxiliary engines for various military applications. The first was a modified 492cc (70mm x 64mm) flat twin motorcycle unit which weighed, fully fitted, 59lbs and developed 5hp in continuous use, peaking at between 8 and 9hp. It was ideally suited to high load generators for signalling stations, portable and mobile searchlights produced by Messers A Lyon & Wrench Ltd and a proposed 'portable' inertia aero-engine self-starter (similar to the Ford Model T mounted 'Hucks' starter developed in 1916 by the pioneer aviator and Blackburn Flying school instructor, Bentfield C Hucks), or coupled to the crankshaft (as on the later Austin V-12) enabling sea-plane pilots, unable to 'swing the prop', to self-start their engines.

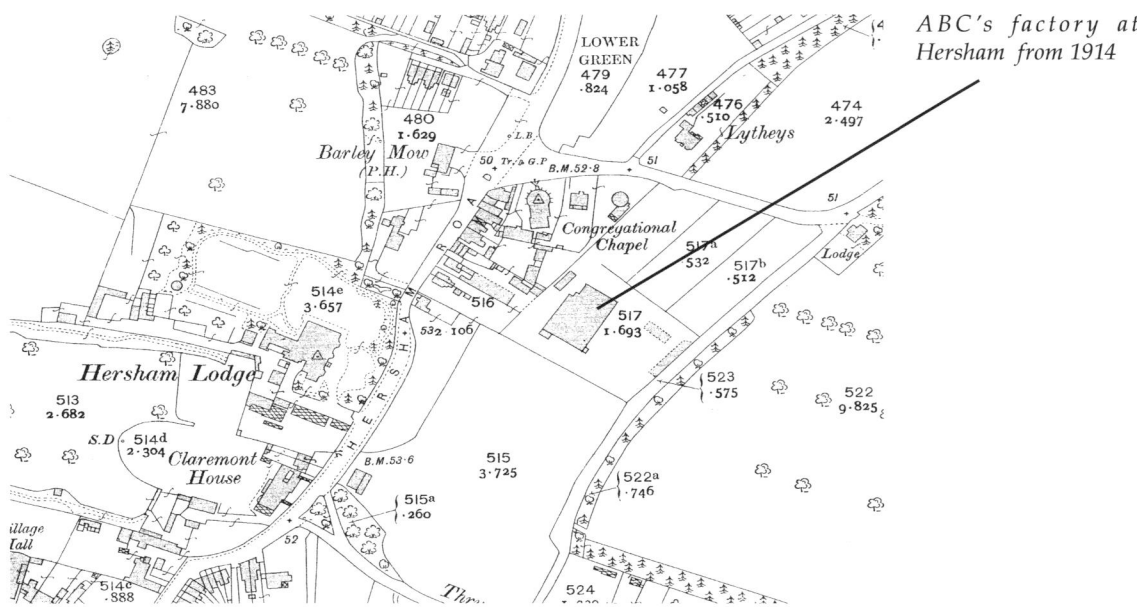

ABC's factory at Hersham from 1914

Brooklands Flying Village circa 1911/12. The large white roofed building, above centre, is the ABC brick built workshop. (Courtesy of Brooklands Museum and Tony Hutchins)

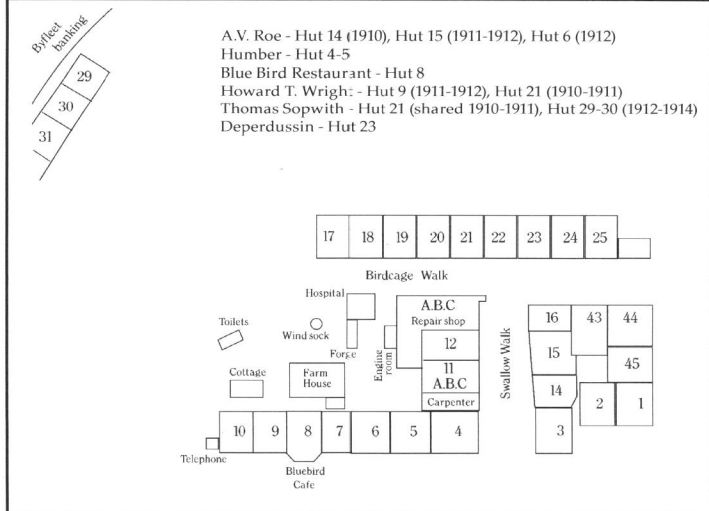

In 1915 Bradshaw was asked to develop a lightweight engine for wireless operations in dirigible observation balloons and later, RFC bombers. Marconi's airborne spark-transmitter wireless telegraphy by Morse code from dirigible airships was first used in the 1912 military trials followed by trials in 1914 by two BE2a observation aeroplanes of No.4 Squadron RFC for guiding artillery gunners. These however required 300watt generators driven off the engine's crankshaft, but as their radio sets weighed 72lbs, it meant they could only be used on 2-seater observation aeroplanes - at the expense of the observer himself! This was resolved in late 1915 by a lightweight Sterling transmitter running off a series of 2.5 volt accumulator batteries; an improved and lighter 20lbs model became standard airborne equipment from 1916. The RFC commandeered Marconi's base at Brooklands for their new RFC School of Wireless in November 1915, this was followed by the School of Wireless Operators at Farnborough in 1916 - the standard to be achieved by wireless-operators was 20, five letter words per minute.

ABC's new 249cc (60mm x 44mm) engine developed between 2 and 3hp in continuous operations between 1,200 and 3,600rpm and, in short bursts, a peak of 4hp at 5,000rpm. This was necessary to achieve the output frequency from the standard Marconi alternator for radio transmission. Out of necessity, these portable wireless generator sets were fitted with a tachometer and fan cooling. Tested on full throttle overnight, these engines developed $3^{1}/_{2}$hp continuously for 150hrs, without fan cooling, before essential maintenance. One engine was tested, briefly, to a claimed 12,000rpm! The bare engine weighed 14lbs, but fully kitted out with magneto, carburettor and cast iron frame, weighed 26lbs.

The 492cc, 5hp engine was also used for inflating dirigibles, blowing air from a 10" output pipe at 3,000 cubic feet per minute, and for the Lyon & Wrench searchlight mounted in the nose of the experimental Supermarine PB.31E 'Night-Hawk' night-fighter, designed in March 1916 - the first such recorded use of an airborne auxiliary engine. The term 'night fighter' was used very loosely for this Anzani radial engined quadraplane of 1916 was crewed by three and was designed to shoot down Zeppelin airships using a $1^1/_2$ lbs Davies recoilless gun and Vickers machine gun. Two prototype Night-Hawks were ordered (s/n 1388, 1389) but only the first was fully tested at RNAS Eastchurch before the project was, wisely, abandoned as the aircraft proved hopelessly slow, even in pursuit of enemy air-ships.

Both ABC engines used machined steel cylinder barrels and ball-race bearings throughout; those on the main shaft were set into steel housings, bolted to the soft aluminium crankcase. A gear driven oil pump was fitted - the cylinder walls and ball races relied on splash lubrication. The two ring pistons were of cast iron (later aluminium) and, as the steel cylinder expanded at a higher rate than the cast iron piston, it was virtually impossible, in theory, to seize the engine under extreme loads. The cast iron cylinder head used an offset manifold arrangement with spark-plug, direct action side inlet valve and, above it, an overhead rocker operated exhaust valve, both driven by push-rods off a one piece, machined camshaft. Uniquely the hand, or kick start, operated on the camshaft giving a 2:1 rotation of the engine for easy starting. The magneto, driven by gears, sat above the engine. A single Claudel-Hobson carburettor was used.

These lightweight units were variously fitted to cast aluminium frames or tubular steel cradles, and were ideally suited to field use for driving generators in the army's mobile workshops and for water-pumps in trench warfare. Granville related in *The Motor Cycle* of 31st March 1938 how ABC almost lost the contract for water-pumps when demand for replacement aluminium pistons suddenly rocketed. Reports were coming in of their ABC trench pump motors failing in regular use; curiously no other ABC engine applications were affected. Inspection of damaged pistons showed suspicious signs of crown damage. Just as the War Office was about to cancel the contract and award it to Douglas, the truth was discovered from a convalescing mechanic who had served on the front and regularly rode an ABC motorcycle. Instructed to keep the trench pumps running at night, their sound merely attracted the Germans' opportunist grenade or machine gun fire in the pitch dark of night. It was then discovered by a canny mechanic that if a certain washer, not normally visible, was removed and dropped through the spark plug hole, the engine would fire up and run normally until the washer had 'holed' the piston, destroying it. By the time this had been discovered, the culprits were well clear and, for once, enjoying a peaceful if uncomfortable, water logged night.

Auxiliary engine production ended prematurely in January 1918 when ABC's Hersham works was turned over completely to ABC aero-engine development. While component parts and spares were still produced, assembly and production of motors for Lyon & Wrench was passed to Fiat Motors in London.

3
Aero-engines

Royal Flying Corps

In 1910, the Army established their Army Aircraft Factory at Farnborough with the sole purpose of designing and building observation aeroplanes to their specification. Their chief designer/test pilot was a young Geoffrey deHavilland who was assisted by a draughtsman (later Assistant Chief Designer), Harry P Folland. Both the Government and its army generals maintained that a static balloon was more than adequate for aerial observation of artillery fire and remained apathetic towards the benefits of an air force despite the growth of such in continental Europe - indeed at outbreak of war, the German air force outnumbered the entire allied forces! When German flyers started attacking the balloons, the generals were proven entirely wrong, but fortunately the first steps towards a rudimentary air force had been taken on 1st April 1911 when the Air Battalion of the Royal Engineers was formed with No.1 Company (Balloons) based at Farnborough and No.2 Company (Aeroplanes) at Larkhill, on Salisbury Plain. In May that year the full potential of aeroplanes in warfare was first displayed at Hendon in front of some 300 Members of Parliament and government ministers.

Until the great war, prospective pilots learnt to fly by their own skill at private schools, tutored by pioneer aviators, or at the newly formed Royal Aero-Club (affiliated to the Royal Automobile Club) at Muswell Manor, Leysdown on the Isle of Sheppey. Co-founded in 1909 by Charles S Rolls, the RAeC moved to nearby Eastchurch the following year and in 1911, three naval officers in the Royal Marine Light Infantry were granted official permission to learn to fly at Eastchurch. Synonymous with aviation's formative years, Eastchurch later became an RAF armaments school and after closing in 1947, was converted to an open prison in June 1950. Several of the original buildings and hangers remain and attempts are now being made to preserve it as a historic airfield.

The Royal Engineers' Air Battalion were transformed on 13th April 1912 into the Royal Flying Corps, adopting as its motto, *Per Ardua ad Astra*, 'Through perils to the Stars'. Those naval officers at Eastchurch were then reformed as the RFC (Naval Wing) and following the formation of an Admiralty Air Department later that year, the naval wing concentrated on coastal defence from bases at the Isle of Grain, Calshot (opened March 1913), Felixstowe (below Felixstowe Docks, August 1913), Great Yarmouth (June 1914) and Dundee (Stannergate, 1914), from where they began experiments on seaplanes, independent of the army. Even the secret Naval Ordnance experimental station at Orfordness, north of Felixstowe (opened in 1912) soon gained an airstrip.

In May 1914, No.1 Squadron Royal Flying Corps moved from Farnborough to Brooklands to begin training on their new aeroplanes. On 14th August 1914, the entire Brooklands site was handed to the Government in the national interest becoming the RFC's Aircraft Acceptance Park and training school. Only those companies directly involved in RFC contracts were allowed to keep their hangars for flight testing. In early 1915, Vickers bought the vacated Itala Motor Works, adjacent to their flying school, and began production of the BE.2 fighter; other Vickers models were produced at Erith, Dartford and, from 1914, Crayford. By the end of 1915, Brooklands had become Britain's principal aircraft manufacturing facility. The RFC Testing Flight, established at Upavon in July 1916, was relocated in January 1917 to Martlesham Heath, near Ipswich as the Aeroplane Testing Squadron, which became the Aeroplane Experimental Unit in October, 1917. Brooklands eventually returned to civilian use, but finally closed as a race course in 1939. It continued as the Vickers' bomber and VC-10 civilian aircraft production facility until spring 1970, when it became British Aerospace's headquarters and component manufacturing base. While the motor-racing club house and buildings survive as the Brooklands Museum, all but the control tower in the Flying Village have been demolished.

The RFC (Naval Wing) was reformed as the Royal Naval Air Service on 1st July 1914 and comprised the Air Department (Admiralty), Central Air Office, Royal Naval Flying School and Royal Naval Air Stations. On 1st April 1918, the RFC and RNAS merged to form the Royal Air Force and on 1st April 1923, an aircraft carrier based wing of the RAF was formed as the Fleet Air Arm. On 24th May 1939, the Fleet Air Arm came under control of the Royal Navy.

Military aero-engines

Britain did produce some successful early aero-engines: the Green in-line, built for them by Aster; the British designed, but French built, ENV 40hp and 60hp V8 engines offered between 1909 and 1910 respectively; the JAP V-engines, and those from Wolseley. Most in-lines were based on motor-car engines; indeed the first Rolls-Royce aero-engine was based on a Daimler-Benz 'Mercedes' car engine. There was also a promising British built Isaacson 50hp water cooled 7 cylinder (and later 100hp 14 cylinder) radial using a double banked cylinder arrangement, designed by Rupert John Isaacson, which is discussed later.

However, unlike France and Germany, the lack of military interest kept British aviation virtually a cottage industry, nudged along by the Royal Aeronautical Society, itself formed in 1866. Aeroplane makers thus relied heavily on imported in-line airship engines such as the Austrian Austro-Daimler, while the French dominated the British market with the Hispano-Suiza V-engines or the Gnôme 'Monosoupape', le Rhône or Clerget rotaries, leaving France's Anzani 3-cylinder radial to dominate the radial market. Most of these engines were also built under licence in Britain.

The close proximity of both rotary and radial engines to the pilot, in a single engined aeroplane, concentrated their combined mass at the front, making them remarkably agile and aerobatic in flight, in comparison to the massive, laboriously slow in-line aero-engined aeroplanes whose length and mass produced adverse momentum in anything other than straight flight. Unfortunately, the Admiralty and the Army Aircraft Factory tended to work against each other, which did little to inspire the fledgling aircraft industries, as demonstrated by the fact that by the end of 1914, Britain had built only 193 aeroplanes and about 130 engines, compared to 30,671 aeroplanes and about 20,200 engines in 1918 - in both eras the balance was about one third rotary engines of which about a third came from France! Yet it was through W O Bentley's offer of his aluminium alloy piston experience to the Admiralty Air Department, that he was commissioned Lieutenant (RNAS) and installed at Gwynnes to supervise production of the licence built Clerget rotary engine, from which evolved the 150hp Admiralty Rotary No.1 engine (later known as the BR1, Bentley Rotary built at Humbers and, later, Daimler and Crossley).

The Admiralty's Air Department (AD) mainly designed 'spotter' (observation) aeroplanes for the navy. Many of their early float-plane designs were based on converted land planes while their AD specifications, such as the 'AD Navyplane' and the 'Porte Baby', were laid down by Lt John Porte RN, Chief Designer at the Royal Naval Air Station, Felixstowe. John Porte had gained his Flying Certificate with the Aero Club de France in July 1911 on a Deperdussin and went on to instruct at the Deperdussin flying school at Hendon before joining Glen Curtiss in America, to test fly his new float-planes. In early 1914, Porte designed a twin engined Curtiss flying boat to cross the Atlantic, but he returned to Britain on outbreak of war.

Most AD aeroplanes took their name from the air station at which they were developed, such as the Grain Griffin based on a Sopwith B1 and produced at the RNAS seaplane station on the Isle of Grain, or the Felixstowe F5 at Felixstowe. These were usually built in conjunction with boat builders such as Short Brothers of Rochester (who held the Wright brother's licence for the UK) or S E Saunders Ltd on the Isle of Wight.

In contrast, the Army and RFC looked to the Army Aircraft Factory for their designs, such as the FE (Farnborough Experimental), BE (Blériot, later British, Experimental), RE (Reconnaissance Experimental), SE (Scout Experimental) or NE (Night Experimental) to meet their quite restricted official needs. At outbreak of war, many aeroplanes and aero-engines were contract built by motorcar makers, boat builders, coach builders and railway carriage makers - these included Messers Dick, Kerr & Co at Preston where Bradshaw had served his apprenticeship, and Phoenix Dynamo at Bradford.

Despite the War Office's earlier response to Bradshaw's proposal to fit an ABC V8 to a Howard Flanders, with the dismissive "no military value in the aeroplane whatsoever", the Government had a change of heart and the first military aeroplane trials of privately built, as opposed to 'approved' Army Aircraft Factory's designs, was held at Larkhill on Salisbury Plain in August 1912. These trials were intended to subject the aircraft to the most searching, rigorous testing and scrutiny so far devised, doubtless to prove the superiority of the RAF's 'approved' designs. Some 31 aeroplanes were submitted; many were experimental airframes. Avro submitted two of their improved enclosed cabin Type G biplanes; one with a 60hp Green to be flown by Lieut. Wilfred Parkes RN, and the other to be fitted with an ABC V8 engine to be flown by Ronald Charteris. The Flanders B2 tractor biplane with the 100hp V8 engine was also entered, but as neither ABC V8 engine was completed in time, the Flanders was evaluated only as a static airframe while Charteris' Avro was hastily reequipped with a spare Green engine. These tests included their construction and compliance with army specification and the rapidity of assembly from packing case to airworthiness (the Flanders took six men, 40 minutes). These were followed by a three hour flying test in which only 10 aeroplanes took part. The Green engine gave an appalling rate of climb and, having completed the flight test, Lt Parke turned to make his approach for landing, but began to spin, recovering only just in time to make a safe landing.

Somewhat disillusioned with the Army's earlier response and the disappointing military trials, ABC continued to offer the 40hp in-line, water cooled aero-engine (now built for them by Armstrong-Whitworth); it still proved a quite popular engine and aeroplanes so fitted, were said to have won 20 contests in or about 1912, including the Shell races at the new Hendon aerodrome, North London which had opened in 1910. Bradshaw also began work on a new rotary engine (believed to have been a 5 cylinder, 100hp job), but that project was abandoned as aero-engine sales slowed through increasing competition. By 1913, ABC had all but ceased as an aero-engine maker as Bradshaw became more closely involved with the many racing drivers and motorcyclists at Brooklands, often being called upon to make improvements to machines at his track-side works for riders such as W O Bentley and S L Bailey, discussed later. But out of this work came a flat twin motorcycle engine which formed the basis of his auxiliary engines (described in the previous chapter) and the ABC Gnat aero-engine.

The war effort

Following the Munitions of War Act of autumn 1915, the British Government finally established in May 1916 several military Boards to bring together common interests. In essence, the Air Board was a 'Ministry of Supply' which combined the respective Admiralty and RFC departments and coordinated all aircraft design and production. Henceforth, aircraft makers were forbidden from building experimental aircraft or engines of their own design without authority. Instead, they were invited to tender to approved Air Board specifications. In 1917, after much pressure from Winston Churchill, the new Air Ministry was formed. Churchill had seen service in India and the Boer War, and had been appointed First Sea Lord, but after the Dardanelles disaster (for which he was not directly responsible) he was reassigned as Minister of Munitions.

The Air Board also comprised eminent men in the fields of aeronautics and engineering. Among these were Adams and Bradshaw. Bradshaw was elected to the committee for ball and roller bearings, which researched their limitations, devising new types of bearing and evaluating lubricating oils and greases. It was this research that led him to develop his oil-cooling principles, for he had observed that several aero-engines ran perfectly for many months in normal usage, but failed within hours under severe bench testing through over-heated bearings, often due to the inadequacy of castor oil as a viscous fluid. His solution was to drill oil journals to the bearings, flooding them with mineral oil recirculated through a cooled sump or reservoir. Bradshaw's later research into the Dragonfly radial engines, then at the limits of technology, led to improved steel-alloys which became the standard in Air Ministry specifications of the 1930s for valves and other critical components.

ABC Gnat powered radio controlled ATs - the Sopwith, above and Royal Aircraft Factory, right.

Bradshaw produced his first deep race bearing in 1916, which did away with the need for thrust bearings at the end of shafts and in 1917, submitted several patented improvements to deep race ball bearings, engine big end bearings, thrust bearings and their housings for rotary and radial engines; these included Pat No. 104483 (laid 27th November 1916) for one piece big-end con-rods which could be slipped over the crankshaft and into which roller bearings could be inserted and retained by spacer rings. This was followed by variations under Pat No 105216 (laid 25th January 1917), Pat No. 129248 (laid 7th March 1917) and in Pat No: 137874 (laid 24th December 1924) 'spacers for roller bearings'. In addition he patented improved thrust bearings (Pat No. 128232 laid 7th March 1917) and the selective heat treatment of crankshaft bearing surfaces (Pat No. 132472 laid 7th March 1917) to allow the crankshaft to remain in a neutral, unstressed state. Most of these applications were to be found in his post-war ABC twin cylinder car and aero-engines, though his Patent No. 142147, laid on 13th April 1918 for lubricated knuckle joints for retaining con-rods, had its application more in rotary engines than radials. A further patent, (Pat No. 133093 laid in February 1918) was for a captive bolt or setscrew, with a milled bolt head, which cut its own seat in a counter bored socket in soft aluminium castings.

Flying Bombs

In late 1914, the German Siemens Shuckert-Werke began development of small, wire guided gliders, fitted with a warhead, and launched from airships. Guided by electric signals, they had a range of up to 5 miles but although their development progressed well, they were not used operationally and the project stopped at war's end, but Siemens had some success with remotely controlled exploding boats. Whether the British were aware of these developments is not known, but if Bradshaw's mind was fertile and inventive, it was nothing compared to his contemporary, Dr Archibald Montgomery Low, a young graduate, who had expanded his wisdom from 2-stroke engines and transmissions, to stoppering of bottles, designing electronic signalling devices, armaments and mine detection. Early in the war, Low was working in Chiswick on electronic range finders for the artillery, based on the principle of what we now know as 'radar'.

Historically, a stop-watch, calibrated for the known velocity of the shell, recorded the time between firing and its landing, thereby indicating the distance (range). Using this effect, the Royal Flying Corps saw great potential in Low's work for developing a radio controlled aircraft which, when packed with explosives, could be crashed into its intended target at a known distance either remotely, by timing device, radio signal or by limited flight endurance (as employed in the V1 'doodlebugs' of World War 2). As a bonus, the aircraft could also be remotely controlled as a target drone for gunnery practice or as aerial defence against the Zeppelins.

Dr Low was duly commissioned into the RFC on 2nd June 1915, and as 2nd Lieut. Low, was made Officer Commanding the RFC's new Experimental Department at Brooklands from where he developed this remotely controlled flying bomb. The Government's Air Board allocated six serial numbers (A8957-A8962) for experimental radio controlled 'AT' airframes, variously described as 'Aerial Target' or 'Aerial Torpedoes'. These serials appear to have been allocated to a small, ski mounted ABC Gnat engined monoplane built in secret at the Royal Aircraft Factory at Farnborough which, it appears, was designed by Harry Folland who later designed the famous SE.5a. Running along a steel guide rail, A8957 lifted off and then promptly crashed on its first flight on 6th July 1915. The second and third prototypes flew on 25th and 28th July but these too crashed and the project was promptly abandoned due to the unreliability of Low's radio control technology. One airframe was rebuilt, but not flown again, and it appears this was finally broken up in 1934.

News of the RFC's requirements prompted Sopwith and others to develop the concept both privately and in conjunction with the RFC. Unfortunately, Geoffrey Bradshaw had only a vague knowledge of his father's involvement, but it is believed that Granville's abandoned patent application (Pat No: 1114 of 23/01/15) "Bombs etc" - may be related to Sopwith's AT, for having earlier used an ABC in-line engine in his Burgess-Wright biplane, Tom Sopwith persuaded Bradshaw to develop a disposable engine of limited flight duration for an 'aerial torpedo'. His first attempt was with a 250cc version of the ABC motorcycle engine to which a 4-bladed propeller was fitted and bench tested. The diminutive engine gallantly fired up and, though designed for running at a maximum 4,000rpm, was tested up to 10,000rpm for some 3 hours without mishap, until suddenly the propeller detached and flew through the roof!

By 1916 Bradshaw had come up with an improved engine design, the 35hp Gnat horizontally opposed twin, which Sopwith duly fitted to an inexpensive and very basic airframe, the Sopwith AT of 1917 which, it is believed, was based on a light biplane, the Sopwith SL.T.BP, built as personal transport for Harry Hawker. It is possible the Sopwith AT was allocated one of the unused AT serials from A8960-62. The AT was assembled in secret in collaboration with the Royal Aircraft Factory - one source says Feltham, but it is more likely to have been Farnborough. It was fitted with servo controls for the electronics mounted in the tail and, to ensure a straight-line take off, a four wheel perambulator undercarriage was fitted. Various accounts have been recorded: In one, the AT was taken to Laffans Plain for flight testing in front of an audience of the RFC's highest command; Bradshaw related that three aircraft were tested at at RNAS Netheravon (Prof A M Low later recalled it was at nearby RFC Upavon). One published periodical account relates that the pilotless aircraft was duly started but as it gathered pace, it flatly refused to obey any signals transmitted to it, and careered around the grassy airfield causing panic among the assembled officers before running out of fuel! Bradshaw relates that one nose dived into the ground due to very lethargic control response; another rose into the air 300ft in such a steep climb that it was starved of fuel and nose dived down towards the assembled observers!

Another account relates that the radio-control system remained obstinately unreliable and the Sopwith AT never did take to the air. After damage sustained in further trials, the project was abandoned but four further airframes (s/n A8970-8973 - from an unused allocation for SPAD S.7s) were built in 1917 and modified as the Sopwith 'Sparrow' for manned flight; these were scheduled for fitting with the Gnôme radial, but at least one was fitted with the ABC Gnat. (In Alex Lumsden's *British Piston Aero-engines,* he refers to a Gnat being fitted to an unidentified Sopwith Bee/Tadpole/Sparrow)

Geoffrey deHavilland, by now at the Aircraft Manufacturing Company at Hendon (which held the British Farman rights), was also working on a light, remotely operated monoplane aerial target of his own design, also fitted with the Gnat. It is thought 'his' aerial torpedo was successfully tested at RFC Upavon on 21st March, 1917 but very few details are to hand. Perhaps we will never know which 'aerial torpedoes' did actually fly!

The aerial torpedo project was finally abandoned in Britain in 1917 due to the impossibly unreliable radio control systems, but in America, development of the Hewitt-Sperry Automatic 'airplane' in 1916, led to an US Army project for 'aerial torpedoes' in which Orville Wright and his Dayton-Wright Company became involved in 1918. The 'Bug', as it was known, was a pilotless biplane with pronounced dihedral wings for directional stability on a tubular steel airframe covered in 'muslin and doped brown paper', powered by a V4, 40hp engine. It used a detachable dolly undercarriage with the intention that a mechanical timer triggered the wings to fold at a predetermined time, causing it to dive at the target! It failed to take off at the first attempt, but a month later it rose into the air and then flew off, out of control "like a thing possessed of the devil", crashing some 21 miles away! A further attempt was made a year later, but the aircraft broke up in the air as it dived uncontrollably. Lawrence Sperry, who designed the Hewitt-Sperry 'airplane' had better success through the Curtiss company with the first controlled flight of a pilotless aeroplane in March 1918 - but the Armistice ended further development.

The first successful radio controlled flight of a pilotless 'target drone' in Britain came on 3rd September 1924, flying from an S Class destroyer; the engine ran out of fuel after 12 minutes controlled flight. This was then developed into the more successful Armstrong-Siddeley Lynx radial powered pilotless aircraft of 1927, again test flown from HMS *Stronghold*, but following flight trials in Iraq, and the loss of four drones, the project was finally abandoned. However in 1949, Bradshaw raised the issue in *Flight* over the future of aerial bombing and reconnaissance by fast jets in the nuclear age, arguing that these offered no advantage over slower ones flown as defenceless, radio controlled drones carrying nuclear weapons. Who, he argued, would wish to destroy an over-flying nuclear bomb above their territory? As he predicted, and as we have seen in the recent Gulf, Afghan and Libyan conflicts, the future truly does lie in radio-controlled, weapon-carrying, reconnaissance drones.

After the war Dr Low held a Chair in the Royal Aeronautical Society for his work at Farnborough and made much play of his 'Professorship'. In the 1930s, he wrote several children's books and established an informative magazine, *Armchair Scientist*, but unfortunately his exuberance and enthusiasm made him unpopular with the scientific fraternity. He was however called in to do work for the Government and was commissioned during the last war as Major in the Pioneer Corps. He became an expert in mine warfare and had much better luck with experimental radio controlled torpedoes for the Navy. After the war he became involved in audio-engineering and, like Bradshaw, television. Low also supported Daphne Oram in her development of the BBC Radiophonics Workshop which many years later produced the haunting 'Dr. Who' theme! A great believer in rocket propulsion, he prophesied manned-rocket missions to the moon and became President of the Inter-Planetary Society. Low was granted 86 patents between 1908 and 1956, his last being 'Improvements to bed pans'!

Gnat horizontally opposed twin

Generally speaking, the operational name of a military aeroplane or aero-engine was determined under an Air Board Technical Department classification system. Aeroplane names were selected from the animal kingdom, while aero-engines were initially named independently by their makers. Many of these predate the 1918 Air Ministry order: Rolls-Royce, for example, had selected birds of prey (Eagle, Hawk, Falcon and most famously, the later Merlin) while Cosmos Engineering chose planets (Jupiter, Mercury). ABC chose flying insects (Gnat, Mosquito, Wasp and Dragonfly), yet their post war engines were named 'Scorpion' and 'Hornet'.

ABC Gnat horizontal twin, front above, rear below

Bradshaw's Gnat was the most basic of designs whose origins date back to the motorcycle engine of 1913. Although it carried the 'ABC' legend, as ABC were fully engaged in auxiliary engine contracts for the army, Bradshaw established a new engineering research workshop, Walton

Above, Sopwith Sparrow. Below, PV8 Kitten

Motors Ltd, where he developed his aero-engines. It appears only about 18 Gnat Mk.Is were produced and these were reputedly built by Selsdon Engineering Ltd of Croydon who, it appears, had also helped produce the pre-war ABC motorcycle engines. Their proprietor, was Gilbert Campling and Works Manager/Engineer, R Brown MIME.

A flat, horizontally opposed ohv twin, 1,600cc ($4^{3}/_{4}$" x $5^{1}/_{2}$") (120mm x 140mm), the lightweight Gnat was originally designed for 30/35hp at 1,800/2,000rpm for the aerial torpedo/target drone for which it employed a reduction gearbox, but the Gnat proved so underpowered when fitted to an aeroplane that military aero-engines were direct drive.

Its first application in manned flight came in early 1917 with two experimental Scout/Zeppelin interceptors for the RNAS: the PV7 and PV8. These were intended to be launched from short runway superstructures atop battleship gun turrets. 'PV' stood for the 'Port Victoria' RNAS Experimental Construction Depot at their seaplane station on the Isle of Grain, Kent. PV7 was appropriately known as the 'Grain Kitten' (s/n N539) while PV8 was the 'Eastchurch Kitten' (s/n N540), the latter named after the Isle of Sheppey RNAS Flying School, established in 1911, at the former Royal Aero Club, Eastchurch.

Port Victoria was established in 1912 as the South East & Chatham Railway's coastal terminus, wharf and jetty on a branch line from Charing Cross for their continental boat services. It also became a RNAS land and seaplane station in 1912, replacing Eastchurch as the RNAS's principal station in 1913. Grain then became a RNAS repair station and in 1915 an Experimental Armament Section. Then, in 1916, Grain became the Marine Experimental Aircraft Depot. After the war, the Orfordness flying boat testing facilities were transferred to RNAS Grain at which point Orfordness concentrated on gunnery. RNAS Grain gradually fell into disuse and decay, finally closing on 17th March 1924. It was demolished in the 1930s to allow expansion of the Isle of Grain oil refinery.

The lightweight 'Kittens' were designed by W H Sawyers as compact biplanes with an 18 foot wingspan and a short undercarriage; the upper wing and Lewis machine gun were at shoulder height to the ground crew. PV7 weighed 272lbs unladen and first flew in June 1917. The test pilot's log recorded: "Preliminary test of ABC 30hp machine. Very favourably impressed (smallest flying machine in world)". However the Gnat proved a major headache as its airframe's designer recorded: "... the only engine that ever survived more that an hour's flying was the hand made prototype originally fitted to it; production engines gave up regularly on their first flight". It appears the engine was very prone to fouled spark plugs which effectively limited flight tests within the airfield's boundary. Whether this fault was due to the engine or the spark plug's heat range - still a common problem today - is not known. Due to its necessary light weight (410lbs all up), tail heaviness and poor service ceiling of 11,900ft (which took over 11mins to reach) and top speed of 74mph at that altitude it meant the Grain Kitten was virtually incapable of attacking the more modern and faster Gotha bombers which had largely superseded the smaller Zeppelins. The PV7 was promptly abandoned in late October 1917.

Development of the slightly larger (and heavier at 586lbs) Eastchurch Kitten PV8 was transferred in March 1917 to Grain; she first flew in September 1917 and was sent to Martlesham Heath for evaluation, but while PV8 suffered poor flight handling and stability, this was much improved through modifications. It had a superior 84mph at 10,000ft and service ceiling of 15,000ft (after 45mins climbing), but though it was promised the more powerful 45hp ABC Gnat II engine, the PV8 was rejected by Martlesham Heath and returned to Grain in November 1917 and the project went no further for the same reasons at the PV.7. However on 18th March 1918, instructions were issued for the Eastchurch Kitten PV8 (N540) to be shipped to America for further evaluation. Whether she was shipped abroad is not known. A replica of N540 exists at the Yorkshire Air Museum.

Martin K.III Kitten

When America entered the war on 6th April, 1917 they had no experience of aerial warfare, so adopted 'proven' British and French aeroplanes, primarily the licence built DH.9 and SPAD S.7 for operations in France. The PV7/PV8 concept of a high altitude Zeppelin interceptor was however taken up by the US Army and a design was submitted by James V Martin in late 1917 as the Martin K.III Scout Zeppelin interceptor. This was an ungainly and slightly larger biplane than the PV7/8 and weighed 582lbs. It had a quite different wing design and ultimately adopted Martin's own design of semi-retractable undercarriage plus oxygen supply and heated flying suit for the pilot. The K.III was powered by an ABC Gnat - whether through the influence of British government is not known, but it soon became known as the Martin 'Kitten'. It seems unlikely the PV.8 Eastchurch Kitten (if it was ever shipped over) had much influence over Martin's imaginative design.

Martin was very optimistic of his aeroplane's performance, predicting 112mph at 10,000ft and a 25,000ft service ceiling, but its maiden flight, scheduled at the McCook Field for December 1918, never took place as the US Army were wholly unimpressed by its structural integrity, and refused permission for take-off until extensive modifications were made, which Martin refused to countenance. The Martin K.III Kitten was promptly abandoned while Martin developed a Liberty engined bomber - which the Army's McCook Field staff immediately rejected as structurally unsafe! But his K.III Kitten did come alive as the experimental K.IV float-plane, with a Lawrence L-3, 3-cylinder radial in 1921 for the US Navy; but it only reached 94mph and had a low ceiling of only 11,400 ft. The Martin K.III Kitten survives!

Gnat Mk.II

Bradshaw had anticipated fuel supply problems at 'high' speeds in manned flight and presented two patents (Pat No. 111681 laid August 1916 and Pat No. 107742 laid 18th December 1916) for adjustable carburettor chokes and main jet positions, to maintain a constant fuel supply under increased venturi action as the air speed rose. These would prove useful in the more powerful 2,280cc Gnat II (120mm x 140mm).

The post-war Mk.II was a well regarded engine and developed 45hp at 1,800 rpm, peaking at 50hp at 2,000rpm - remarkably powerful for its day. A rotary plunger pump drew oil from a tank and fed the crankcase and crankshaft bearings, using splash lubrication for the piston and machined steel cylinder walls. Like the Gnat Mk.I, it weighed 115lbs and its push rods to the exposed valves were driven by roller tappets. An ML Magneto and 48mm ABC carburettor was fitted.

The Gnat II was fitted to the Gosport Aviation Company's 'Shrimp' prototype of 1919. One of six new designs, this single-seat open cockpit, 23ft span biplane-floatplane was 'ideal for the man residing on an island or upriver'. Three were eventually built with Beardmore and Puma engines. The company was formed during the war by Charles Allom, a racing yachtsman, and Charles Nicholson of the famous Camper Nicholson Ltd boatyard at Gosport. The Gosport company also built one of the two flying boat hulls for the aborted Fairey Atalanta of late 1918 (s/n N118). Their chief designer was F P Hyde-Beadle who had already designed the unique, but unsuccessful, Perry-Beadle biplane flying boat of 1913 which had employed a single, but insufficiently powerful, ENV engine with chain drive to the twin air screws. It used a 'Consuta' sewn-mahogany panelled hull, built for them by S E Saunders Ltd at Cowes.

Another user was the rather ugly Blackburn 'Sidecar' mid-wing monoplane of 1919 (G-EALN). This was designed in anticipation of post-war civilian flying for the well-to-do for get-away breaks to the country and sporting events. She had a deep, triangular shaped fuselage with a side-by-side two seater arrangement. Entry to the cockpit was by a hinged door, motorcycle-sidecar fashion. The Gnat Mk.II however proved underpowered even for this 'lightweight', and the only 'Sidecar' made was exhibited at the famous Harrods store in March/April 1919, priced at £450. It did eventually sell and was later successfully flown in 1921 with an Anzani 100hp radial by its new owner, Blackburn's London based manager.

It is also reported that in 1919, BAT proposed resurrecting the pre-war Santos-Dumont 'Demoiselles' design as the BAT FK.28 'Crow' fitted with an ABC Gnat. This 14ft long 'motorcycle of the air' had a 15ft span and 150 mile duration. The pilot sat in an underslung nacelle-cum-undercarriage with the propeller spinning perilously close to his nose. For garaging, the aerofoil was simply removed by releasing 12 bolts, but the project was abandoned and BAT ceased trading in 1920. The Blake Bluetit of 1930 also had the Gnat fitted. Others included the Bristol Babe, Farman Moustique, Maachi 16, the aforesaid P8 Kitten and the Martin K.III, tentatively offered for the 1920 civilian market, as was M. E de Marcay's 'Passe-par-tout' biplane.

Gnat aero-engine production appears to have ceased around 1920/21 after only around 20 engines had been built.

4
ABC Radial Engines

Rotary or radial?

During the war, Bradshaw returned to an idea he had in late 1914 for a new, air-cooled, rotary or radial engine with which to win the war.

A static radial engine is similar to a motor car engine. Its cylinders radiate from a static crankcase fixed to the aeroplane's fuselage. The exploding gases force the pistons down causing their connecting rods to turn the crankshaft, as in a normal car engine. The propeller is fitted to the end of crankshaft; the rotating propeller draws cooling air over the cylinders and radiator as well as providing thrust for take off and flight.

A rotary radial engine is one whose cylinders also radiate from the crankcase. However on these engines, the crankshaft is fixed rigidly to the aeroplane and does not turn. The crankshaft has a single, offset crank about which revolves a connecting rod 'ring' from which radiate the connecting rods to each piston. As with a 'normal' engine, these con-rods are of equal length, but as the engine fires, it is the cylinders which are forced to move against the static piston causing the cylinders to rotate about the crankshaft which then pull the eccentrically mounted connecting rod ring around the crank pin causing the fixed pistons to move, relatively, up and down the cylinder creating compression in the usual Otto cycle. The propeller is fixed to the crankcase and the engine is cooled as its cylinders 'windmill' through the air, but therein lie several major problems!

Rotary engines

The majority of early British aero-engines were in-line or V-pattern automobile engines developed for airship use, but in order to extract the maximum power to attain flight, every component had to be made as light as possible. Even the forged I-section connecting rods were often drilled along their length to reduce weight without losing strength. But with their primitive knowledge of metallurgy and stresses gained from over-engineered steam engines, mechanical failure at the most inopportune moment was virtually guaranteed and engine service life was a measured in a matter of hours rather than days - typically 16 hours before a major strip-down and rebuild of a Gnôme rotary engine - though a competent engineer could strip a Gnôme rotary within 30 minutes. Compare this to a service life of 70 hours for a conventional in-line aero-engine. It should be remembered that the strong, lightweight aluminium alloys, such as Y-alloy universally used in later engine castings, was only developed by the National Physical Laboratory as the Great War was ending.

But in Britain, an inventive Charles Benjamin Redrup had developed a unique 2-cylinder "Reactionless" airship aero-engine by 1906. In this, the cylinders rotated via gears in the opposite direction to a rotating crankshaft; he later developed a 3-cylinder and proposed a 2-row, 10-cylinder rotary engine. This contra-rotating action reduced the gyroscopic effects - a principle later adopted in the powerful Siemens ünd Halske engines, but none of Redrup's designs found success. Despite many setbacks, Redrup's automotive and aviation career closely parallels that of Granville and the reader is commended to Redrup's biography, *The Knife and Fork Man*.

However, it was to be the French Seguin brothers who in 1906 developed the first successful lightweight 5-cylinder radial aero-engine. Laurent Seguin then improved on this with his 7-cylinder Gnôme 'Omega' 50hp rotary in 1908, setting the standard for the, mainly, French rotary aero-engine designs of the day, such as the Clerget of 1911, or the more complex le Rhône of 1912 designed by former Peugeot engineer, L A Verdet. Many of these French engines were fitted to private venture and military aeroplanes and were licence built in Britain, Italy and America.

The original Gnôme rotary used nickel-chrome steel alloy barrels. These were machined from a solid billet of steel leaving an incredibly thin 1.5mm wall which made it very prone to distortion in the extreme heat of combustion. It tolerated only one rebore. Even the larger Salmson radial only had a 2mm thick wall. While many V-twins were of similar thickness, most in-lines had ¼" (6mm) thick walls. The choice of machined billets of steel was never questioned, for they were considerably lighter than cast-iron and, being machined, inherently stronger; the Americans later developed tougher, de-stressed, drop forged steel-alloy cylinders which, being tougher, required more time to machine. Machined barrels could also carry many more, and much thinner, cooling fins around the critical combustion chamber, doubling the heat dissipating surface area. These cylinders had a thick, blanked off, 'Poultice' type combustion chamber end which was drilled for valve ports and to which was bolted a cast iron cylinder head, with its single mechanical push-rod exhaust valve. The Gnôme also had a hollow crankshaft, which greatly reduced weight. As rotary engines did not employ a conventional carburettor, the air and fuel mixture was drawn into the crankcase, picking up on its way any excess of the immiscible castor oil cylinder and bearing lubricant; this was then drawn into the cylinders

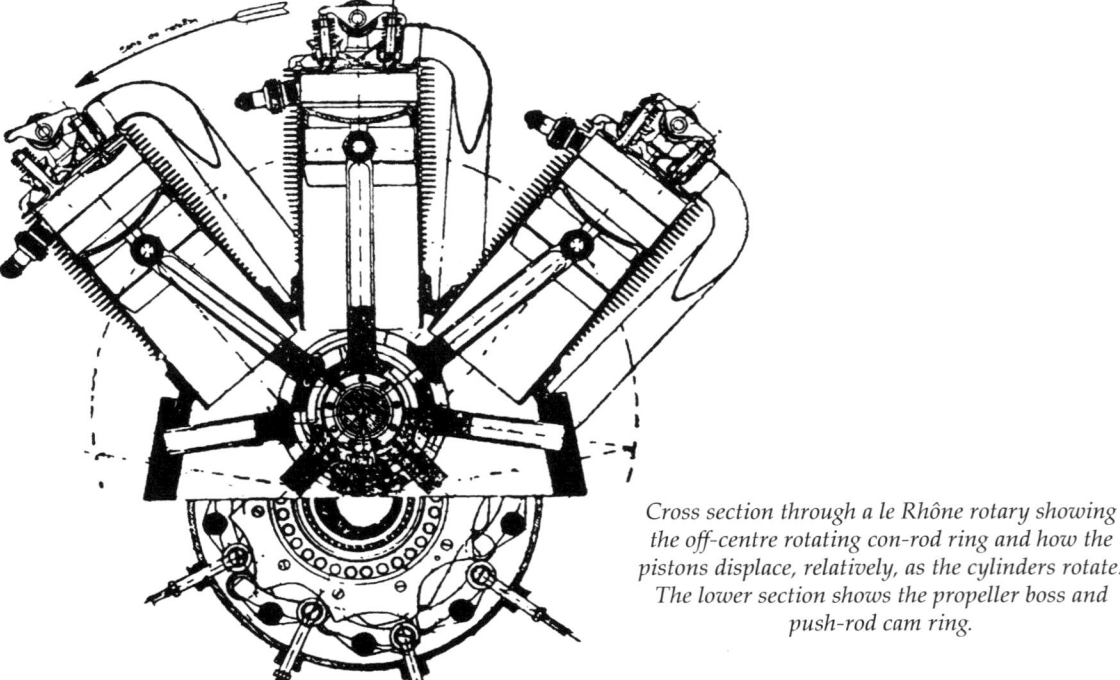

Cross section through a le Rhône rotary showing the off-centre rotating con-rod ring and how the pistons displace, relatively, as the cylinders rotate. The lower section shows the propeller boss and push-rod cam ring.

through a simple spring loaded clapper inlet valve in the cast iron piston crown as the engine rotated; oil consumption of 2 gallons/hour was quite normal!

The automatic inlet valves were a weak point in the engine design, because they required regular maintenance to get a uniform 4lbs loading on each piston's valve spring. However they often proved more reliable than later designs even though they had limited operational altitude. But in 1913 the Seguin brothers redesigned the inlet arrangement by using a two-stroke inspired inlet port design which was exposed by the piston at bottom dead centre; now having only the one exhaust valve, it became known as the Gnôme 'Monosoupape' (single valve) engine. Conventionally, exhaust valves and ports were at the front of the engine for maximum cooling. The, usually incompletely burnt, oil enriched exhaust gases were exhausted into the air through short, stubby exhaust ports which, being hot, often ignited the exhausted oil thereby engulfing the hapless pilot and aeroplane in flames, fanned by the propeller's draught!

In contrast, the le Rhône of 1910, 80hp Clerget of 1911 and the Clerget inspired 150hp Bentley Rotary had an external fuel supply from the crankcase to the cylinder head. These engines differed primarily in that the le Rhône used cast iron cylinder barrels to overcome the problems of cylinder distortion and used a single rocker arm with a single push-pull 'push-rod' with fixed valve timing. The Clerget however used thin, machine turned steel barrels with fragile bronze 'obturator' rings which moulded themselves to the distorted barrel. The le Rhône company was taken over by Gnôme in 1915 and became Gnôme et Rhône.

With the rotary's in-built inertia from its spinning mass, it dispensed with the need for a flywheel so necessary for slow running on an in-line or V-engine. But the absence of a regulating throttle and carburettor meant rotaries could only satisfactorily run when flat out; the engine speed had to be controlled by 'blipping' (cutting off) the ignition to allow the engine's compression to slow it down. The Bentley rotary employed a selective ignition system which allowed it to cut-out alternate cylinders, which better controlled the engine and made it safer to fly. The last Gnôme Monosoupape 9B engines were equipped with a single cylinder setting for a 'rough' tick-over.

As their axial compactness made all rotary and radial engines inherently very stiff, they had no need for additional structural mass or bracing and thus further helped reduce weight. However, the spinning cylinders meant that each cylinder had to be balanced within half an ounce to prevent the engine disintegrating through vibration and as the extremely thin cylinders needed replacing at regular intervals, routine rebuilding could be a very tedious process. Further weight reductions came from the air-cooling as all the cylinders were equally exposed to the air flow, even on those having two banks of cylinders offset on the same axis. Weight savings over a conventional in-line or V-engine were huge, hence the rotary's superior power to weight ratio.

However, rotaries suffer some major disadvantages. As the rear of the cylinder was not exposed to the oncoming air flow, it ran considerably hotter than the air-cooled front. This reduced its endurance through barrel distortion and unequal wear. The gyroscopic effect of 400lbs or so of iron spinning at some 1,500rpm seriously affected the aeroplanes's handling during a turn, especially on take-off, requiring full left rudder. However, the gyroscopic effect in light fighters was of great benefit in allowing immensely fast right hand, diving turns but it made them horrendously slow in a left hand turn as the aeroplane first tended to rise into the air making it easy pray to attackers. In both turns, full left rudder was always required to counter the gyroscopic effect.

Furthermore, the air-resistance caused by the spinning cylinders varied as the square of the engine's speed and was, clearly, dependent upon the diameter and power of the engine - which effectively restricted these engines to around 1,600-1,800rpm. The air resistance from the spinning cylinders also sapped the engine of between 10%-20% of its power. These problems could be reduced by having more cylinders of shorter stroke, but smaller barrels present even less surface area for dissipating the considerable heat. A further limitation to power was found to be in the limit of natural aspiration of a rotating engine at around 2,000rpm. The German Siemens ünd Halske rotary overcame these problems with a much slower running, stubbier barrel engine which used a large, gear driven propeller, although at its maximum speed this was insufficient to cool the spinning cylinders. This most powerful of engines only saw service towards the end of the war in the Siemens-Shuckert D.IV and D.V Scouts, but its phenomenal power frightened the Minister of Munitions into calling for the mass production of a more powerful British radial aero-engine - they hastily selected the ABC Dragonfly.

Radial engines

The advantages of static, radial engines, include controlled, pumped lubrication systems and conventional metered carburettors. Both made the engine less prone to exhaust gases catching fire and considerably improved fuel consumption through better control of the engine's speed. However external carburettors are very prone to icing at high altitude, whereas the rotary's induction system draws warmed air through a hot crankcase, regardless of altitude. Perhaps the greatest advantage over the rotary is that a radial directs its full power into the propeller for performance rather than losing some 10%-20% in spinning the rotary engine; also the small, gyroscopic effect of the spinning crankshaft had no affect on a turn, though that from the propeller remains the same.

The French Salmson company produced the first large, water-cooled, static radial in 1911 but it was an over complicated design of con-rods and crank-pins driving an epicyclic gear train. The smaller, 3-cylinder air-cooled Anzani radial of 1909 was a simple design of quite low power and had cast iron barrels, but while Anzani had produced some reliable 5 and 6-cylinder radials before the war, their two row 10 and huge four row, 20 cylinder 200hp units were not a success. There was also a promising, but short lived Viale 70hp 7-cylinder radial designed by the Italian engineer Spirito Mario Viale who manufactured them at Boulogne-sur-Seine, France between 1910-1914. Viale offered 3, 5, 7 and 10-cylinder designs, before leaving to work for the French Government, in a similar capacity to Bentley.

One of the first successful British static radials was the Isaacson 50hp water cooled, 7-cylinder of 1909. It was designed by Rupert John Isaacson of Leeds in association with Horace and Henry St John Sanderson of Ben Rhydinng, Ilkley, who were steam engineers with experience in quarry equipment, but little is known of the Sanderson brothers. It appears Isaacson was, or later became, a works manager for Manning Wardle & Co at their Boyne Engine Works, Jack Lane, Leeds, where the Isaacson radial was made. The company was more famous for steam locomotives and several of their 0-6-0 engines are still in use on preserved railway lines. Manning Wardle went into liquidation in 1927 and were bought by Kitson & Co, while part of their Boyne works was bought by their neighbour, the Hunslet Engine Company (in whose works Redrup had built his radial aero-engine).

Isaacson's first patent of 1905 relates to friction clutches followed by vaporisation of fuels for internal combustion engines. By 1908 he was working with the Sanderson brothers on blast nozzles and valve arrangements for steam engines, but by 1909 his attention had turned to a 6-cylinder radial internal combustion engine which comprised two rows of 3-cylinders on a common crankcase. The crankshaft had ball-race big end bearings and cooling was by a common, annular, water jacket for the enclosed cylinder heads which itself was cooled by air flow from the propeller. It appears two further versions were produced: a 50hp, 7-cylinder or, when double banked, a 14-cylinder rated at 100hp - both ran at 800rpm. Isaacson's radial engine was first fitted in 1910 by Robert Blackburn into his second Antoinette inspired monoplane and later, into the Blackburn Mercury Mk.I of 1911, but the engine was prone to bearing failure and did not prove a very successful design.

Isaacson then introduced a patented 2:1 epicyclic co-axial reduction gear within the air-screw hub (Pat No. 19110083 of 1911). An example of this improved engine was fitted to James Valentine's Bristol-Prier monoplane of 1911 for the Michelin Cup (restricted to all-British aeroplanes and pilots) and into the 7th (and final) Avro Type D biplane of 1912 for Avro's flying school, which had recently moved from Brooklands to Shoreham. One 45hp example was fitted to a Blackburn Mersey Mk.III monoplane for the 1912 Military trials and proved most promising, until a fatal crash.

Between 1912 and 1913 Isaacson developed a cylinder lubrication and fuel induction system, mixing carburated fuel with the oil mist within the crankcase. An automatic inlet valve was to be set in the cast iron piston crown, (Pat No. 191219268), much like the Clerget, with a spring loaded valve. He reckoned his improved automatic valve would offer a noticeable weight reduction greatly improving the engine's power to weight ratio. His redeveloped 7-cylinder engine, now developed 68hp at 1,080rpm, weighed a mere 196lbs and consumed only 3.8 gallons of fuel per hour. In an effort to secure greater sales, the engine was designed to be made fully interchangeable with the contemporary Gnôme rotary. One Isaacson 7-cylinder was successfully fitted to a Howard Flanders biplane.

A 200hp 14-cylinder Isaacson radial was developed in early 1914, but this together with a proposed 9-cylinder 250hp rotary and an 18-cylinder 465hp rotary, were all abandoned when war was declared - at which point Isaacson returned to steam engineering for the war effort. His final patent application was in 1918, just before he died, for a mirror attachment to a motorcycle oil drip feed lubricator allowing the rider to observe its action on the move. The Patent was granted posthumously.

After his rotary engine experiments, and now keen to see British aero-engine production increased to meet an increasingly likely war, Charles Redrup co-founded the Hart Engine Co of Leeds in 1913 to develop a new 9-cylinder 'reactionless' rotary but he soon realised the severe limitations in the rotary engine's design with ever larger capacities, and that the future lay in a radial design. He duly built a 150hp 9-cylinder radial for the 1914 military trials at which it performed well, but it failed to meet the strict criteria for ease of manufacture. Nevertheless, it prompted an approach by the Air Department to Vickers for them to help redesign the engine in an attempt to ensure Britain had the capacity to produce sufficient aero-engines for the war; but despite much enthusiasm, the redesigned Hart engines were not a success. In 1929, Redrup developed a unique, axial radial engine followed by much work at Avro.

Odd or even number of cylinders?

Bradshaw strongly upheld the engineering convention that an odd number of cylinders ensured no two pistons were ever 'stationary' at the top or bottom of their cycles and thus dynamic balance was ensured. This resulted in a constant and smooth torque, thereby reducing inertial loads and stresses on the pistons and con-rods. In a wager against his challenger, Harry Hawker, to disprove the widely held theory that an even number of cylinders would not work as efficiently as odd, Bradshaw made a one-off 8-cylinder radial engine with a consecutive firing order (some sources have erroneously suggested this was the 6-cylinder Mosquito.) It worked, but the discussion which followed is not recorded!

In fact there were as many 'even' cylinder rotary and radial engine designs offered by the end of the Great War as 'odd', though to be fair most of these 'even' designs comprised double row 'odd' engines, aligned between the forward cylinders to expose them to the air-flow, reduce the overall diameter and minimise cross sectional area to reduce drag. A further factor was the contemporary convention that the largest practicable air cooled cylinder should not exceed 5" bore; greater power therefore could be achieved only by increasing the number of cylinders. In fact the Bentley BR.2 and ABC Dragonfly were the only war-time rotary and radial engines to break this convention at

5.5" bore and it was only in post war radial designs that they tentatively crept up to 6.1" bore by 1930. Double row engines usually required a forward support for the propeller bearing, affecting the airframe design.

Overheating

Contemporary rotary and radial engines were notorious for overheating. The front of the cylinder, exposed to the full force of the air-flow, ran much cooler than the sheltered rear, causing barrel distortion, and requiring a rebuild every few hours. It was estimated that about 70 cu.ft. of moving air was necessary for each brake-horse power/minute: a huge volume which required a large frontal surface area increasing drag and thus requiring extra power - reinforcing the desire for double row engines.

To overcome the problem of cylinder distortion, Clerget adopted special L-shaped bronze 'obturator' piston rings, but though these rings would mould themselves to the distorted barrel, they had a very limited service life of less than 15 hours and, ultimately, failed to solve the problem.

Bradshaw firmly believed that a copper coating on an air-cooled cylinder would more evenly distribute the heat and better prevent distortion. The contemporary 120hp, 6-cylinder Beardmore, the Green overhead cam in-line and the 150hp Sunbeam V8 water-cooled engines had begun to use electrolytically deposited, extremely thin copper water-jackets formed over wax coated cast-iron barrels, but Bradshaw's ideas for a copper coating on an air-cooled engine raised more than a few eyebrows and led to many a heated debate. Yet, Bradshaw's own tests revealed that a plain, machined nickel-steel barrel used on a conventional rotary engine with an air flow of 150mph, was six times hotter (around 200°F) at the back than at the front, while in contrast a $1/16$" thick copper-coated steel barrel was only two and a half times hotter (down to around 80°F) at the back. He was later granted a patent (Pat. No: 102186) in November 1916. Yet though copper coated steel barrels were fitted to the ABC Gnat without comment, only the later ABC Dragonfly radial was to suffer the stigma of alleged failure 'due to its copper coated steel cylinders'! In 1944, American engineers proved Bradshaws' theories of more uniform heat dissipation to be correct and duly developed a 'top secret' technique of fusing a thick copper coat to machined steel barrels.

Aluminium alloys have a similar dissipating effect to copper - indeed, Orville Wright's engine of 1903 used thin aluminium water jackets. Aluminium also had the added advantage of reduced weight. Many engineers later suggested that Bradshaw's choice of copper was a silly choice given the availability of aluminium; but the quality of aluminium alloy castings were very variable indeed, particularly their porosity, causing Hispano-Suiza to enamel both sides of their castings: so Bradshaw chose to play safe!

As the Air Department's Technical Liaison Officer seconded to Gwynnes of Hammersmith, W O Bentley proved the benefit of aluminium alloys in the Admiralty Rotary (AR.1) and the production version, the Bentley Rotary BR.1, with its steel-lined, cast aluminium cylinder barrels and alloy pistons.

At the time, Gwynnes were the British licensee for the Clerget rotary and already had seconded to them Spirito Viale who also had experience of aluminium alloys. But Gwynnes resisted Bentley's new rotary designs and under Air Department orders, Humber were then contracted to build the Bentley design as the 130hp AR.1. Prototype AR.1/BR.1 engines suffered badly from fractured valves; these were cured by a new lightweight design - ironically it was Bradshaw who was later accused of always designing everything too light for commercial purposes! Such was the BR.1's success that Bentley got permission to develop his 200hp BR.2 in the summer of 1918; this engine remained in service with the RAF until 1928.

The Royal Aircraft Factory had for some time also been formulating designs and specifications for RAF pattern air-cooled in-line and V-type aero-engines. Their two senior development engineers, Professor A H Gibson and his assistant Samuel D Heron, called for the adoption of aluminium alloy cylinder heads, with widely spaced and angled overhead valves for maximum heat dissipation and that the combustion chamber be machined with an internal thread so it could be screwed onto the end of the cylinder barrel. However, cast alloy heads require hardened steel valve inserts to prevent damage as the valves slam shut against the extremely hot and relatively soft aluminium alloy. Bradshaw was certainly well aware of these different coefficients of expansion for steel and aluminium alloys, so he probably played safe once again by choosing a cast iron head for his ABC radial engines.

War

At the outbreak of war, William Weir, a Scottish industrialist, was appointed by the Liberal Prime Minister, Herbert Asquith as Director of Munitions in Scotland. On Lloyd George's appointment as the Prime Minister of a coalition government in 1916, Weir was appointed Controller of Aeronautical Supplies at the Ministry of Munitions and given a seat on the Air Board. Realising the Royal Aircraft Factory was failing to keep up with German technology (hardly surprising given the Government's indifference to military aviation in 1912), one of Weir's men was now installed as Superintendent at the Royal Aircraft Factory to transform them into an efficient production unit.

Meanwhile two speculative, privately sponsored, prototype fighters were under development; the Vickers FB.12 tractor biplane fighter using the Vicker's sponsored, 150hp, Hart 9-cylinder radial of 1916 and the BAT 'Bat' with ABC's 6-cylinder 'Mosquito'. Due to major development problems, the first Hart 9-cylinder was not delivered until late 1916 and was not flown in the FB.12 until early 1917. The Hart also proved a major disappointment in the new Vickers FB.16 Scout of late 1916 and was duly superseded by a 150hp Hispano-Suiza powered FB.16A which crashed in December 1916 during aerobatics, killing its test pilot. No further development was undertaken on the Hart engine.

ABC Mosquito

Relying on his copper-plated, machined steel barrels, Bradshaw hoped to solve both the cooling and gyroscopic problems of rotary engines by developing a new six cylinder, ABC Mosquito radial, which it appears may have been commissioned by Samuel Waring for his new BAT aeroplane. As ABC were fully committed to producing the ABC auxiliary engines, Bradshaw duly set up a subsidiary

BAT prototype with ABC Mosquito

research and development company, Walton Motors Ltd. How much he was influenced by the Isaacson will probably never be known, but the Mosquito was a two row, 6-cylinder, 120hp radial using many components from the Gnat. It had a single overhead inlet and twin, side-by-side overhead exhaust valves at the front, operated by cast alloy rockers from exposed push rods, in line axially, in front of the cylinder; the foremost operated the single inlet valve at the rear while the innermost actuated the forward, side by side exhaust valves. The inlet ports at the back were connected by cast alloy tubes to an annular-ring manifold, fed by two carburettors. The engine was duly tested on the private venture BAT prototype in 1917, whose construction had been approved under Air Board Licence No. 11, but it proved greatly underpowered and though the Mosquito engine was rejected outright by the Royal Aircraft Factory, a government contract was placed for six BAT airframes with the improved 7-cylinder, 170hp ABC Wasp radial and the 100hp Gnôme Monosoupape (later replaced by the 110hp le Rhône 9J).

ABC Wasp

By April 1917, the Air Board had settled on their requirements and drew up Specification A.1a in October 1917, for a high speed, high performance single seat fighter. They duly invited tenders for powerful aero-engine designs. Hopefully their superiority would allow them to out-manoeuvre the likes of Manfred von Richthofen and his 'Flying Circus' of Fokker triplanes, or Oswald Boelcke's formidable Albatros D1, bringing the war to an early end. Air Board Specification A.1a prompted the development of three prototype Wasp 7-cylinder radials for the Sopwith 8F.1, BAT FK22/2 Bantam and Westland Weasel prototypes, and the later Westland Wagtail. A proposed Martinsyde twin engined fighter was abandoned early on.

Whereas the Mosquito was a two-row radial, the 170bhp, 10.9 litre Wasp (114 x 150mm) was a single row unit of greatly improved design. The majority of magneto ignition systems pre-war were of German manufacture. With outbreak of war, supplies soon dried up. Sadly, early British units proved

Left, Wasp engine (GB) and above, installation in a Sopwith Snail

unreliable, so war time stipulation for military engines required twin spark plugs per cylinder and twin magnetos to ensure the engine could be kept running on at least one set! A handful of Wasp engines were said to have been built at Gilbert Campling's Selsdon Aero and Engineering Company at Sanderstead Road, Croydon but it soon became clear that the Wasp was seriously under-developed and on two occasions the experimental airframes were grounded, by order, so that the engine could be sorted out. In one of his "Jottings from a designer's notebook" articles for *The Motor Cycle* in 1952, Granville confessed to his over enthusiasm by over stating the Wasp's potential performance. On paper it would develop 150bhp, but they had managed to extract 175bhp in bench tests. Rather optimistically then, he stated 175bhp as "Maximum bhp" on the official Government questionnaire. But in military trials, these engines regularly fell short of the maximum rating by some 4hp to 6hp. As Bradshaw later confessed, "it was my undoing", for had he have declared 150bhp, he would still have secured the contract, but instead had to waste a lot of energy and money on specially tuning each engine to extract the declared maximum power! He soon learnt to adopt a tactic used by a fellow engineer, well versed in Government contracts, who confided in him; "if you design a boat to do 18 knots, you should quote its designed speed at, say, 16 knots, then you will probably get a glowing report saying 'Designed speed exceeded by $1^1/_2$ knots' and in the eyes of the powers that be, you are a made man"! That philosophy still holds true today.

But despite his error of judgment, and its lack of exhaustive testing, the Wasp powered Bantam offered a far superior performance over the Gnôme (125mph v 110mph at 10,000ft; climb to 10,000ft in 9 mins v 16 mins 50 secs). Sir William Weir, Controller of Aeronautical Supplies was so impressed that when news reached the Ministry of Munitions that the Germans had developed more powerful engines, and with the Government keen to bring the war to an early end, Bradshaw was called to a Ministry of Munitions conference in late 1917 and instructed to immediately build the largest and most powerful version he could of the 'impressive' Wasp. Within 28 days, Bradshaw's team of six machinists at his Hersham works had built a 9-cylinder version which was duly christened 'Dragonfly'; a second prototype quickly followed.

Meanwhile, work on the Wasp continued and a much improved and more powerful Mk.II appeared boasting 200hp at 1,800rpm. It weighed 320lbs complete with a pair of 48mm ABC carburettors and a pair of P.L Type 7 Magnetos. It retained the usual thin wall machined steel poultice, closed-end, barrels with a cast iron cylinder head bolted to the barrel. This more reliable Wasp Mk.II arrived in September 1918 but in October the Royal Aircraft Establishment (created out of the Royal Aircraft Factory in April 1918 to avoid confusion of their initials with the Royal Air Force), declared the Wasp obsolete and all further work was cancelled.

But the Wasp did see limited use post-war in the Saunders Kittiwake (G-EAUD) twin engined biplane land/flying boat of 1920; this was designed by F P Hyde-Beadle who had years earlier been involved with Norman Thompson of White & Thompson Ltd, Bognor. Thompson had designed their No.2, two-seat side by side pusher of 1913 using an ABC 120hp V8, which gave a most promising performance when flown by both Charteris and Lt Porte. But success only came on the outbreak of war with their No.3 two-seat anti-submarine patrol flying boat and their Hyde-Beadle designed monocoque fuselage 'Bognor Bloater' tractor biplane of 1914, fitted with a Renault 70hp engine, which saw limited RNAS service.

Hyde-Beadle then joined the Gosport Aviation Company before moving to S E Saunders Ltd in 1920 and, inspired by government trials for post war civilian aircraft, designed the massive Saunders Kittiwake twin engined biplane flying boat (G-EAUD) with a wing span of 68ft and a massive two tier cabin for 7 passengers which filled the gap between the hull and upper wing. It was an ungainly

Wasp engined Saunders Kittewake

looking aeroplane and unjustly described as the "Isle of Wight ferry with wings"! Powered by a pair of ABC Wasp Mk.II 200hp radials, trials of G-EAUD proved fraught with problems relating to the variable geometry wing, hull and retractable undercarriage. However once finally airborne and at its maximum speed of 116mph, it was soon discovered that one of the engines was failing to deliver sufficient power. This was fast becoming a familiar problem with ABC engines in which the good engine was a hand build pre-production ABC job, and the faulty one was contract built! But alas, the Wasp proved insufficiently powerful and recurring problems with the airframe caused the project to be abandoned following a crash in the summer of 1921 when, being flown by an experienced pilot who was unaware of the Kittiwake's foibles, the hull was holed on rocks in the Solent during take off. The flying boat was scrapped in July 1921 and the hugely disappointed Hyde-Beadle left the company.

An American prototype biplane, the Huff-Daland HD-4 of 1920, also used an ABC Wasp Mk.II. From this was developed a military prototype TA-2 of which three airframes were built in 1920; the first, a non flying model, was fitted with the Wasp Mk.II while the other two flying examples had the Curtiss OX-5. Huff-Daland became better known for their later crop sprayers.

ABC Dragonfly

The Air Board's revised 1918 programme included the Type A.1a (later RAF Type I) long distance high speed single seat fighter; the type A.2a two seater fighter; the later RAF Type III short range fighter reconnaissance; RAF Type IIIB artillery observation and the RAF Type IVB long range photo-reconaissance. These nomenclature have a complex hierarchy.

The Type A.1a high performance, high altitude fighter specification required a minimum speed of 135mph at 15,000ft which demanded powerful new aero-engine designs of 300hp, but of under 600lbs weight and smaller than 42" overall diameter, to minimise drag. It resulted in submissions from Brazil-Straker (14-cylinder Mercury radial), Siddeley-Deasey (RAF No.8 radial inspired Jaguar), and ABC Ltd.

The Dragonfly was a 22.8 litre, 9-cylinder (140 x 165mm) 315hp engine based on the Wasp and was, at the time, claimed to be the world's largest air-cooled radial aero-engine. On paper, this giant of an engine was a most promising (if so far unproven) design and, being relatively light at 600lbs, could develop a theoretical 340hp at 1,650rpm - an unparalleled power to weight ratio for its day at 0.57hp/lb compared to the contemporary 200/230hp Bentley BR.2 or 200hp Clerget 11-cylinder rotaries. 300+hp was deemed essential to meet the Ministry's specification for the RAF's high performance fighters to combat Germany's 240hp Siemens ünd Halske S.IIIa powered Siemens-Shuckert D.IV and D.V Scouts.

The Dragonfly retained the Wasp's 3-valve layout and twin spark plugs per cylinder arrangement but was now fired by a pair of AK9 magnetos, while a pair of ABC carburettors (later Claudel-Hobson) fed the engine at the rate of 0.56 pints per bhp-hour at 1,650rpm. It retained the Wasp's rotary pump lubrication system, friction-less roller tappets and compact coiled strip-steel 'Volute' valve springs which offered a more compact solution to wire-coil springs and tended to maintain their 'spring' for far longer as the area most prone to overheating and softening was also the smallest diameter and thus, inherently, the stiffest. However, at high speed, the Volute spring's coils were prone to rubbing against each other causing heat spots and loss of temper allowing the spring to fail, causing the valve to drop resulting in loss of compression. This was resolved in later designs by using three concentric coil springs, to form a fail safe system, albeit requiring longer valve stems and guides.

In his book *An Account of Partnership - Industry, Government and the Aero-engine*, Major George Purvis Bulman of the Engine branch, Aeronautical Inspection Department of the War Office, describes how the ABC Dragonfly won a spontaneous order by Sir William Weir in late 1917. During autumn 1917, Bulman been had been assessing development of the promising Wasp and its likely adoption in trainers, subject to its 100 hour trials. He also saw the drawings of the 9-cyl Dragonfly engine which he reported to his superior, Lt.Col J G Weir (Sir William's half-brother), Controller of Technical Branch, Air Board.

Of major appeal to the government was the fact that the Wasp (and Dragonfly) had been designed for simplicity in manufacture and assembly, greatly enhancing its scope for high rates of production at shadow factories by a variety of engineering companies, many of whom had little or no experience of aero-engine manufacture. Bulman then records that "from out of the blue came news that, on account of its amazing simplicity of construction no less than 10,000 Dragonflies were to be made" - despite the fact that it had not been built - and how in vain he exploded in fury at this 'crazy decision'. On 27th April 1918, Sir William was appointed Secretary of State for the Royal Air Force and ennobled that summer.

In April 1918, a prototype Dragonfly, having already recorded 72 hours bench test running-in at ABC, was fitted at the Royal Aircraft Factory with a calibrated propeller. After a ten minute warm up, it registered 365 bhp. Later tested on a Sopwith Bulldog and subjected to flight evaluation by Harry Hawker, in terms of speed and manoeuvrability, it 'ran rings around' a captured German fighter fitted with 'the latest engine' (believed to have been an Albatros D.V). Bradshaw later claimed the Bulldog had unofficially taken the British height and speed records during these trials.

To understand the Dragonfly's allure, we need to compare its power to weight ratio with its contemporaries and against the unproven Cosmos Jupiter prototype of October 1918.

Figures for bhp can vary quite widely depending on source and author (see * in table on p54), these figures should therefore be viewed as a fair comparison. For example, tests at Martlesham Heath on a Sopwith Snark and a Siddeley Siskin recorded 340bhp and 320bhp at 1,650rpm respectively as shown below, and that of the contemporary Jaguar was measured at only 336bhp, compared to later measurements of around 400bhp.

Left, Dragonfly engine in profile, propeller fits to the right. Front view on the right. (GB)

Engine	Power hp	Weight lbs	Power to weight ratio hp/lb
Cosmos Jupiter:	395	662	0.59
ABC Dragonfly:	340	600	0.57
- RAE trials	365	600	0.60
- production engines	340	656	0.52*
	320	656	0.49*
ABC Wasp:	170	320	0.53
Brazil-Straker Mercury:	315	587	0.53
Bentley BR.2 rotary	250	500	0.50
Siddeley Lynx:	215	512	0.42
Siddeley Jaguar:	336	850	0.39
le Rhône J rotary	120	323	0.37
Clerget 9B rotary	130	381	0.34

While both the Brazil-Straker Mercury and Dragonfly met the Air Board's 1918 specification A.1a (later RAF Type I), so impressed was William Weir at the Dragonfly's performance and superior power to weight ratio, that he instructed Bradshaw to immediately issue sets of drawings to thirteen selected manufacturers as well as Wrights in America. However, the newly formed RAE demanded the engine first be thoroughly examined and exhaustively tested, whereupon the prototype promptly disappeared into their workshops! But, without waiting for their results, Weir went ahead with his 1918-1919 programme and ordered the immediate production of 8,580 Dragonflies along with 8,278 Bentley BR.2 rotaries and 8,295 Liberty, 6,891 BHP-Siddeley Puma, 5,861 Hispano-Suiza and 4,564 Rolls-Royce Eagle V-12 engines. The Dragonfly order would eventually rise to 11,500 with delivery of 4,135 engines scheduled for June 1919. With these orders firmly in place, Weir then held back all further work on the promising Mercury and RAF No.8 14-cylinder inspired Jaguar.

Meanwhile, the RAE carried out their own production-engineering analysis of the Dragonfly and gradually improved upon its design and serviceability, but this resulted in 'a 70lbs increase in weight and a 20bhp drop in power after 2 minutes of full load', which left Bradshaw, characteristically unamused and blaming the RAE for the engine's ultimate

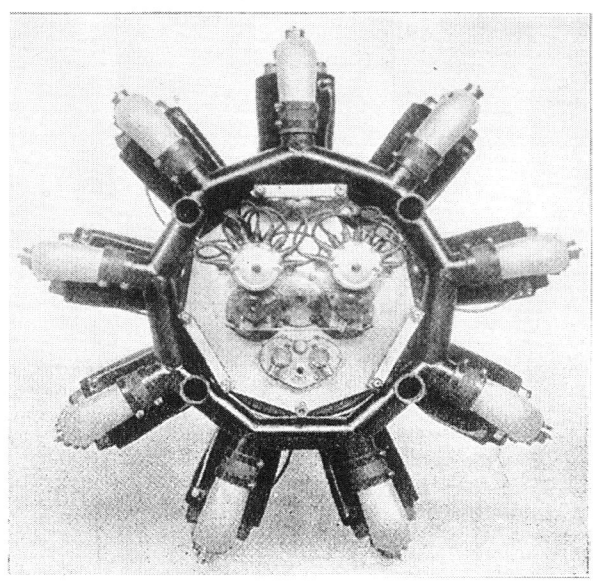

Rear view of Dragonfly engine - note the twin magnetos

failure. Bradshaw's many critics often maintained that he all too easily blamed others for the failure of his designs, but to his credit, Bradshaw readily confessed that he was merely the conceptual designer and the problems of manufacture are those of the production engineer and specifier, who invariably have the final say in design and performance. This was well demonstrated by earlier comments on the difference between the performance of the hand-built ABC and contract built Gnat and Wasp engines in the PV8 Kitten and Saunders Kittiwake. Even Phelon & Moore's own draughtsman reflected that interpreting Bradshaw's ideas into production engineering drawings was often a near impossible task, but that problem is equally true with today's high technology Computer Aided Design lead 'draughtsman' - as any experienced production engineers will confirm.

Nevertheless, the first Dragonfly engines had reached the aeroplane makers by late autumn 1918, but it was soon discovered at Nieuports that the engine suffered bad vibration and overheating and only gave between two and 17 hours service life; this was fine for fast interceptor work but of no use to long range or reconnaissance fighters! The root cause of these problems was only fully analysed after the war through extensive research by both Bradshaw and the RAE.

Shadow factories

While orders for Wasps were few and mainly placed with established aero-engine or motor car makers, the huge Dragonfly orders required many more companies not conversant with aero-engine manufacture. To that end, the Government adopted the practice used in the mass production of the BR2 rotary, appointing William Shackleton, Chief Engineer of the Sheffield Simplex Car Company, working with the AID, as overseer for the Dragonfly. All production details, and the many modifications, were now controlled centrally and duly despatched to the factories. In addition to a few of their own hand-built Dragonfly prototypes, ABC/Walton Motors were presented with an official order, AS2223/1, for three prototype Dragonflies, two of which were completed by 31st December 1918. Production orders were sent to 'thirteen' companies in the United Kingdom. The list of companies varies from source to source, but under the official Air 1/2301/215/12 notice, these included: Beardmore Ltd; F W Berwick

& Co; Belsize Motors; Clyno Engineering; Crossley Motors; J B Ferguson; Guy Motors; Humber Ltd; Mather & Platt Ltd; Maudsley Motors; North British Locomotive Company (Ruston Proctor); Ransomes, Simms & Jeffries; Sheffield Simplex; Vickers, Vulcan and Wolseley. Of these, only ABC, Clyno, Ferguson, Maudsley, North British and Wolseley actually supplied pre-production Dragonfly prototypes, to a total of 21 engines.

Beardmore of Glasgow were well known shipbuilders who had already developed a variant of the Austro-Daimler in-line aero-engine, contract built for them by Arrol-Johnston in Scotland. Their subsidiary company, Galloway Engineering Ltd, completed their order for twelve Wasps with delivery of the first eight by March 1918, but none of the Beardmore order for 1,500 Dragonflies was met.

Only one of the order for twelve Dragonfly prototypes were produced by J B Ferguson Ltd of Belfast. Ferguson were a bodybuilder and machine tool company. Their 'Fergus' car in 1915 was the first ever to use rubber engine mounting blocks, but war-time restrictions caused its manufacture to move to Newark, New Jersey. Joseph Bell Ferguson was the elder brother of Harry Ferguson of tractor fame.

Belsize were motor-car makers of Manchester. Of the 1,000 Dragonfly orders placed, only 48 had been built by 1919; a reduced balance of 300 engines remained on order, but was cancelled in 1919. Berwick (plant engineers) eventually produced 231 out of two orders of 500 each, but nothing else is known of these.

Crossley Motors were major contractors supplying the RFC with motor lorries and staff cars as well as assembling several aeroplane types. In addition they were contracted to build 900 Beardmore 160hp 6-cylinder in-lines and 600 Bentley BR.2 rotaries of which only 83 were delivered. They also received an order for twelve Wasp Mk.Is of which only eight appear to have been built, but they did carry out further work on the Wasp. However, none of their order for 1,000 Dragonfly radials was met. Maudsley, another motor car maker, had produced five Dragonfly pre-production prototypes by July 1918 out of their order for twelve prototypes, but none of the 500 production engines were produced.

Clyno Engineering, more famous for their Vickers machine gun motorcycle-sidecar platform, was ordered to produce eight prototypes and 500 production engines. The first prototype was delivered on 22nd April 1918, the second in June and by the end of the war, six prototypes and four production engines had been delivered, with the final two prototypes and 47 production engines being delivered in early 1919, at which point their contract was cancelled.

Guy Motors received an order for a prototype Dragonfly engine which they produced in a record time of 24 days from receipt of the blueprints through continuous day and night toil! The company duly received two telegrams, one from a Col J G Weir stating "Congratulations to everybody concerned on the building of the Dragonfly in record time"; the other, dated 15th February 1918 from William Weir, Director-General of Aircraft Production: "Reference to your telegram of yesterday regarding Dragonfly engine, I heartily congratulate you on your magnificent performance, the result of which, it is hoped will mark a new milestone in progress." These were followed on 19th April with an order for twelve Wasp prototypes and then an order for 600 Dragonflies, but while the Wasp order was fulfilled (some sources suggest only one Wasp was built), the entire Dragonfly order was cancelled before production commenced. While in later years many dismissed the Dragonfly as one of Bradshaw's many follies, Guy Motors Ltd was more than proud to take credit, in their impressive Silver Jubilee commemorative booklet of 1939, for "the versatility of the Company's designers" in producing the Wasp and Dragonfly radials.

The Humber car company was a pioneer in the manufacture of bicycles, motorcycles, motorcars and indeed aircraft. They also produced their own engines and had been contracted by the Government to build the Avro 504K and, following Gwynne's reticence, built the new Bentley BR.2 rotary engines under Bentley's supervision which put Humber's works at near capacity. It is not known how many, if any, of the 850 Dragonflies ordered to be produced at Humbers' Coventry works were completed. Gwynnes only built eleven out of an order for twelve Wasp radials.

Sheffield Simplex, another famous motor car maker, produced seven (some sources say eight) out of an order for twenty Wasps and six Dragonfly prototypes but only 300 out of the 500 Dragonfly production orders were completed by 31st December 1918. The Vulcan Motor Manufacturing and Engineering Company are said to have been contracted to build 600 Dragonflies, but none were assembled, nor is it believed any of the 500 orders placed with Ransome, Sims & Jeffries were built. Mather & Platt Ltd also failed to produce any of their order for 750 engines though North British produced four of their twelve orders.

Ruston Proctor Ltd of Lincoln became the largest sub-contractor for Sopwith aeroplanes; their extensive engineering facilities allowed them to also build aero-engines and half a million horseshoes! They merged in September 1918 with Richard Hornsby & Sons to form Ruston-Hornsby. However, though they were ordered to produce 12 Dragonfly prototypes, it appears only four were built by May 1918 and of the 1,500 ordered, none are believed to have been built. Likewise, Wolseley, a Vickers subsidiary. Production was to be at Vicker's Maxim gun works at Crayford in Kent, rather than at Wolseley's works, which was already fully committed to Hispano-Suiza engines. It appears that only one Dragonfly prototype out of an order for twelve from Wolseley was built. Vickers also only produced four out of an order for 212 Wasps and none of their order for 1,000 Dragonfly engines.

Production orders for the Wasp are not very clear but it appears only around 50 Wasp prototype and production engines may have been built - but how many of these were Wasp Mk.I and how many Mk.IIs, in unclear. Of the 11,050 Dragonfly engines ordered, only 1,147 are said to have been completed. Technically, the Dragonfly remained in production until the summer of 1919 for, although the engine was troublesome, like most contemporary aero-engines, it was not yet considered incurable. Besides, it was still out performing all other radial engines until the arrival of the redesigned Siddeley Jaguar and Cosmos Jupiter in 1920.

Wright radial

America's own experience of building their first radial engines well illustrates how unjustly Bradshaw's reputation was painted in the post-war years.

Bradshaw always firmly maintained that in addition to the many British aero-engine makers, he was also instructed to give his Dragonfly drawings to Wrights in America. Support for his claim can be found in Public Records Office records, (see Appendix 1 Walton Motors), yet many suggest this was another of his 'false' claims so he could take credit for the hugely successful Wright 'Whirlwind' of the 1930s which became one of the world's most loved and reliable radial aero-engines of all time. Yet even if his claim was false, Wright's prototype radial was uncannily similar in design and specification to the Dragonfly.

Although the Wright brothers had begun commercial production of their pusher type bi-plane in 1909 at Dayton, Ohio, they soon lost their lead to the Glenn Curtiss and Glenn Martin tractor type aeroplanes, favoured by the US Navy. After Wilbur Wright died in May 1912, Orville concentrated on aero-engines with their trusted 4-cylinder in-line engine, but although America didn't enter the war until April 1917, Orville had already secured a major share in the Simplex Automobile Co of New Jersey in 1915 and, to reinforce their position, merged with their rival aeroplane makers, Glenn Martin in September 1916, forming the Wright-Martin Corporation.

It was through Simplex that Orville secured, in 1916, the US rights to the manufacture of the superb French Hispano-Suiza V8 aero-engines, but the 'Hisso' 180hp V8 proved to be one of the most difficult of all engines to master by their licensees (as Wolseley found to their cost!) and it was not until late 1917 that Wrights were finally producing trouble free engines, albeit de-rated to 150hp which, though they proved underpowered for combat use, were considerably more reliable than the Wolseley built 200hp Hispanos. America's only other aero-engine of worth was the lumbering Packard/Hall-Scott designed 'Liberty' in-line engine of late 1917. There was only one 10-cylinder radial developed in America between 1913-1917, the 14.3 litre Smith Static, in which our Admiralty showed great interest, but after the six engines built under licence by Heenan & Froude in England proved wholly unsatisfactory in the AD Navyplane, the Smith radial was rejected. Herschel Smith, in his *Aircraft Piston Engines*, describes this engine as an 'atrocious piece of engineering' and is quite sympathetic to the ABC Dragonfly!

America's success at designing war planes was little better; all they had available were basic training aeroplanes and spotters. The US Government duly selected the deHavilland DH.4 light bomber, Bristol F.2B fighter and French built SPAD S7 Scout for war in France. Some 4,800 deHavilland DH.4s were built in America by the hastily created Dayton-Wright Company between late 1917 and the Armistice; these mainly used the 12-cylinder Liberty engine. But of the 2,000 F.2Bs ordered, and intended for Wright-built Hispano engines, the first American built example was tested with a most inappropriate, and badly fitted, Liberty engine - and promptly crashed! The US Army blamed the aeroplane and promptly cancelled the order; however a small batch of Hispano powered F.2Bs were built and successfully flown after the war.

By 1917 Wright-Martin were the principle American aero-engine makers - they built over 5,000 Hispano engines during the war - which would have made them the natural choice for licence building the 22.8 litre (5.5" x 6.5") ABC Dragonfly to meet British demands. Wright's first radial prototype was the 23.8 litre (5.6" x 6.5") 9-cylinder assembled during 1919, but this remained a very problematical engine. It developed 350hp, weighed an incredible 884lbs and had poultice headed barrels, but very little development work actually took place for, at war's end, Glenn Martin left to concentrate on his own aeroplane design. His place was taken by Fred Rentschler who, having worked on the Hispano development had no interest in radials. However, like the RAF, the US Navy had settled on air-cooled radials for their post-war requirements and selected a small capacity, 12.9 litre, Lawrence Aviation J-1 180hp, 9-cylinder radial, designed in 1919, which used cast-iron heads on poultice ended barrels; this was developed from Charles Lawrence's successful 3-cylinder radial design of 1912.

However, Lawrence's business was undercapitalised and, desperate to see their order for 200 J1, 9-cylinder radials in production, the US Navy virtually forced Wrights to merge with Lawrence to increase production capacity. Wrights' board were not at first interested, but given the fact the US would no longer buy their Hispano engine, they had little choice! With Lawrence's expertise on hand at the newly formed Wright Aeronautical Corporation of May 1923, they were able to develop their Wright prototype turning it into a reliable R-1 engine with direct help from Samuel Heron who, with Professor Gibson, had undertaken pioneering work at the Royal Aircraft Factory on cylinder head designs.

Siddeley-Deasey were also well aware of Gibson and Heron's work, for in January 1917 Major Frederick Green had joined them from the Royal Aircraft Factory having designed the BHP 8-cylinder engine which Siddeley-Deasey transformed into the Siddeley Puma 6-cylinder in-line engine. They then began work in late 1917 on a new short stroke, two bank, 14-cylinder Jaguar radial with screwed on aluminium heads, developed around Green's RAF No.8 radial engine in which Samuel Heron had also been involved. There was also a 7-cylinder Lynx version, but despite the Jaguar being successfully fitted to the Siddeley SR.2 Siskin, development of both Lynx and Jaguar engines was delayed until after the war due both to perfecting the Puma and by the Government's decision to commit to the ill-fated ABC Dragonfly. At the time, the Minister's decision was quite sound for although both Siddeley engines proved relatively reliable and trouble free, neither prototypes matched the crucial power to weight ratio of either Bradshaw's Wasp or Dragonfly engines.

In the less fraught post-war years, the Jaguar was greatly improved, although it still suffered torsional vibration. After a serious difference of opinion over John Siddeley's 'interference in (Heron's) cylinder head design', Samuel Heron left what was now Armstrong-Siddeley in 1921 and moved to America to work as a civilian research engineer at the McCook US Army engineering air base, in Dayton. His place at Siddeley's was taken by Spirito Viale who had joined Armstrong-Siddeley as Chief Designer of aero-engines in 1919. Viale later returned to his native Italy but as Mussolini's power grew in the 1930s, he returned to Britain and joined Rolls-Royce.

With Heron on hand at McCook, he was able to introduce his head technology to the US Army and Navy. The breakthrough came in September 1924, when Wright's President, Fred Rentschler left the company to develop even larger radials. At this, Lawrence became Wright's new President. Rentschler had secured financial backing from the Pratt and Whitney machine tool company and in July 1925 established their aero-engine division in direct competition with Wrights. Ironically, his first successful radial, with machine turned cast steel barrels, was the 425hp 'Wasp'. The prototype was completed in December 1925, within four months of conception, and tested by the US Navy in May 1926. Unlike Bradshaw's ill-fated Dragonfly, at least the Pratt & Whitney 'Wasp' had undergone proper development and been subjected to the new, gruelling, 50 hour proving trial at 90% rated output.

In 1926, Samuel Heron joined Wrights full time and, using his alloy 'Heron head' principle, fully redeveloped the Wright R-1. However it was the smaller Lawrence J-1, 9-cylinder radials which, as the Heron head equipped J-5, evolved into the Wright 'Whirlwind' - one of the world's most loved and reliable radial aero-engines of all time. Heron was also closely involved in the J-5 radial used in Charles Lindbergh's epic trans-Atlantic flight on 21st May, 1927.

The Wright Aeronautical Corporation and the Glenn Curtiss Aeroplane Company merged in 1929 to from Curtiss-Wright. In the 1960s they showed interest in Bradshaw's 'Omega' toroidal, reciprocating piston engine - described later.

ABC Gadfly

Granville Bradshaw still had high hopes for his radial engines and a 14-cylinder, two-row prototype was proposed towards the end of the war, but it doesn't appear as if it was built. However, he did develop a 5-cylinder radial engine in the late, or immediate, post-war years. Here again it is not known if a prototype was built. This 60hp radial is referred to as the ABC Gadfly, of which very little is known, but there are two references to its intended use in a new, experimental, 'Wind-waggon' motorcar of 1919 and in the Bristol Babe light aeroplane.

Impressed by the performance of the ABC Gnat in the PV7 and PV8 Kitten, Bristol's chief designer, Captain Frank Barnwell, saw great potential in a post-war civilian, 'owner pilot', single-seat biplane and duly designed the Bristol Babe (originally called 'Bobby') which was eventually sanctioned by the Bristol board in April 1919. Two ABC Gadflies were duly ordered, but soon afterwards, following their acquisition by Harper-Bean, ABC abandoned aero-engine work and the engines were never completed. Bristol then fitted a prewar, French made, 5-cylinder Viale radial of the design that Frank Barnwell had earlier helped install into an A V Roe Type F, but though the Viale gave a good performance, serious overheating limited its flying endurance to a mere half an hour. The Viale was duly replaced by a Gnôme et Rhône 60hp lightweight rotary, but such was this engine's horrendous vibrations at over 45hp, it was immediately rejected and the entire Babe project was abandoned.

Propeller fixing

One of Granville's patent applications, No. 106944, laid on 27th November 1916, was for an improved, stress relieving propeller blade mounting in which the blade root was fitted into a metal cup, or socket. This was keyed to the propeller boss to ensure its correct position and was retained by a single bolt passing through a deep, resilient washer which served a double purpose of easing stresses placed on the boss when fully tightened, and acting as a shock absorber cushion to minimise propeller 'flutter' under adverse conditions. Though a full specification was lodged in 1917, it appears the patent application went no further and it is not known if this device was ever tested by ABC on their 'Wind-Waggon' or used on an ABC aero-engine.

5
ABC Powered Aeroplanes

In November 1914, the RFC established its Experimental Flight at the Central Flying School (formed in 1911) at the former Army training camp at Upavon. Testing was then transferred to the new Aeroplane Experimental Unit at Martlesham Heath, Ipswich in January 1917. Farnborough was the home of the Air Battalion Royal Engineers No.1 (Airship) Company which became the Royal Flying Corps in May 1912. The camp's Royal Aircraft Factory built several famous aircraft, such as the BE.2 and SE.5, but by 1916 they had increasingly become a research establishment, building their first wind tunnel in late 1916. To avoid confusion with the new Royal Air Force (RAF), formed on 1st April 1918, the Royal Aircraft Factory was renamed in July 1918 as the Royal Aircraft Establishment (RAE). By 1920, the RAE Farnborough had become the principle research establishment for the Air Ministry, absorbing Martlesham Heath's flight testing and on 20th March, 1924 Martlesham Heath became the Aeroplane & Armament Experimental Establishment; this was moved to the relative safety of Boscombe Down in September 1939.

Bradshaw claimed the first Dragonfly powered aircraft had entered RAF squadron service in France on 10th November 1918, the day before the Armistice, but he is clearly wrong about this. Alan Lake's *Flying Units of the RAF*, shows no record of any war-time, or post-war, squadron service of any Dragonfly powered aeroplane. However, there were certainly several Dragonfly powered aeroplanes undergoing evaluation at both Martlesham Heath and the Royal Aircraft Establishment, Farnborough before, and after, the Armistice. There was certainly a BAT FK.23 Bantam undergoing field evaluation in France during the summer of 1918 and it could be this accounted for Bradshaw's misunderstanding. Most RAF squadrons were disbanded in early 1919 and few were reequipped until late 1920, long after the Dragonfly and many other engines, had been abandoned, and consequently, no ABC powered aeroplanes entered RAF Squadron service.

The following installations are well documented. It should be noted that an individual Mosquito, Wasp or Dragonfly engine was often used experimentally in more than one air-frame.

Airco (deHavilland)

George Holt Thomas, owner of the Aircraft Manufacturing Company, already held the British licences for the Farman designs. In 1914 he secured the services of Geoffrey deHavilland, the former chief designer of the Army Aircraft Factory, and now adopted new works in Colindale Avenue, Colindale, close to the Hendon aerodrome from where they tested their designs. He also secured the licence to build the Gnôme rotary aero-engine at Walthamstow. Three Dragonfly powered Airco DH.11 Oxford twin-engined, biplane bomber prototypes were ordered in early 1918 (s/n H5891-H5893) as a replacement for the similar DH.10, but only one, H5891, was completed by August 1918. Delays with the Dragonfly development caused a redesign in November 1918 for the 6-cylinder in-line, aluminium block, Siddeley Puma, however in March 1919, as the airframe was nearing completion, the ABC Dragonflies were finally fitted, even though there were known problems with the magnetos. H5891

DH11 twin engined bomber

first flew in January 1920, but after a few successful flights, a conrod bearing seized on take off, but the test pilot, F T Courtney made a safe landing. The ongoing problems caused deHavilland to consider an Oxford Mk.II variant (s/n H5891-H5893) powered by the Puma, but these mass produced versions of the BHP based, cast iron 8-cylinder Galloway 'Adriatic' engines, suffered endless teething problems, particularly with their porous aluminium alloy castings and poor valve gear - many a Puma powered DH.9 was lost through engine failure during active service. The Puma's problems were mainly due the production engineer's haste in increasing production output causing the 300hp Puma to be de-rated to 230hp, making it grossly under-powered for the, otherwise very successful, DH.9 bomber of 1917. As a result, both Oxford II airframes (H5892 and H5893) were cancelled leaving Airco to undertake a frustrating search for more powerful and reliable engines. They again settled for the ABC Dragonfly in a DH.11 variant, the DH.12, but this design got no further than the drawing board.

Geoffrey deHavilland's proposed post-war DH.20 single seat sports biplane, not dissimilar to the BAT Bantam, but with folding wings, was to have been powered by the ABC Wasp Mk.II 7-cylinder radial, but this project also got no further than the drawing board. With the loss of military orders, Holt-Thomas turned to air transport and motor car bodies but, on 1st March 1920, sold his Airco group to BSA. However, so perilous were Airco's finances that all but the air transport operations were put into liquidation, at which point Cpt. Geoffrey deHavilland (Airco's Technical Director) stepped in to acquire the aeroplane design and construction side, reconstituting the former Aircraft Manufacturing Company as the deHavilland Aircraft Co. Ltd on 25th September 1920, based at the nearby Stag Lane aerodrome. Airco's former office at The Hyde became the head office for Aerofilms Ltd and was renamed Aerial House.

Armstrong-Whitworth

Although Armstrong-Whitworth (A-W) had taken on production and development of both Adams' and ABC's V-8, V-12 and in-line aero-engines, none of these featured in any of Armstrong-Whitworth's military aeroplanes. However these engines did impress A-W's new designer, Frank Murphy, who had replaced Frederick Koolhoven when he left in 1917 to join BAT.

A late entry to an Air Board Type A.1a specification, (Long distance, high altitude single seat fighter) the stumpy and compact A-W Ara was a Dragonfly powered biplane, modelled on their BR.2 rotary engined Armadillo, then under development. Three Ara prototypes were ordered (s/n F4971-F4973); the first (F4971) being near completion in June 1918, but delays in Dragonfly production meant it was not fitted with an engine until September 1918, but then in October 1918, the Air Ministry cancelled the Ara in favour of the Nieuport Nighthawk and only F4971 was ever flown. Fitted with a Badin venturi fuel system, F4971 achieved 150mph in level flight and climbed to 10,000ft in $4^1/_2$

Armstrong-Whitworth Ara

minutes where she attained 145mph. Its excellent performance was aided by an aerodynamic nacelle which exposed only the upper cylinders, but this merely added to the engine's unreliability! In stark contrast, the lighter BR.2 powered Armadillo only reached 125mph.

The second Ara prototype (F4972) was finally completed in 1919 and was fitted with an enlarged rudder and revised upper wing. Although A-W recognised its high speed would serve it well as a post-war sports aeroplane, they decided to close their aeroplane department in late 1919 and the third prototype, F4973, was abandoned. These were the last aircraft to be built by Armstrong Whitworth at Gosforth and, having acquired Siddeley-Deasey in 1919, Armstrong-Whitworth gained the contract for the Siddeley 'Siskin' chosen by the RAF as their first post-war fighter (see Siddeley, below).

Austin Motors

The Austin Motor Company developed their advanced and well armed Greyhound 2-seat fighter/reconnaissance biplane for the Dragonfly engine in response to the RAF Type III specification, in competition with the Bristol Badger and Westland Weasel. Designed by John Kenworthy, who had earlier designed the RAF BE.2, three Greyhound prototypes were ordered in mid 1918 (s/n H4317-H4319). H4317 was built in the summer 1918 but delays in Dragonfly engine supplies meant it too was not completed until just after the war. She was eventually flown to the Royal Air Force's Aeroplane Experimental Unit

Austin Greyhound

at Martlesham Heath on 15th May 1919 for evaluation. As with the Ara, the Greyhound had exceptional performance, reaching 130mph at 6,000ft and 121mph at 15,000ft. H4319 was also completed with a modified tail plane and fin while H4318 went to C-Flight, Royal Aircraft Establishment at Farnborough where she underwent extensive tests on fuel systems, cylinder head design and engine cowling design. Tests at Martlesham Heath ended in 1920, but H4318's last recorded flight was on 12th June 1922. Plans by Austin for an air-postal service variant of the Greyhound were aborted.

AVRO

Alliott Verdon-Roe adopted the Dragonfly for their ambitious three-man, twin engined, biplane photo-reconnaissance bomber, the AVRO 533 Manchester which evolved from the Sunbeam pusher-powered AVRO 523 Pike of 1916 and the improved Rolls-Royce powered AVRO 529 long-range bomber version which first flew in March 1917. Three Pike prototypes were built for the RNAS in 1916 but the first two crashed at Martlesham Heath. Only two AVRO 529s were built.

The AVRO 533 Manchester Mk.I was built at their new Hamble works and featured both a novel king-post aileron operation and an aerodynamic engine nacelle design. Work on the three flying prototypes (s/n F3492-F3494) began in the summer of 1918. The first, F3492, was completed in October 1918 but delivery delays with the Dragonfly engines gave AVRO the opportunity to refit it with the 'Siddeley Puma' 6-cylinder in-line, 300hp engines; this became the AVRO 533A Manchester Mk.II and first flew in December 1918, remaining on trial until September 1919. It reached 125mph in level flight. This aircraft was then returned to Manchester to be fitted with Napier Lion engines, but the project was abandoned.

The second airframe (F3493) was fitted with the Dragonfly engines in late December 1918 as a Manchester Mk.I, and undertook extensive tests at Hamble before flying, in October 1919 to Martlesham Heath, taking only 90 minutes. Evaluation flights recorded 128mph. Both Manchesters Mk.I and Mk.II had staggering performance for a large biplane which could, incredibly, loop the loop! F3494 was completed as a Mk.III, but it never received its intended 400hp in-line Liberty engines for, with the war now over, there was no further demand for heavy bombers.

A proposed AVRO 532 two seat, short range, reconaissance/spotter biplane fighter fitted with a Dragonfly, was designed in April 1918 to meet RAF Type IIIA, IIIB and IVB specifications, but this got no further than the drawing board. A single seat AVRO 531 Spider biplane prototype was built in the spring of 1918 and fitted with the Clerget 130hp rotary, with optional ABC Wasp II or BR.2 engines. Intended as a replacement for the AVRO 504K night-fighter variant, the RAF showed no interest as

AVRO 533 Manchester

they had by then selected the Sopwith Snipe for that role.

At least two AVRO 504K biplanes were fitted experimentally with the ABC Wasp Mk.I and Mk.II in 1919: the civilian registered AVRO 'work's hack' (K-147) first flew with a Mk.I and then acted as a test bed for AVRO's proposed 504L seaplane, but the Wasp proved unsatisfactory and the BR.2 radial became standard fitting for the 504L. K-147 was later converted to an AVRO 548 and sold to the Welsh Aviation Company (re-registered G-EAFH) with whom it won all three races at the Croydon meeting in September 1921 with, it appears a Renault 80hp in-line engine. When the Welsh Aviation Company went into receivership, G-EAFH was bought privately by a Dr Whitehead of Bekesbourne, Canterbury and used on his doctor's rounds! She was later sold for acrobatic displays, but crashed in 1935.

AVRO 504K

The other 504K was s/n D9068, built by Grahame-White at Hendon, and fitted with the latest ABC Wasp II in 1919. This was used by RAE Farnborough to test new fireproofing methods using asbestos and aluminium bulkheads, asbestos insulated pipe work and fuel tanks. In its modified form she first flew in February 1920 but in June 1920 suffered crankshaft failure which delayed further flying until September 1922.

B.A.T

The British Aerial Transport Company was formed at Willesden, North London in 1917 by Samuel Waring, of furniture fame (see also Appendix 1 Walton Motors). He chose as his designer the Dutchman, Frederick Koolhoven (formerly Chief Engineer of the British Deperdussin Syndicate and from 1913 chief designer with Armstrong-Whitworth). Koolhoven had co-built the 'Gordon Bennet' Deperdussin racer of 1913 - the first aircraft to attain 120mph in level flight.

The prototype BAT 'Bat', later designated FK.22, was built as a private venture in 1917 under Licence No.11, and carried no official serial number. First flown in that September with the ABC Mosquito engine, she had a cigar shaped, monocoque ash and birch ply-wood fuselage, with upper and lower wings fixed direct to the fuselage, rather than by struts. The pilot's head was above the upper

BAT FK22 Bantam with Mosquito Radial

Top, BAT Bantam FK23. Below, BAT Baboon and bottom, BAT Basilisk

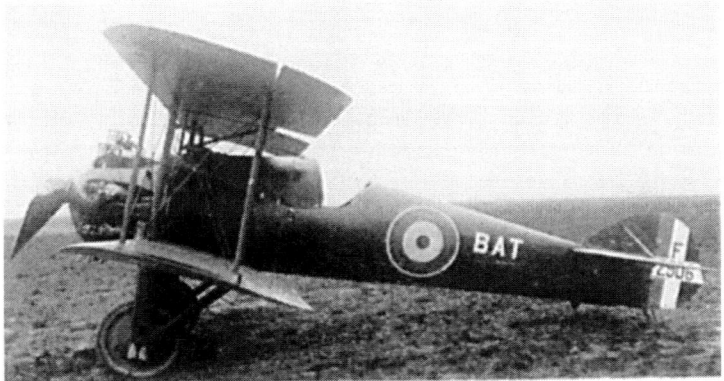

wing. Although the engine was known to be underdeveloped, it so impressed the Ministry of Munitions that an initial order was placed under contract AS 25314 for six BAT Bantams (s/n B9944-B9949), of which only four were built and flown in 1918.

Conflicting reports suggest that B9944 was originally fitted with the ABC Mosquito 120hp radial as an FK22, but as the underpowered engine and frame proved unsatisfactory, after close inspection, the Mosquito was rejected by the Royal Aircraft Factory and B9944 was cannibalised in December 1917 for spares in favour of the improved FK.22/2 'Bantam' Scout (s/n B9945) and, now fitted with a 100hp Gnôme engine, went for evaluation at Martlesham Heath in January 1918, proving a most satisfactory aircraft to fly. In February 1918 she was transferred to a Training Squadron and thence to the Central Flying School and was later fitted with a Gnôme et Rhône Monosoupape 110hp unit as a 'Bantam Mk.II'. The other FK.22 airframes all differed slightly; B9946 was to be fitted with a 170hp Wasp, but it appears was never completed. The Wasp powered B9947 was flying at Martlesham Heath by the March 1918 as an FK.23 but was returned for modifications to both wings and engine on several occasions.

In July 1918 trials of all Wasp powered BAT Bantams, Sopwith Snails and Westland Wagtails were suspended 'until engine trials had been overcome'; it appears these were soon resolved and B9947 finally went to Farnborough in September 1918. It also appears that both B9948 and B9949 were cancelled in early summer, but with the improved 200hp Wasp Mk.II engine becoming available, they were resurrected in 1919 by BAT and re-registered J6579-J6580 as working prototypes for the Wasp Mk.II engine. Both were sent to Martlesham Heath in February 1921 for comparative tests against the Westland Wagtails, but persistent (undisclosed) problems with the Wasp Mk.II meant only J6579 underwent brief trials; the Wasp Mk.II proved no more reliable than the Mk.I.

The FK.23 Bantam's improved and streamlined body and its 138mph level flight (compared to only 100mph with the Gnôme) made it one of the fastest of its type! An order for twelve Wasp powered FK.23 Bantams was placed in spring 1918, (s/n F1653-F1664) and all but F1662-F1664 were completed. F1653 was being evaluated at Farnborough in July 1918 as part of C-Flight and was extensively tested in looping the loop and engine cooling, for which she was fitted with a new ducted cowl in November 1918. These Wasp powered Bantams proved fast and manoeuvrable but were prone to difficult spin recovery, although that was soon overcome by revised aerofoil controls. Officially regarded as an abandoned project in January 1919, test flying of F1653 continued at RAE Farnborough until summer 1921 when she was flown by Flt. Lt Paul 'George' W S Bulman at the July 1921 Hendon air display after which she remained at Farnborough until late 1921. F1654 first flew in August 1918 and went to Martlesham Heath and thence to Orfordness for armaments trails.

F1656 was also test flown post war by C-Flight, RAE Farnborough in extensive engine and fuel cock testing and was also fitted with a 50% larger, experimental rudder and elevators, but having established the criteria for stability in flight, she became surplus to needs and was disposed.

One of this batch of nine Wasp powered FK.23s was flown to France by BAT's test pilot, Peter Legh in the summer of 1918 for evaluation at France's principle aerodrome/test station at Villacoublay near Versailles, south west of Paris. This most probably accounts for Bradshaw's erroneous claim that Dragonfly powered aeroplanes were in RAF service in France at the time of the Armistice. After the war most of these Bantam FK.23s entered into the Civilian Register. F1654-F1657 became, respectively, K123 (later G-EACN), K125 (later G-EACP), K154 (later G-EAFM) and K155 (later G-EAFN).

BAT now offered their Wasp II powered Bantam as their 140 mph 'Sport'. F1657 (K155), which had a specially modified lower wing for racing, was flown by BAT's new test pilot, Cyril Turner and won the May 1919 Whitsun cross country event. F1661 was fitted with a Wasp Mk.II and tested at Martlesham for civilian service; she reached 146mph at 10,000ft, taking 17 minutes to reach 17,000ft. Her controllability and manoeuvrability were described as better that any existing fighter. Severe vibrations from the engine were soon cured by additional stays. F1661 then entered the civilian register as G-EAYA and, after BATs' collapse, was bought back from its private owner by its designer, Frederick Koolhoven and taken to Holland in 1924 where she was registered as H-NACHS and, after extensive modifications to the airframe, was fitted with the 200hp Siddeley Lynx engine, achieving 153mph in 1924.

Of the remaining three FK.23 airframes, one is said to have later been fitted with a Gnôme et Rhône Monosoupape 100hp radial; another entered the United States Air Service as a P-167, s/n AS/94111, going into storage in 1922. A post-war airframe was registered as K143 (G-EAFA) and was converted to a side-by-side FK.27 two seater aerobatic aeroplane in 1919; she flew at Hendon in 1920 powered by a Wasp Mk.II.

The Bantam was developed into the FK.25 Basilisk experimental scout/fighter powered by an ABC Dragonfly engine. Three prototypes were built in the late spring 1918 (F2906-F2908) and, once again due to engine delays, first flew in September, reaching 162mph - one of the fastest aircraft of its day! F2906 crashed on 3rd May 1919 following a fire during an altitude test flight from Hendon, when

BAT's test pilot, Peter Legh jumped to his death just feet from the ground. F2907 was completed in 1919 and then evaluated at Martlesham Heath in October 1919. F2908 was not completed until 1919. Due to a design flaw, the Basilisk airframes suffered from weak undercarriage and only a few tests were carried out before abandoning their evaluation in mid-1920.

BAT also developed the FK.24 Baboon experimental advanced tandem trainer under the A.2d specification. This was intended for the ABC Wasp Mk.I engine and six were ordered (s/n D9731-D9737), of which only the first three airframes were completed in 1918. In the event, only one was fitted with a Wasp engine (D9731), the other two with ABC Dragonflies. BAT entered D9731 into the Civilian Register as K-124 (G-EACO) in July 1919 and won the July 1919 20 mile Hendon cross-country event. She had a flying speed of only 90mph. BAT intended offering the Baboon after the war as a flying school trainer, but no orders came.

An ABC Gnat powered BAT FK.28 'Crow' prototype was built and flown in 1919, (this has been described earlier). Ambitious plans were also laid for a plywood bodied FK.26 'Commercial' bi-plane freighter/enclosed four-seater cabin aeroplane, with the pilot in a rear mounted open cockpit. Having a 46ft span and 1,000-lbs payload, it was powered by a single Rolls-Royce Eagle VIII engine. The prototype was completed in early 1919 and, with a pair of Bantams, was entered in the 1919 Aerial Derby. BAT then started a regular London-Birmingham air service with the Commercial in September 1919 and flew several commercial flights in October 1919 between Hounslow, Amsterdam and Paris, with plans for more regular commercial services. But with no military or civilian orders forthcoming, BAT's aircraft manufacturing business closed and after a major fire at the Central Aircraft Company's works in March 1920, BAT let a major part of their Willesden works to CAC. Frederick Koolhoven left in 1920 and returned to Holland with Bantam K123, which survives in the Aviodrome Museum, Lelystrad, Holland.

Boulton-Paul

More famous as ironmongers, wire-workers and woodworkers, Boulton-Paul survive today as a major joinery business. During the Great War, they assembled under contract many Sopwith Camels, but Boulton-Paul also produced several aeroplanes of their own design including three prototypes of their twin engined P7 Bourges biplane fighter-bomber of late 1918 (s/n F2903-F2905). These were designed for the ABC Dragonfly engines and to have the performance and climb of a fighter; indeed they reached 124mph and became the first aerobatic twin engined aeroplane to loop the loop, but only after the wicker-work seats had been shown to be too weak for attaching any safety harnesses! Each prototype differed slightly for Boulton-Paul were so desperate to get the Bourges airborne, that the first prototype, F2903 (built in November 1918) flew with 200hp Bentley BR.2 9-cylinder rotaries and was only later fitted with Dragonfly radials and then flown to Martlesham Heath for evaluation in August 1919. The second prototype, F2904, was a Dragonfly powered Bourges Mk.IA with a special gull wing top-plane to increase the rear gunner's field of fire. The third, F2905, a Mk.II, had a single, nose mounted 450hp Napier Lion engine with its unique W-12 cylinder arrangement and a straight upper wing.

The Boulton-Paul Bourges

All three Bourges entered the civil register: F2903 as K129/G-EACE (scrapped in May 1920) and F2905 as G-EAWS. The second prototype, F2904, crashed in early 1919 but was salvaged, transformed into the 'P8 Atlantic' and refitted with twin Napier Lion engines in the hope it would take Lord Northcliffe's £10,000 trans-Atlantic prize, but it crashed on take-off!

Bristol

The Bristol & Colonial Aeroplane Company was most famous for their F.2B fighter of early 1917, designed by Frank Barnwell and affectionately known as the 'Brisfit'. But even as this was entering production, Barnwell was already working on its successor, the F.2C to meet RAF Type IIIA and IIIB specifications.

Alas, neither the existing 230hp BR.2 rotary nor the prototype 9-cylinder Salmson water cooled 260hp radials were able to meet the Air Board's specification. But on the announcement of the 320hp Dragonfly, Barnwell redesigned the F.2C in April 1918 and three F2.C Badger RAF Type III short range fighter-reconnaissance prototypes were ordered (s/n F3495-F3497). The first air-frame was ready by June 1918, but delays in the Dragonfly production caused the Air Board to revise the order and allow the fitting, experimentally, of one of the new 9-cylinder 400hp Cosmos Jupiters then under development, into F4396, redesignated Badger Mk.II. With the war now ended, only F3495 was fitted with the improved Dragonfly II.

This airframe provides a convenient direct comparison between the two engines. The Dragonfly powered F3495 first flew on 4th February 1919, but crashed due to a fuel airlock. She was rebuilt with a larger rudder and flown again on 13th May 1919 achieving 135mph at sea level and 122mph at 15,000ft. In contrast, the more powerful Cosmos Jupiter powered F4396 of April 1919 managed 145mph at sea level. Major engine cooling problems with the Dragonfly were resolved by using different nacelle designs, but a major, and unresolved fault remained with crankshaft failure, a problem shared by the Sunbeam Arab and only later discovered to be due to synchronous torsional vibration. However F3495 was still flying at Martlesham Heath in September 1919 but appears to have by then been fitted with a Cosmos Jupiter engine.

Though the Cosmos powered F3496 was completed in October 1918, it first flew on 24th April 1919. She was evaluated at Martlesham Heath until 1920 and then Farnborough, where she showed great promise, although the airframe did require further modifications. Parts from the partially built, but cancelled, F3497 were used for the Siddeley-Puma powered civilian Badger X (K-110 G-EABU) which first flew in May 1919 but subsequently crashed and was rebuilt for civilian use. Impressed by

Bristol Badger

the Jupiter powered Badger II, a further Badger II was built to Air Board order (s/n J6492) for post-war testing of the Jupiter radial engine, but the engine's continuing development had by now cost Cosmos dear and forced their liquidation. J6492 was later fitted with the new Bristol Jupiter II in 1921 and was then evaluated at Martlesham Heath and later Farnborough.

English Electric

The company drew up plans in 1919 for a two-seat, twin float amphibious fighter under the P10 project, which included six proposals for a biplane or triplane design, using a Rolls Royce Falcon V-12, Cosmos 9-cyl Jupiter or ABC Dragonfly radial. The project went no further than the drawing board. The company had greater success with the post war ABC powered Wren, discussed in the next chapter. During the last war, English Electric built Halifax and Hampden bombers under contract as well as early examples of the deHavilland Vampire jet fighter. They later became part of the British Aircraft Corporation.

Nieuport

The Nieuport & General Aircraft Company of Cricklewood had the greatest potential demand for Dragonfly engines for their Nighthawk single seat fighter. Highly praised by its pilots, the Nighthawk was widely held at the time as being one of the finest aerobatic aircraft ever built and, being one of the most advanced fighters of its day, entered RAF service after the war as their first operational aircraft fitted with a radial, as opposed to rotary, engine.

Nieuport's chief designer was Henry (Harry) Philip Folland who designed the Nighthawk specifically for the Dragonfly engine to meet the Ministry specifications. As he had also designed the Royal Aircraft Factory's SE.4 and SE.5 fighters, the Nighthawk incorporated many SE.5a parts and, being very similar to the Sopwith Snipe, facilitated ease of maintenance in the field.

Work began in May 1918 on the three Nighthawk prototypes (s/n F2909-F2911). So satisfied were the Ministry with the airframe that they once again took the rash decision to place huge orders before the prototype, F2909, had even flown! Production thus got under way in late 1918 with F2909's unflown airframe now being modified for shipborne use. After the Armistice, the Ministry revised an order for 150 Sopwith Snipes, placed through Nieuport, to be built as Nieuport Nighthawks (s/n H8513-H8662), however only 40 of these (H8513-H8553) were completed during 1919.

Nieuport Nighthawk

It appears F2910 was the first to fly with the improved Dragonfly II engine and both she and F2911 were tested at Martlesham Heath in July 1919 but F2911 later suffered a collapsed undercarriage and was scrapped in early 1920. The Dragonfly powered Nighthawk proved to have superb handling and a magnificent performance of 151mph at sea-level, even if the engine's unresolved problems did limit its service life to a few hours.

The Ministry's faith in the Nighthawk airframe however proved sound and five were converted for the RNAS as Nieuport 'Nightjars' (s/n H8535-H8536, H8538-H8540) with H8535 being evaluated at Farnborough while H8553 was modified for ship-borne service with an undercarriage 'hydrovane', designed to prevent an aeroplane tipping nose-first when making a landing at sea for recovery by the mother-ship. A further 15 were ordered in 1919 (s/n J2403-J2417) however many of these are recorded as 'reallocated'. J2403 achieved 43 flying hours at Martlesham Heath between February and May 1920 before being scrapped due to its particularly troublesome Dragonfly engine; J2405 and J2416 with Dragonfly Mk.IIs were used in general equipment and parachute evaluation between late 1920 and 1922 along with H8533 which had been damaged on arrival and never flew again. J2416 then became a Gloster Mars VI 'Nighthawk' with the Siddeley Jaguar radial engine. This order was followed in 1919 by a peace time order for 48 Nighthawks (s/n J6801-J6848) to be built by both the British Caudron Company and the Royal Aircraft Establishment who were still carrying out extensive work on the Dragonfly to bring it up to muster, but the order was eventually cancelled.

Both J6925 (Jaguar radial) and J6927 (Jupiter radial) were briefly tested at Martlesham Heath in early 1923 but both showed poor performance at high altitude and were exceptionally noisy; indeed, these early post-war Jaguar engines proved very troublesome though Martlesham Heath conceded they were more reliable than the Dragonfly!

Meanwhile, H8534 had been re-engined and became a Gloster Mars VI, while H8544 had become a civilian racer. After the war, Nieuport built three Nighthawk variants as demonstrators for the civilian racing market or for fast air-mail services. The first was a Nighthawk, K151 (LS1 - later G-EAEQ) which flew at the 1919 Hendon Aerial Derby and went to India in 1920. The second, G-EAJY was a Nieuport 'Nieuhawk' and first flew in July 1919; she was then entered into the 1920 Aerial Derby. The third was a Nieuport 'Goshawk' (LS3 G-EASK) fitted with a Dragonfly engine and which set a British Air Speed Record at Martlesham Heath in June 1920 of 166.5mph in the hands of Maj. J H Tait-Cox. G-EASK was then loaned by Harry Folland to Harry Hawker for the 1921 Aerial Derby; tragically Hawker was killed flying it in a practice session.

With so few post-war orders at war's end, the Nieuport & General Aircraft Company was forced to close in November 1920. Their designs and assets were bought by the Gloucester Aircraft Company who had secured settlement with the Government in late 1919, as compensation for the cancelled Nieuport Nighthawk assembly contracts by securing sufficient quantities of surplus Nighthawk parts for them to continue the Nighthawk's development under their 'Mars' programme with Harry Folland, who had joined Gloster as their Chief Design Engineer. But, interesting though it is, as the Mars does not involve the ABC Dragonfly engine, the reader is referred to Derek James' authoritative work on Gloster. Suffice to say, the Nighthawk evolved into the Siddeley Jaguar powered Mars VI Nighthawks and the Bentley BR.2 powered Mars X Nightjar, of which only a handful entered RAF Squadron service in 1921. The design was declared obsolete in 1923, but one Nightjar, J6969, became the Gloster Grebe prototype from which evolved the Gloster Gamecock of 1924 and thence the Gauntlet and finally the famous Gladiator.

How fitting then that the ill-fated Dragonfly radial should have inspired both the first and last radial engined biplane fighters in RAF service! Harry Folland left Glosters in 1935 to establish the British Marine Aircraft Company at Hamble which became Folland Aircraft Ltd in 1937.

Nieuport London bomber

In addition to the Nighthawk, Nieuport also developed the London triplane bomber of 1919, powered by two Dragonfly engines and probably to the same RAF specification as the Sopwith Cobham. Six prototypes were ordered in 1918, (s/n H1740-H1745) with H1740 first flying on 13th April 1920 and H1741 flying in July 1920. Both were evaluated at Martlesham Heath. The remainder were cancelled in November 1918 before assembly had begun.

With Nieuport's closure, their Cricklewood factory, at the junction of Temple Road and Cricklewood Broadway, was taken over by S Smith & Sons for their motor car instrument works. Opposite them, across the marshalling yard, was Handley-Page's Somerton Road works.

Siddeley

The Siddeley-Deasey Motor car company began aero-engine design in 1917 when Major F M Green joined them from the Royal Aircraft Factory (he had no connection with the Gustavus Green engine company). Their in-line Siddeley Puma and RAF No.8 inspired Siddeley Jaguar radial was described in the last chapter. However, the Jaguar was intended for their speculative, experimental, SR.2 (later called the Siskin), based on their SE.7 proposal and a batch of six prototypes was duly ordered in mid 1918 (s/n C4541-C4546), but Siddeley were concentrating so much on their troublesome Puma in-line engine that, though C4541 was not completed until May 1918, it was still awaiting an engine. Then, following the placing of a huge order for the Nieuport Nighthawk, the Siddeley contract was cut to just three SR.2 airframes and instructions were issued that they all be fitted with the ABC Dragonfly. Although C4544-C4546 were cancelled, it appears work on these was resumed, though was never completed.

Dragonfly production delays meant that C4541 did not fly until April 1919. She and C4543 went to Martlesham Heath for evaluation. Both Siskin's Dragonfly engines proved trouble-free and with sufficient endurance. The C4541's performance and handling was highly praised, yet these two Dragonfly engines yielded only 320hp compared to those pushing out 340hp in the Sopwith Snarks. Despite their acclaim, little interest was shown in the Dragonflies and in late 1919, C4541 was refitted with the new Armstrong-Siddeley Jaguar engine. Now known as the Armstrong-Whitworth 'Siskin', it was ordered in large numbers for the RAF, along with the Gloster Grebe, to form their new post-war fighter squadrons. Having proven her worth, C4541 was returned to Siddeley in August 1921, albeit dismantled. C4542 was flown from Coventry to Martlesham Heath in January 1920, covering the 120 miles in 50 minutes, but severe engine vibration had caused damage to the airframe and she was duly scrapped. C4543 underwent extensive trials between March 1920 and June 1921 with a modified Dragonfly II engine.

Sopwith

Despite their close friendship and Tom Sopwith's personal admiration for Bradshaw, the ABC Dragonfly was not much liked by the company and ABC's association with Sopwith is rarely mentioned in official Sopwith histories, yet the Dragonfly was fitted experimentally to several Sopwith airframes including a proposal for a Wasp in a Sopwith Pup.

Designed to Air Board Specification A.1a, the Sopwith 8F.1 Snail high wing monoplane of late 1917 was their first plywood, monocoque construction airframe and was destined to be fitted with the new ABC Wasp radial. An order was placed under contract AS37484 for six Snails in November 1917 (s/n C4284-4289). Only the last two had a plywood skinned monocoque fuselage. As so much trouble had been experienced with the Wasp in the BAT Bantam, the next available Wasp engine, which was originally allocated to the second Snail, was diverted to BAT for their Bantam. It was the third Wasp engine which finally arrived at Sopwiths in March 1918 to be immediately fitted to Snail C4284. A further engine soon arrived for C4285 and they both flew at Brooklands on 4th April 1918. C4284 attained 124mph at sea level and 121mph at 15,000ft. Both she and C4285 then went to Martlesham Heath, but high altitude flying caused the carburettors to ice up, delaying their tests, so two of the exhaust ports were redirected through an inlet muffler to eliminate icing.

The Wasp powered C4284 and C4288, having been tested at Brooklands, went to Martlesham Heath on 9th May, 1918. The stiffer, monocoque C4288 was faster, at 116mph, than the Sopwith Camel

Right, Sopwith Snail. Below, Sopwith Snipe

Dragonfly powered Sopwith Dragon

but far heavier, less manoeuvrable and controllable; she was duly diverted to RAE Farnborough for further tests in June 1918. The Snail also proved inferior in flight to the Sopwith Snipe and, with Wasp production being abandoned in October 1918, the now incomplete airframes of C4285 and C4286 were sent to Farnborough as spares for the two flying prototypes before they were all then broken up for firewood in November 1919.

The Sopwith 7F.1 Snipe of late 1917 was intended to be the Camel's successor. Six prototypes (B9962-9967) were built under contract AS31668. Most used the Bentley BR.1 or BR.2 rotaries, but the sixth (B9967) was fitted with the first production Dragonfly engine (No.1) as a Snipe Mk.II and then test flown at Brooklands in April 1918 alongside the Dragonfly powered Sopwith Bulldog Mk.II (H4423) - also said to have been fitted with the first production engine - but despite its phenomenal performance of 147.8mph at 10,000ft (compared to a miserable 121mph from the BR.2 radial engined Snipes), B9967 suffered persistent magneto failure together with misfiring due to carburettor icing at 18,000 ft, as did the Snail, when being extensively tested by the RAE to cure the Dragonfly of its ills. She was refitted with a Clyno built Dragonfly (No. WD48204) in September 1918 but problems with the oil circulation on the Clyno engine caused bearing failure; this was quickly cured by Bradshaw at Walton Motors, but then a magneto shaft broke, shattering the engine. A Dragonfly Mk.II, built by Sheffield-Simplex in early 1919, was then fitted with a modified exhaust heated carburettor manifold muffler to resolve carburettor icing, but it also suffered magneto shaft failure.

Despite Martlesham Heath's reservations about the Snipe's airframe, production models were fitted with the readily available BR.2 rotary, but so impressed was the Ministry by the Dragonfly powered B9967's earlier performance that one Snipe, E7990 was modified in July 1918 to a new specification as the Sopwith Dragon; she first flew in February 1919. A further thirty Dragonfly powered Snipes (F7001-7030) were ordered as Sopwith Dragons with F7001 being completed in July 1918. F7003 flew with C-Flight RAE on endurance tests while F7017 was sent to Martlesham Heath and remained there for testing until October 1919.

A further batch of 500 Snipes (J2542-J3041) was diverted from an order of 950 for conversion to Dragons. This was followed by an order for 300 Dragons just after the Armistice (s/n J3617-J3916) with

Sopwith Bulldog

the aim of forming a new squadron of high performance fighters, but only four of this large batch were delivered by Sopwiths (J3628, J3704, J3726, J3809); the rest were cancelled. J3726 was sent to Martlesham Heath and remained there for testing until October 1919. J3809 joined C-Flight, RAE Farnboorugh and underwent extensive engine testing including rate of ascent, slow running and climbing tests with pilots using oxygen.

There is a report of a Dragon being prepared for the US Army. It was fitted with the only Wolseley built Dragonfly engine, which promptly broke down during its inaugural test on 28th June, 1918! It appears this was J3628 which was then sent to the McCook Field, Dayton, Ohio, for evaluation by the US Army's aeronautical engineering establishment where she became P-149 (s/n 94106). She was still flying in 1926, but with which engine is not known.

When flying with a perfectly running Dragonfly engine, the Dragon was a very highly regarded aeroplane with a phenomenal 150mph performance, precise handling and a service celling of 25,000ft, but the engine's faults were never fully cured and though the Sopwith Dragon was officially adopted by the RAF in September 1921, it never entered squadron service and was declared obsolete in April 1923.

The Sopwith Bulldog of 1918 was designed as a private venture in response to an Air Board Type A.2 specification for a two seat fighter to replace the Bristol F.2B 'Brisfit'. The Bulldog was very similar to the Snipe and four prototypes were ordered as X2-X5, although later records variously show them as X2-X4 or X3-X5! X3 and X4 became H4422 and H4423 Bulldogs Mk.I and Mk.II respectively. ('X' signifies a licensed civilian prototype). Intended to be fitted with a Hispano-Suiza 200hp engine, supply problems meant the first airframe, X3, was fitted with a Clerget 11-cylinder rotary engine which proved hopelessly under-powered. X4 (H4423) was ready by April 1918 and was fitted with the first production Dragonfly built by Clyno (WD48204) in June 1918, making it a Bulldog Mk.II. She attained 101mph at 13,000ft, but after 20 hours flying, was refitted in June 1918 with the improved Clyno built Dragonfly Mk.II (Clyno 50008) while Clyno WD48204 went into a Sopwith Snipe. Once again we have a direct comparison of performance between two different engines when tested at Martlesham Heath: the Dragonfly powered X4 achieved 15,000ft in 16 minutes compared to 38 minutes for the Clerget rotary powered X3. H4423 remained at Farnborough until March 1919, when all further work on the Dragonfly was abandoned. X5 was cancelled in May 1918.

The Cobham was Sopwith's only attempt at a massive, twin engined, triplane bomber. Three prototypes were built in August 1918 (s/n H671-H673) to a general specification for the RAF. These were designed for the Dragonfly engine, but production delays lead to H671 being fitted with the Siddeley Puma in-line as a Cobham II in July 1920; she was completed in spring 1919 and was evaluated at

Sopwith Cobham bomber

Right, Sopwith Snark, below Sopwith Snapper

Martlesham Heath in August 1919. The other two airframes were eventually fitted with Dragonflies: H672 arrived at Martlesham Heath in May 1920 but crashed on take off in August 1920 due to her undercarriage collapsing, ending further trials. H673 was grounded in July 1919 following its maiden flight when the project was abandoned. All were disposed of in late summer 1920.

Three prototype Sopwith Snark Triplanes were ordered in April 1918 (s/n F4068-F4070) against an RAF Type 1 high altitude fighter specification. Problems with the Dragonfly engines temporarily halted production of the Snark until June 1918. Only F4068 was completed and first flew in April 1919, however persistent magneto problems temporarily grounded the aircraft until September 1919. She was then evaluated at Martlesham Heath in November where she remained until 1920 when her Dragonfly engine gave 340bhp at 1,650 rpm but her top speed of 130mph was considered too slow and along with reservations over the airframe design, her trials were not completed. The other two Snarks were finally built in late 1919 with F4070 first flying in 1920 but she undertook very few trials between October 1920 and March 1921. These three Snarks remained at Martlesham Heath until late 1921 and although said to be one of the best performing triplanes of all types, they were not thought as satisfying to fly as the earlier Sopwith Triplane.

The Sopwith Snapper Scout was an exceptionally fast biplane built by Sopwith to RAF Type 1 specification; three were ordered in May 1918 (s/n F7031-F7033). The first to be assembled (F7031) was held back awaiting an engine and was not completed until April 1919. She was first flown in July 1919 at Brooklands and thence to Martlesham Heath in September and finally to Farnborough in January 1920. She reached 140mph at 3,000ft and 133mph at 15,000ft. The other two were flying at Farnborough by June 1920. F7032 was then taken back by Sopwiths, re-registered as K-149 (later G-EAFJ) and entered in the first post-war Aerial Derby of June 1919 to be flown by Harry Hawker, but they were forbidden to fly it as a private entry for the Dragonfly engine was still under development for the RAF and, thus, on the Official Secrets List! F7032 was scrapped in August 1920. F7033 was completed in June 1920 and then sent to Farnborough for research work.

Westland

The Westland company was formed in 1915 by twin brothers, Ernest and Percy Petter as an offshoot of their famous marine engine company, Petters Ltd of Yeovil, Somerset. In 1915 Westland was contracted to assemble Short seaplanes followed by Sopwith Strutters and in 1916, went on to design the Westland Wagtail biplane Scout and Weasel biplane fighter. Ironically the Wagtail was known at Westland as the 'Wasp' and was due to be called the 'Hornet' in official circles, but under the Air Board's new classification system of 1918, all Westland aeroplanes would henceforth be named after birds.

The Wagtail prototype was designed in 1917 to Air Board Specification A.1(a) as a solo fighter. Although six prototypes were ordered (C4290-C4295), this was promptly reduced to just three (C4291-C4293). C4291 was assembled in January 1918 and completed in April 1918 with the fourth prototype 170hp Wasp Mk.I engine; she even looped the loop on her maiden flight, but problems with the engine saw C4291 grounded on 20th April and the cylinders returned to ABC. In May 1918 she was tested at Martlesham Heath against a Martinsyde F3 in a dog-fight; the Wagtail outturned the F3 on each occasion, but its 170hp Wasp was no match in level flight and climb to the F3's 300hp Hispano-Suiza. An unfair test, perhaps. But the Wasp's reliability problems remained and at the end of May, all Wasp powered aeroplanes were grounded until the Crossley-built Wasp engines arrived in September 1918. C4291 flew again in October 1918 and then went to Orfordness for gunnery trials in November. The Wagtail reached 125mph at 10,000ft but though an improved Wasp Mk.II was duly fitted, the RAE had by now declared the Wasp's faults were incurable and

Top, Westland Wagtail, below, Westland Weasel

Wasp production ceased in October 1918. C4292 survived until 1920, still with the original Wasp Mk.I engine while a modified C4293 began its trials in April 1918, flew at Martlesham Heath in May and thence to Farnborough on 18th May.

The Wagtail was widely held to be the best to fly in its class and a further two Wagtail airframes were ordered in 1920 (s/n J6581-J6582), along with a Bristol Badger and a Westland Weasel as test beds for the new generation of radial engines. J6581 was fitted with the Wasp Mk.II radial in January 1921 and flown to Farnborough for trials, but in September 1921 she was refitted with a 160hp Armstrong-Siddeley Lynx. She then flew to the Experimental Establishment at Grain before returning to Martlesham Heath; she was still at Farnborough in 1922. J6582 also had her intended Wasp Mk.II replaced by the Siddeley Lynx in November 1921 and successfully passed the 50 hour endurance trials. She was still flying at Martlesham Heath in October 1922.

The Weasel was a 2-seat fighter developed from the Wagtail to RAF Type IIIA specification. Three prototypes were built in summer 1918 (s/n F2912-F2914). F2912 and F2913 were fitted with Dragonfly engines in December 1918; F2912 was evaluated at Martlesham Heath in April 1919. F2913 first flew at Yeovil in February 1919 before going to Farnborough. These Dragonfly powered Weasels achieved 147mph at sea level and 122mph at 15,000ft. F2913 was later fitted with the Cosmos Jupiter 9-cylinder radial and remained in RAF service until 1924. F2914 was also flown at Martlesham Heath in November 1919 but was later fitted with a supercharged Armstrong Siddeley Jaguar engine and was used as a flying test bed at RAE Farnborough until 1925. One further, much modified, Cosmos Jupiter 9-cylinder powered Weasel was ordered in August 1919 (J6577) but was not completed until the autumn of 1921 when it was flown at Farnborough on very rare occasions, as the new 375hp Jupiter engine was reportedly considerably heavier than the Dragonfly, for which the Weasel airframe had been designed. This so severely reduced the structural safety of the airframe that only straight and level flying was permitted. She barely reached 130mph at 10,000 feet and only then, by fitting heated muffling to the induction pipes to prevent icing, was the improved Jupiter passed fit for flying. However, a few days after its $3^{1}/_{2}$hr endurance flight, the Jupiter powered Weasel (J6577) caught fire in mid air and crashed on 29th July, 1922 after only 46 hours flying time; the pilot was lucky to escape.

Synchronised machine-guns

A major turning point in aerial warfare came with the synchronised firing of machine guns through the fast spinning propeller blades giving the pilot of a single engined fighter far greater accuracy. Firing was triggered by an engine driven cam through mechanical action and, later hydraulically in the superior post-war Constantinescu system.

Many of the ABC radial powered fighters were so equipped and at least one ABC auxiliary engine was employed at RAF Uxbridge in 1918 for a rudimentary gunnery-simulator which gave fighter pilots the confidence to fire their synchronised guns through the fast spinning propeller blades without shooting them to pieces!

6
Post-war ABC Aero-engines

Dragonfly problems

Everyone, bar none, had high hopes for the ABC Dragonfly. Its incredible power to weight ratio and proven, quite outstanding performance in prototype aeroplanes, as reported in the previous chapter, was far in excess of others and, potentially, would give the RAF just what they needed to end the war. It is no surprise then that much of Martlesham Heath's work immediately after the war was spent on flight testing approved Dragonfly powered aeroplanes, but alas, the Government's confidence in unproven prototypes - not just that of the Dragonfly - and their urgency in getting it into production without a thorough pre-production programme, meant that many early engines exhibited serious faults, some examples of which had a remarkably short service life of only a matter of hours.

But against this, it has to be remembered that those engines hand-built by ABC/Walton Motors consistently proved to be more reliable than contract built, volume produced examples. And therein lay Sir William Weir's blunder for his aim by 1917 was to rationalise Britain's war-time aviation industry by having simple, standard designs which could be manufactured by all manner of industries.

Bradshaw's Dragonfly fully met that criterion; Roy Fedden's Mercury did not and to reiterate this point, Rolls Royce knew that only under their supervision could engineers with skills matching Rolls-Royce's own standards, build good aero-engines to the extent that their General Manager, Claude Johnson was quoted as saying he "would rather go to prison before he allowed anyone else to build Rolls engines, war or no war"! In fact, Rolls-Royce only licensed Brasil, Straker & Co to build the Falcon engine and components for the Eagle. During the last war, only Packard in American and Ford at Manchester were licensed to build the famous Merlin. Several stories are told of resentment at Rolls-Royce's 'exacting' standards: noting that Rolls-Royce engines required much hand fettling to get a perfect fit, the Ford Motor Company told Rolls-Royce that they couldn't possibly build to Rolls-Royce's standards; Fords worked to much finer tolerances with a guaranteed fit, first time! Even Phelon & Moore, who were contracted to make special nuts and bolts told Rolls-Royce to take their contract elsewhere after too many false complaints about P&M's workmanship.

As events unfolded, the reliability and service life of these early contract produced Dragonfly engines was not that much better than the early Gnôme rotary, Sunbeam Arab or Wolseley built Hispano-Suizas V-8s; even early examples of the Rolls-Royce Falcon and Napier Lion engines had major, but easily cured, problems - and all these had been built by professionals in the field. Yet the Dragonfly's problems were nothing in comparison to the Siddeley-Deasy Tiger V12 of 1918, ordered by the Government towards the end of hostilities, when a pair of hardly bench tested Tiger engines were installed in the Siddeley Sinaia bomber prototype. It was a struggle to get the engines to run for any length of time, but worse, so poorly designed and built was the airframe, that the fuselage buckled in two when later being wheeled out of the hanger.

Curing most faults on both the ABC Wasp and Dragonfly prototypes proved to be a relatively easy task without requiring a major redesign. As there was no indication at the time that these promising engines were incurable, the Royal Air Establishment, Martlesham Heath and Granville Bradshaw all continued with their development programme throughout 1919, but in January 1920 an Air Ministry communique advised that the Dragonfly was not now considered sufficiently satisfactory for adoption by the RAF for their post war programme and that all further manufacture and flying with these engines would cease, unless authorised by the Director of Research. One month later, the RAE finally abandoned work on the Dragonfly in favour of the new generation of radials, such as the Armstrong Siddeley Jaguar and Cosmos Jupiter whose development had been held back by the Government in preference for the Dragonfly.

Only in the retrospective post-war calm was the ABC Dragonfly heavily slated. Major George Purvis Bulman of the AID later expressed: "Had the War dragged on into 1919, the Dragonfly would have lost it for us in the air. We would have been beaten out of the sky; it was a hideous episode illustrating the effect of a single unwise decision, based on 'cleverness', and should never be forgotten by any future arbiter of high responsibility". While W O Bentley states in his biography: "That the war ended in November, before this radial (Dragonfly) reached the squadrons in any numbers was merciful... It still succeeded in killing several good men, among them the brilliant test pilot, H G Hawker. His engine caught fire and he went straight in". In fact Bentley's claim is unfounded, for while both Peter Legh and Harry Hawker were flying when fire broke out, Legh had jumped from his Basilisk at low level, being killed instantly on hitting the ground, while Hawker survived the crash but died soon after. No blame was ever attributed to the Dragonfly engine in the post mortems or Air Accident Investigation, as is related later.

Bill Gunston states in his brief *World Encyclopaedia of Aero-engines* that his "long talk with Cpt Norman Macmillan and Major Oliver Stewart MC left him in no doubt that the ABC Dragonfly would have necessitated a frantic re-engining programme for thousands of aircraft had World War I lasted into 1919". If the Dragonfly engine was such disaster, one would reasonably expect both Macmillan and Stewart, as WW I fighter pilots and post-war test pilots who had both test flown a Dragonfly, to have reported this calamitous engine in their several books on the development of aviation - but they make no mention of it!

Who knows though what damage the fast, Dragonfly powered fighters could have inflicted upon the enemy had the Ministers allowed it the proper development programme (just as they had the Bentley rotary), for everyone admitted that when running properly - and they certainly were by 1920 - the Dragonfly was a truly impressive engine.

So, what went wrong?

It has to be remembered that the Dragonfly was one of the very first serious attempts at a truly large, high powered, static, air cooled radial aero-engine and was, potentially, liable to major teething troubles. In the final analysis, the problem was put down to scaling up the under-developed Wasp radial to generate double the power, causing it to seriously overheat at the constant, high revolutions required by an aero-engine at all altitudes. The lack of a proper development programme and, more seriously, the then unknown phenomenon of destructive 'synchronous torsional vibration' also played a part. Yet, to the uninformed, the problem lay entirely with Bradshaw's copper plated machined-steel barrels and his silver plated tongue!

Steel cylinder barrels

There was nothing fundamentally wrong with Bradshaw's choice of machined steel barrels nor poultice head design, for these were the standard choice for many WW I aero-engines even into 1920. Indeed, Roy Fedden's Cosmos engines used the same principle of turned, finned steel-alloy, close-ended barrels, but his were later fitted with a composite aluminium alloy head to the Gibson and Heron specification. Copper plating was also proven by others to be beneficial.

Cylinder heads

Although aluminium was known to be very efficient in dissipating heat, Bradshaw stuck to cast iron heads mainly due to the proven unreliability and porosity of contemporary aluminium-alloy castings (Hispano-Suiza stove enamelled inside and outside of their aluminium alloy engine castings to overcome porosity problems) while major problems existed in retaining hardened steel valve seat inserts in alloy cylinder heads. The more reliable Y-alloy was not developed until after the war. Roy Fedden used alloy heads in his Cosmos radial engines, but these were prone to severe warping under the intense heat of full power, and even in normal running his engines showed a noticeable drop in power output over a quite short period due to head distortion. It was only in the relaxed postwar years that, under Bristol and with research at the RAE, that they were able to address this problem. As owners of the Hillman Imp cars of the 1960s will be all too well aware, even much more modern water cooled all-aluminium engines suffer head distortion, but at least Granville Bradshaw couldn't be blamed for those.

The Dragonfly's head also suffered from a poor inlet manifold porting design which caused uneven running and, especially at high altitudes, carburettor icing. The latter was partially resolved by the RAE using exhaust heated induction manifolds. The post-war Jupiter prototype also suffered badly from icing and required heated induction muffs. As specifiers of aircraft and engine designs to the RAF, the Royal Aircraft Establishment had a vested interest in curing the Dragonfly, and other engines, of their ills and throughout 1919 they tried several variations of cast aluminium, double inlet valve head designs on the Dragonfly which eventually raised full throttle power output from 290hp to an impressive 350hp, equalling the original hand-built prototype, but with greater reliability.

Cooling

Cooling was a major problem on all aero-engines; even water cooled engines had their faults. When the Handley Page V/1500, powered by two Rolls Royce Eagles V12s, arrived at Martlesham Heath for testing, it was found they had consumed 50 gallons of coolant just on their short delivery flight - but that was not entirely Rolls-Royce's fault! Relying on adjustable radiator louvres to control air flow, these were often forced shut by the propeller's thrust when they should have been kept open.

The Dragonfly's cast iron cylinder heads reputedly glowed dull red in use - presumably under full power. Much of the problem here lay in fuel to air ratios which were weakened in an attempt to extract maximum power at high altitude without running the engine at too weak a mixture which would exacerbate overheating and valve failure at low altitudes. Consequently, the engine was designed to run rich, which proved most uneconomical - but as a post-war racing engine the Dragonfly excelled! Many contemporary aero-engines also ran rich, whether by accident or design, but unbeknown at the time, the unburnt fuel actually helped cool the exhaust valve. In the later war years Heron and Gibson both experimented with hollow stemmed valves filled initially with water, which proved useless, and then mercury, but with its high surface tension, mercury proved equally useless. A solution was ultimately found with sodium.

Cooling was made worse by the aeroplane designer's obsession with streamlined engine nacelles and cowls to reduce drag; but these only added to overheating problems as there was now less of the cylinder exposed to the cooling air flow. Bradshaw was well aware of this conflict and presented three post-war patents relating to engine cooling.

The popular belief that the speed of the air flow cools the engine, based on the cooling effect on the skin through evaporation of sweat, does not apply to a 'dry' engine. What matters is the air's ability to carry away, in a controlled manner, heat dissipated over as large a surface area as possible, which meant more, thinner fins, rather than fewer thicker ones. In this respect, machined steel barrels proved far superior than cast alloy ones. Bradshaw's own tests also showed that fast moving air directed into exposed fins merely caused turbulence and created heat pockets, trapping heat at its hottest spot where the fins joined the barrel. But carefully directed and controlled air flows eliminated such hot spots.

In August 1917 he lodged a patent (granted as Pat. No: 139827 in 1920) for a ducted, streamlined nacelle which directed cooling air to each cylinder barrel. This was followed by Pat No: 142516 laid on 21st January 1918, for elliptical cylinders with corresponding elliptical pistons and fins to increase surface area for a proposed 14-cylinder Dragonfly prototype which was, in essence, a double row Wasp. When ABC Motors (1920) Ltd went into liquidation in 1924, Bradshaw recovered (and renewed) this particular patent from the receivers. In July 1919 he lodged a provisional patent for a spacer ring of a highly heat conductive material, such as copper or aluminium, to be inserted between the barrel and cast iron head to help carry away heat from the exhaust valve; this was granted as Pat No: 152383 on 7th October 1920.

Bradshaw's theories on controlled, cooling air flows were supported by H C H Townend who in 1929 designed his famous 'Townend ring' - an aerodynamic ring surrounding a radial engine which not only reduced drag and improved airspeed by some 10mph, but also helped control air flow around the cylinder heads. It was fitted to the Gloster Gladiator, Fairey Swordfish and many others. It was also discovered during the last war that much of the success of the powerful German BMW radial aero-engine was due to its directly driven cooling fan mounted in front of the engine, which not only supplied a cooling air flow when taxying, but also reduced and controlled the air flow when diving at high speed.

Crankshaft vibration

Engine vibrations were a major problem for all aeroplane designers. Usually the airframe's wooden structure and wire stays absorbed them, but sometimes engine vibration destroyed the airframe! The early post war Napier Lion W-12 engines were prone to vibration, oil leaks and failure but later versions proved extremely reliable.

As to the Dragonfly, it was discovered that excessive crankshaft vibration occurred only at certain engine speeds, causing main bearing failure and severe charring by high frequency vibration of the wooden propeller where it was secured by bolts. In fact heat generated by high Radio-Frequency vibration is used today to weld plastics. This however severely limited the Dragonfly's service life to between 1 and 17 hours - only marginally better than the similarly afflicted war-time Sunbeam Arab V-8 which had been ordered into large scale production in January 1917, and which had also begun to exhibit this serious defect as early as May 1917. The problem was not fully understood at the time, for while the V-8 Arab had a long crankshaft for its eight cranks and was thus potentially liable to torsional vibration, the Dragonfly's was as short as possible with only one crank. But nevertheless it was analysed at the end of the war as 'synchronous torsional vibration'. Quite unwittingly, Bradshaw's Dragonfly developed its designed power output at the peak periods of the crankshaft's natural resonance (or frequency) causing it, and other components, to fail - as an analogy one must think of soldiers marching in step on a suspension bridge or a wine glass shattering as an opera singer sings a certain high frequency note.

Once the problem was fully understood from experiments at RAE Farnborough, they were able to cure it by building in dynamic damping. This lesson led to a major redesign of both the Cosmos and Jaguar radials, then still in their prototype stage, but to cure the Dragonfly, would have meant a radical redesign which, with the war now over, was considered rather pointless.

Indeed, much good came out of the 'disastrous' Dragonfly and the extensive research work carried out under Major B C Carter at the RAE for, with the luxury of a leisurely peace time development, both the Air Ministry and US Government demanded henceforth that all prototype aero-engines undergo a 50-hour pre-production endurance test at 90% rated engine power, as against the pre-war practice of 10hrs continuous running with no more than 3% variation in horsepower. The Dragonfly at first struggled to reach the 50-hour mark. The first engine to pass was the redesigned, Bristol built, Cosmos Jupiter 9-cylinder radial in 1920, yet even the Armstrong Siddeley Jaguars had problems and one, fitted to a Nieuport Nighthawk, seized after $50\frac{1}{2}$ hours. These endurance trials also showed up failings in the ignition systems of many engines.

The opposition

To be fair to Bradshaw, none of the rotary engines, other than the Bentley, was particularly reliable. The Gnôme had a notoriously short service life and both it and the Clerget were famous for overheating and often catching fire. W O Bentley overcame many of these problems when seconded by the Air Department as their Technical Liaison Officer at Gwynnes of Hammersmith; while his Humber built 130hp AR.1/BR.1 was successfully developed, after several teething troubles, into the hugely successful 200hp BR.2 in the summer of 1918. This was the last rotary engine to be developed; it remained in RAF service until 1928.

The much vaunted Sunbeam Arab V8 also proved a disaster with a poor crankcase design, weak cylinder attachment and the destructive 'synchronous torsional crankshaft vibration'. The Arab was finally declared unusable in late 1917, nine months after its acceptance and after 1,800 engines had been built, of which only 81 were delivered to the aeroplane makers.

The 200hp Hispano-Suiza V8 with its geared propeller drive also proved problematic when built under licence by Brasier et Cie and by Wolseley; indeed so serious were these defects, and those of the Wolseley Viper variant, that they caused a crisis in 1917 with SE.5a fighter production when the Viper crankshafts lasted barely 4 hours. This serious design flaw was quite fairly blamed on a poorly written Government specification!

Although held back by the Government in favour of Bradshaw's Dragonfly, both the Cosmos and Siddeley radials benefited greatly from the research undertaken at the RAE on Bradshaw's engines which helped transform them into the great success they later became.

Cosmos radials

Brazil-Straker & Co of Fishponds, Bristol was part of the Straker-Squire car company. In 1914 their Fishponds factory came under direct Admiralty control after its Chief Design Engineer, Roy Fedden, had greatly improved the American Curtis OX-5 engine to the Admiralty's satisfaction. As a result of this, Brazil-Straker built many Rolls-Royce aero-engines for the war effort. Their 14-cylinder Brazil-Straker Mercury radial engine was designed by Roy Fedden and L F G Butler to meet a 300hp Admiralty requirement of late spring 1916. Seeking an airframe in which to test the Mercury, Frank Barnwell of Bristol & Colonial readily offered Fedden a Bristol Scout, for he too was desperately seeking an alternative engine to the troublesome Sunbeam Arab, but by July 1918, the Admiralty order for 200 engines had been overturned by the new Air Board in preference for the Dragonfly.

With the Armistice, all military aeroplane orders were immediately cancelled and the entire aviation industry now faced a very uncertain future forcing most to explore other avenues for their survival. Brazil-Straker was duly bought by an Anglo-American shipping company, Cosmos Ltd, and the aero-engine side became the Cosmos Engineering Company, leaving Fedden to continue research and development of the Mercury and Jupiter radials to meet new RAF specifications. The first Mercury was bench tested in October 1918 and fitted to a Bristol Scout F.1 (B3992) which first flew in December 1918 and attained 145mph. Fedden then persuaded Barnwell to develop a new airframe for his new 9-cylinder Jupiter; this he did and the Bristol Bullet (J6492) was first exhibited with a mock-up Jupiter engine in August 1919 but did not fly until June 1920 when it attained 155mph. When several Nieuport Nighthawks were fitted at the RAE with the modified Dragonfly II and the, alternative, Cosmos Jupiter radial, it was once again the ABC Dragonfly which offered the superior performance and unmatched power to weight ratio demanded by the RAF. Even the 14-cylinder Cosmos Mercury could only maintain maximum power for a very brief period causing it to be temporarily abandoned in favour of an improved Jupiter which benefited from the full and thorough pre-production testing in late 1919, denied to the Dragonfly.

But following a disastrous shipping contract by their parent company to White Russia, when the goods were impounded during the Bolshevik revolution in late 1918, Cosmos went into receivership and funding for Cosmos aero-engine development was suddenly withdrawn. By February 1920, Cosmos was forced into liquidation. Such was the post-war economic uncertainty, that even Armstrong Siddeley declined to take Cosmos over and kill the Jupiter in order to protect their own Jaguar radial. The Bristol & Colonial Aeroplane Company also stayed well clear for, like Sopwiths, Bristol had gone into voluntary liquidation in March 1920 to avoid the Government's Excess Profit Tax on war-time production and had reformed as The Bristol Aeroplane Company.

It was only at Fedden's insistence, knowing the importance of the Jupiter engine, that the receivers allowed him to keep his design team and the Jupiter project intact. Now keen to see the Jupiter enter production for the RAF, the Air Ministry placed great pressure on the reluctant Bristol Aeroplane Company to take over its development, and so in July 1920 a new Bristol aero-engine division was formed with Roy Fedden as Chief Engineer. The Jupiter project finally got properly under way in April 1921. Ten years later, the now fully proven Jupiter, became one of the most famous radial engines of the inter-war years. It was license built by Gnôme et Rhône and 26 other companies throughout the 1930s, powering some 262 aircraft types including many from Bristol and Gloster. In the 1950s, Sir Roy Fedden became R&D Director at Leyland Motors.

ABC - post-war

The delays in sorting out the Dragonfly's problems meant that by the Armistice, production had reached only 1,147 engines. The majority of these were held in storage before their eventual reduction to scrap metal.

At war's end all companies employed in Government munitions and contracts work were required to submit their accounts for compensation claims: W O Bentley received £8,000 from the Royal Commission on Awards to Inventors for his BR rotary engines; Guy Motors £65,000 compensation for cancelled ABC Wasp and Dragonfly orders, while Selsdon Aero & Engineering submitted a claim for £11,000 for ABC Gnat, Wasp and auxiliary engines.

Granville submitted his accounts, under Walton Motors, between 1918 and 1921 for his work on the Wasp and Dragonfly, plus royalties and the supply of copyright designs to other aero-engine makers in Britain and America. In addition there was a 'hardship claim' for the cancelled auxiliary engine contract. This netted a pre-tax award of 'around £43,000' described in Appendix 1 Walton Motors. The only regret he had over the Dragonfly was being given a fine for driving his Rolls Royce at 15mph through Hyde Park when returning from one of many meetings with the Men from the Ministry!

Bradshaw became an Associate Fellow of the Royal Aeronautical Society in February,1917. In recognition of his prompt and valuable services to the war-effort with the Dragonfly engine through Walton Motors (not, one notes, ABC), Winston Churchill, Minister for Munitions of War, recommended to the Prime Minister that he be awarded the Order of the British Empire. The Order was sealed on 3rd June 1918 and at 11pm that night, Churchill telephoned through his personal message of congratulations. Meanwhile, the OBE had been despatched by messenger who arrived at Bradshaw's house at 3am in the morning! Granville Bradshaw was later offered a knighthood, but declined.

In 1920, at a celebratory dinner to honour the 'Pioneers of British Aviation' and the first 100 certificate holders, Granville was seated between two (unnamed) world record breakers. Peter Masefield (later head of Vickers), who knew Bradshaw well, praised his immeasurable contribution to the development of the internal combustion engine; indeed few other designers have had five permanent exhibits at the Science Museum.

Although the Dragonfly was rejected by the government, Walton Motors (based within ABC's works at Hersham) offered it alongside the ABC Gnat Mk.II, Wasp Mk.II, and the venerable 4-cylinder 40hp for the post-war civilian market, but all these engines were now to be built by Selsdon Engineering.

ABC Scorpion advert from 1924

This allowed ABC to concentrate on a new lightweight cycle-car while Bradshaw turned his attention to a new ABC motorcycle for Tom Sopwith. The cycle-car attracted the attention of the empire building John Harper-Bean, prompting him to invest heavily in ABC and in March 1920, ABC's directors (Bradshaw, Cpt. Charteris and Maurice Yorke) agreed to put the company into voluntary liquidation. Its assets were taken over by a new company, ABC Motors (1920) Ltd made up principally of Harper-Bean's directors, with the intention of producing some 5,000 cycle-cars a year.

While Charteris and Yorke remained on the board, Bradshaw took no further active part and rented offices in central London with Gilbert Campling as his distributing agent, at 90, Jermyn Street (formerly German Street), from where he was retained as a consulting engineer to ABC Motors (1920) Ltd, allowing him to concentrate on a new oil-cooled engine and the Belsize-Bradshaw motor-car. However, by 1922 Harper-Bean's empire was in serious financial difficulties and both Charteris and Yorke resigned in July 1922, after which Bradshaw severed all formal links with ABC Motors (1920) Ltd and formed his own company, Granville Bradshaw Ltd in 1922. Ronald Charteris remained at Broom House, Horsley until 1922 before moving with Louisa to his mother's, Lady Charteris' home at 80 Culverden Road, London SW.12; she died aged 93 in 1930. Ronald died at Chobham on 26th November 1950.

In July 1923, ABC Motors (1920) Ltd, went into receivership. The factory and major assets of its car and engine designs were bought from the liquidators by a consortium lead by Thomas Andrew Dennis, who had been ABC's company secretary since 1915. A new ABC Motors Ltd was formed in December 1923 with Dennis as Managing Director. Their telephone number changed to Walton on Thames 774/775. Walton Motors Ltd, which was associated with Bradshaw and ABC aero-engines, was wound up in January 1925 and the final remnants of ABC Motors (1920) Ltd were disposed of in October 1925.

Although T A Dennis tried to keep the ABC cycle-car in production, it was clear from entries in the 1923 *Daily Mail* Light Aeroplane Competition (limited to 750cc capacity) that ABC's future now lay in flat twin aero-engines and not with motorcars or motorcycles.

Light Aeroplane Competition

The glider was to play an important part in the clandestine development of Hitler's Luftwaffe, in direct contravention of the Versailles Treaty of 28th June 1919 which prohibited Germany from developing military aircraft. In keeping with their earlier promotion of aviation, *The Daily Mail* offered a £1,000 prize for powered gliders at their first Gliding Competition to be held in October 1922 at Iford, on the South Downs, overlooking Lewes, East Sussex. This quickly inspired the development of ultra-lightweight powered gliders leading to a new Air Ministry requirement for glider-trainers and, now with cooperation from the Air Ministry, *The Daily Mail* announced the next motorised glider competition to be held at Lympne in 1923.

W O Manning, then English Electric's chief designer, promptly drew up plans for a flat-twin motorcycle engined monoplane which evolved into the S1 Wren built in February 1923, (s/n J6973). Though designed at their Phoenix Dynamo offices in Bradford, it was built at the Dick, Kerr works in Preston, and featured a high revving 398cc ABC engine (69 x 54mm, 3.8bhp @ 2,000 rpm, weight 35 lbs) which peaked at $7^1/_4$hp at 4,500rpm. When fitted with a reduction gear, which brought the revs down to 2,700 rpm, it developed a more refined 3hp and was first flown on 8th April 1923. Though its maximum air speed was 50mph, it remained perfectly controllable at only 20mph.

The S1 Wren had cost only £600 and taken eight weeks to build, which admirably met the Air Ministry requirements, but after evaluation at Farnborough and Martlesham Heath in August 1923, it

ABC motorcycle engine Wren S1 No 4

was rejected by the RAF. Nevertheless, so successful were its test flights that English Electric decided to build and enter two Wren Mk.IIs for the October 1923 *Daily Mail* Light Aeroplane Competition to be held at Lympne, Kent. These were entered as Nos.3 and 4. Flights of over 180 miles and heights of 1,200ft were recorded and No.4 went on to share first place having consumed a miserly 87.5 miles per gallon of petrol. No.4 was then exhibited at the famous British Empire Exhibition, Wembley in 1924. She was restored to flying condition at the Shuttleworth Collection of Historic Planes, Old Warden, Bedfordshire in 1956, using parts from No.3 which had been stored at Preston in 1926, until being sold privately. It was intended that No.3 be fitted with a rebuilt engine using new high compression pistons, but over the passage of time its airframe had deteriorated beyond repair. A proposed S.2 Twin Wren, two seat trainer/tourer, powered by a single Bristol Cherub flat twin or a Villiers 9hp two-stoke motorcycle engine came to nothing.

The Wren's competitors at Lympne were to have included P W Kingswell's ABC motorcycle engined tandem-wing monoplane, but problems with his advanced airframe design prevented him from taking part. However, a pair of deHavilland DH.53 'Humming Bird' monoplanes, especially built for the event, were entered at the last minute. The first and second Humming Bird prototypes, (G-EBHX and G-EBHZ) were fitted with Douglas 750cc twin-cylinder motorcycle engines, but their valve gear proved so troublesome that they were replaced by the 26hp 700cc Blackburne 'Tomtit' engine and in this guise, six DH.53s were ordered by the RAF in 1925 for communications and experimental work; this including launching from a dirigible airship, and remarkably, returning to a dock with it. Six more DH.53s were sold on the civilian market - the first production model, G-AUAC went to Australia and was later fitted with an ABC Scorpion before going to Samoa in 1937. G-EBHZ passed to the Seven Aero Club and remained flying until its Scorpion engine seized through lack of oil; she was scrapped in 1937. The sole surviving DH53, G-EBHX, having been flown privately in Kent until the 1930s with a Bristol Cherub III engine, was rediscovered in 1955 and underwent restoration at the Shuttleworth Trust in 1960 with extensive assistance by apprentices at deHavilland's Hatfield works and at ABC Motors who refurbished a suitable ABC Scorpion to replace the missing Cherub engine. Rejuvenated, she took to the air again on 4th August 1960.

There was also a Handley-Page HP-23 Sayers motor glider, jointly designed by W H Sayers, technical editor of *The Aeroplane*. This too was fitted with a 349cc ABC flat twin motorcycle engine, but it proved insufficiently powerful to gain flight. Even a Douglas 500cc flat twin proved little better and despite repositioning the engine above the wing and making a special rubber catapult assisted take off, the HP-23 failed to rise into the air and was duly scratched from the race.

Another entrant at the October 1923 Lympne event was the rather ugly Salmon Tandem monoplane (G-EBHQ) designed by Percy Salmon, chief draughtsman at the RAE. This had a 3½hp Bradshaw motorcycle engine - almost certainly an oiled cooled single - but it never managed to take off.

ABC Scorpion

Inspired by the Lympne trials, ABC launched a modified, reduced bore version of Bradshaw's 24hp, 1.2ltr (91.5mm x 91.5mm) 4:1 CR ABC light-car engine in June 1924 as the 'Scorpion' aero-engine (87.5 x 91.5mm 1,100cc) to fall within the 1,100cc Light-plane class; it weighed 91lbs. The car engine was reversed for aeroplane use with the flywheel replaced by propeller boss, a ball race rear and roller front bearing plus thrust bearing for propeller. The inlet and exhaust valves were now the same size allowing the head to be reversed so the exhaust port bend was well away from the propeller, following the same technique employed on the Wren's engine. It cost £80 (£85 with twin carburettors). The first engine was fitted to a German Caspar C.17 low-wing monoplane and the second to Hawker Cygnet (G-EBJH) for the 1924 race, but it proved underpowered. Nevertheless, these early engines were also fitted to the Noel le Parmieter Wee Mite and Westland Woodpigeon.

Under development by ABC's Mr Elliott, an entirely new large bore version appeared in June 1926 as the Scorpion Mk.II which, at 1.5ltr (102mm x 91.5mm) developed 35hp at 2,300rpm, peaking at 38.3hp at 2,550rpm. It weighed 93lbs. Permissible five minute bursts at 2,750rpm produced 40hp. As usual, its cylinders were machine turned from a single billet of cast steel-alloy and were fitted to a one piece cast alloy crankcase with integrally cast induction manifold to both heat the fuel:air mix and help cool the engine oil. The deeply finned cast iron cylinder head had twin spark plugs fed from a single BTH magneto. A Zenith double choke carburettor fed the engine at 0.5 pints per bhp per hour in normal flight. Uniquely, the drop forged one piece steel crankshaft was case hardened at its bearing surfaces, leaving the rest of the shaft in its strongest, unstressed state.

The Scorpion soon found many friends among the light aeroplane designers in the late 1920s-1930s and powered the following aeroplanes on the British register:

Hawker Cygnet

ABC Robin (G-AAID); Boulton & Paul Phoenix P.41 (G-AAIT); Comper CLA.7 Swift prototype (G-AARX); Heath Parasol (G-AJCK) Henderson-Glenny HSF.2 Gadfly Mk.I (G-AAEY) and Gadfly Mk.II (G-AARJ), Hendy 281 Hobo (G-AAIG); Luton LA.4 Minor prototype (G-AEPD) and LA4A Minor (G-AFBP, G-AFUG, G-AHMO); Mignet HM-14 Pou-de-Ciel 'Flying Flea' (G-AEDN, G-AEHM, G-ADZS); Navarro Chief; Noel le Parmentier Wee Mite (G-ACRL); Short Satellite (G-EBJU); Westland Woodpigeon (G-EBJV), and the Wheeler Slymph (G-ABOI, which is preserved).

In 1929, ABC developed their own 100mph light aeroplane, a single seat high-wing monoplane ABC Robin (G-AAID); it was one the few cabin monoplanes of its day. Designed by A A 'Tony' Fletcher (who had worked for Martinsyde and the Central Aircraft Company of Kilburn, designing their Centaur biplanes), the Robin weighed 739lbs and was powered by a Scorpion Mk.II engine. With a proposed price of £395, the sole distributors were to be National Flying Services Ltd of Trafalgar Square, London. The prototype was built by S E Saunders on the Isle of Wight and first flown at Brooklands in June 1929, then again in November 1929 when she gave an impressive performance. Evaluated at Martlesham Heath in May 1930, she proved underpowered and difficult to land while grave concerns over its construction required several modifications before an airworthiness certificate was issued. Alas, only one airframe was built but, without any orders, she was scrapped in 1932. The Centaur IIA biplane airliner of 1919 had twin 160hp Beardmore engines - both prototypes crashed. The two seater, Anzani powered Centaur IV biplane of 1919-21 was more successful with eight built before sales slumped in the post-war depression causing the company to close in May 1926.

The Comper Swift was a well designed aeroplane. Developed by Nicholas Comper from his earlier CLA.3 racer, it weighed only 600lbs which allowed the relatively underpowered Scorpion to give it a very lively performance, but its true agility came only with the larger and more powerful 75hp Pobjoy, R-type 7-cylinder radial in 1931. The Hendy Aircraft Company's single seat cantilever 'Hobo' monoplane of 1929 was originally fitted with the 35hp Scorpion, but this too was replaced by a Pobjoy 90hp radial in 1934.

Hawker's Cygnet lightweight biplanes were entered in the 1924 competition; G-EBMB with a British-Anzani V-twin and G-EBJH with the horizontally opposed 30hp Scorpion Mk.I, flown by Fred Raynham. The Anzani engine suffered magneto and valve problems and came 4th overall, while the Scorpion behaved impeccably until the penultimate lap when a rocker arm fractured from over strong valve springs; she came in third overall. G-EBJH was then sold to the RAE Aero-Club but was refitted in 1926 with a Bristol Cherub III engine entering the 1926 competition in which she came fourth, while G-EBMB, also refitted with a Cherub, came sixth. The Cherub remained the major competitor to the Scorpion.

ABC Robin solo cabin monoplane

Short Brothers also built an ultra lightweight twin-seat monoplane for the 1924 Air Ministry competition held at Lympne. The Short Satellite monoplane used a fabric coated light duralumin, monocoque fuselage frame and was originally fitted with a 32hp Bristol Cherub I twin which proved underpowered and failed to pass the test. She was refitted with a Cherub II in April 1925 and then entered in the 1925 August Bank Holiday meeting at Lympne, but failed to take off due to structural damage. The Satellite monoplane was presented to a syndicate of RAF officers who had formed the Seven Aero Club at Eastchurch in 1926 and was duly fitted with an ABC Scorpion Mk.II with Fairey-Reed duralumin propeller. She was entered in the 1926 trials, but though the new engine greatly improved its performance, the Satellite was disqualified following undercarriage damage. The club also entered a Scorpion powered Westland Woodpigeon in the 1926 trials, but she retired early with a seized rocker arm.

Designed by C H Latimer-Needham, the first Luton ultralight monoplane was built in 1936. Their prototype high wing monoplane Luton LA4 Minor of 1937 was powered by a 40hp ABC Scorpion Mk.II and proved a very popular kit-build aeroplane, fitted with a variety of engines. In 1943, a fire destroyed Luton's Phoenix Works at Gerrards Cross, but in 1958, Latimer-Needham formed a new company, appropriately called the Phoenix Aircraft Company at Cranleigh, Surrey to develop further the Luton Minor and Major with assistance from Arthur W J Ord-Hume, who later showed great interest in Bradshaw's revolutionary toroidal reciprocating piston engine for a new ultra-lightweight Phoenix monoplane.

Top, Navarro Chief Trimotor and bottom, ABC Scorpion MkII flat twin

Three Scorpions were fitted to J G Navarro's 'Navarro Chief' of 1931. Joseph had worked in aircraft production at Brush Electrical on Farmans, at Pemberton-Billing and Whiteheads before setting up as The Navarro Aircraft Company as subcontractors at Burton-on-Trent in 1916 (renaming it the Burton Aircraft & Manufacturing Company in 1917); the business was declared bankrupt in 1921. After the war he operated as the Navarro Aviation Company at Rochford (Southend) airfield offering joy-rides in Avro 504s between Whitstable, Southend, Skegness and local coastal resorts, but put this into liquidation in June 1919 amalgamating it with the Wellesely-Brown Aircraft Company of Kingston-on-Thames, to became Navarro-Wellesley Ltd (which went into liquidation in 1924), but he left that July to concentrate on setting up a new glider and flying club association. Then in 1923, he became involved in ambitious plans for a trans-Atlantic sea-plane service using ships moored at sea as refuelling stations, for which he laid plans for a new cargo flying-boat. This project came to nothing.

In early 1929 he proposed an advanced high wing monoplane with an interesting hinged butterfly type air-brake rudder, and a diagonally split tail plane which also acted like an aileron/air-brake to combat spin. Even the main plane, with its dihedral wing-tips, folded back for small hangers. Satisfied with a scale model, he established his Navarro Safety Aircraft Ltd, at Heston in October 1929 and assembled a full scale 3-seater, plywood bodied, tri-motor 'Navarro Chief' whose three ABC Scorpions would propel it at 110mph but which was designed to stay aloft, fully loaded, on just two engines. Weighing only 1,600lbs gross, it was declared the world's smallest tri-motor aeroplane. The prototype was completed in March 1931, with its test flight 'imminent', but nothing more was heard of it. The company had closed by 1933.

The Scorpion also found a limited market abroad - Poland, Romania, Australia, America and Germany. It was fitted to the 1928 RWD-1 sports monoplane, designed by Jerzy Drzewiecki in 1927 of which two examples were built: a static test bed and a flying version from which were developed a range of lightweight monoplanes. In America, the Curtiss-Wright company built the CW-1 in 1929; a lightweight, sports high-wing monoplane with skeletal fuselage which, like the Flying Flea, soon proved a highly popular and cheap to buy machine which saved Curtiss-Wright from the depression. Later production models used the American 3-cylinder Szekely SR-3-0 radial engine.

In France, the pioneer of 'flying wing' gliders, Charles Fauvel, chose the lightweight Scorpion for his 1928 Peyret-Mauboussin PM-10 single seat powered glider with which he took several altitude and duration records in 1929.

Yet one of the most intriguing applications was in the early days of helicopters. It will be remembered that Bradshaw's original 40hp Star engine was used in Bartelt's Ornithopter of 1911, but in 1921 a M Passat borrowed a Scorpion from ABC for his 'Helithopter' experiments in vertical lift using four axially rotating and flapping 'ornithopter' wings mimicking the action of a bird, while rotating like a helicopter. His experiments continued into 1923, but without success. Another weird application of the Scorpion was in Jos Louis Sanchez-Besa's experimental 'Multiplane' aeroplane of late 1921 whose rack of 21 cork filled Venetian blind aerofoils were of fixed or variable pitch. A prototype was built, but not flown. In 1929, two Austrian engineers, Bruno Nagler and Raoul Hafner, both working in Britain, developed the R1 'Revoplane' helicopter which briefly took off, albeit rather clumsily, powered by, it is said, an ABC Scorpion. Nagler returned to Austria and became a leading light in Germany's helicopter technology. Hafner remained in Britain and worked on helicopter development in competition with Juan de la Cierva. Hafner became most famous for his war-time 'Rotachute' towed gyro-glider and the Jeep based 'Rotabuggy' towed gyro-fieldcar.

ABC Hornet

Although Darracq, Dutheil-Chambers and Geoffrey deHavilland had all developed a horizontally opposed flat four aero-engine by 1910, the 75hp, ABC Hornet is regarded as the first commercially successful of this type. Development began in 1927 under T A Dennis for light aeroplanes and airships and after 80hrs testing, was submitted for RAE acceptance trials. It suffered very little vibration and proved a very smooth running engine.

Launched in June 1928, it was in essence a long stroke 3,990cc (102mm x 122mm) 'double-Scorpion', which developed 75bhp at 1,875rpm, peaking at 82bhp at 2,150rpm. Weighing 225lbs, its centre of gravity was dead centre between the four barrels. Though the machined steel barrels and many components were shared with the Scorpion Mk.II, it had a unique, three section, three bearing cast alloy cylindrical crankcase and was fitted with thrust bearings for pusher or tractor propeller mounting. Dual spark plugs per cylinder were fired by a specially designed 'Watford' magneto with integral advance/retard, set for full advance at one third throttle. A single Zenith twin-choke carburettor was fitted.

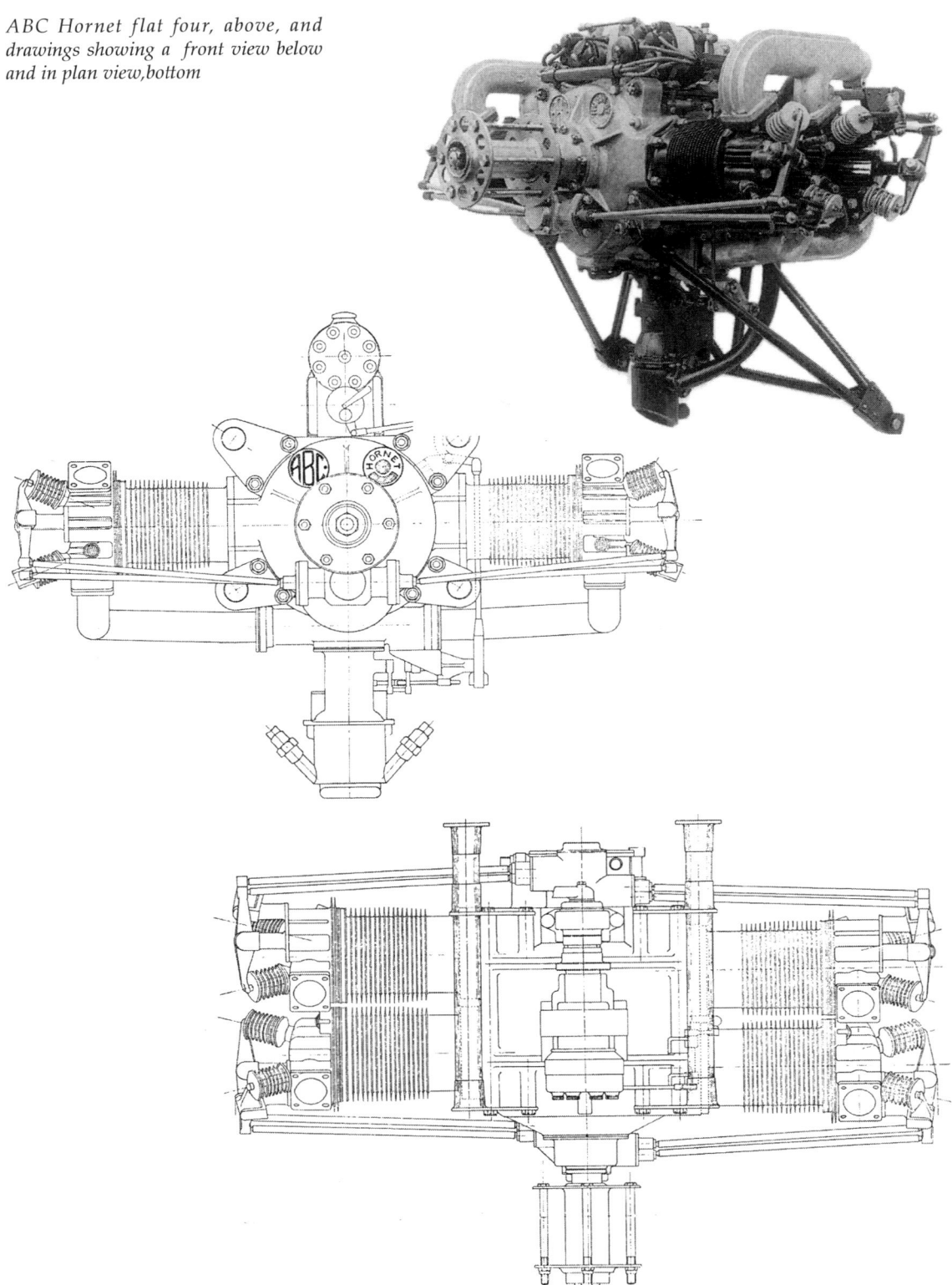

ABC Hornet flat four, above, and drawings showing a front view below and in plan view, bottom

Robinson Redwing

The Hornet was first fitted, but not flown, in the prototype Westland Widgeon III civilian high wing monoplane of 1927 which was developed from the Widgeon Mk.I entered in the 1924 Lympne trials. Production models of the Mk.III adopted the ADC Cirrus 4-cylinder in-line engine. The Hornet was also fitted to a handful of other prototypes: CAC Mk.1 Coupe, Robinson Redwing Mk.I and the Southern Martlet (G-AAII).

Harold Boultbee had formed his Civilian Aircraft Company at Burton-on-Trent in 1928. His prototype high wing, two-seat CAC Coupe monoplane was powered by the 75hp Hornet engine, but Boultbee experienced many problems in fitting the Hornet into the air-frame. First flown in July 1929, he also encountered major vibration problems, but these were resolved in conjunction with ABC and, with an improved crankcase allowing additional frontal support, resulted in an extremely smooth running Hornet appearing in spring 1931, but after these 18 months of trials it was decided to adopt the more powerful 100hp, Armstrong-Siddeley Genet Mk.II 5-cylinder radial for the five production Coupe Mk.IIs built at their new works at Hull.

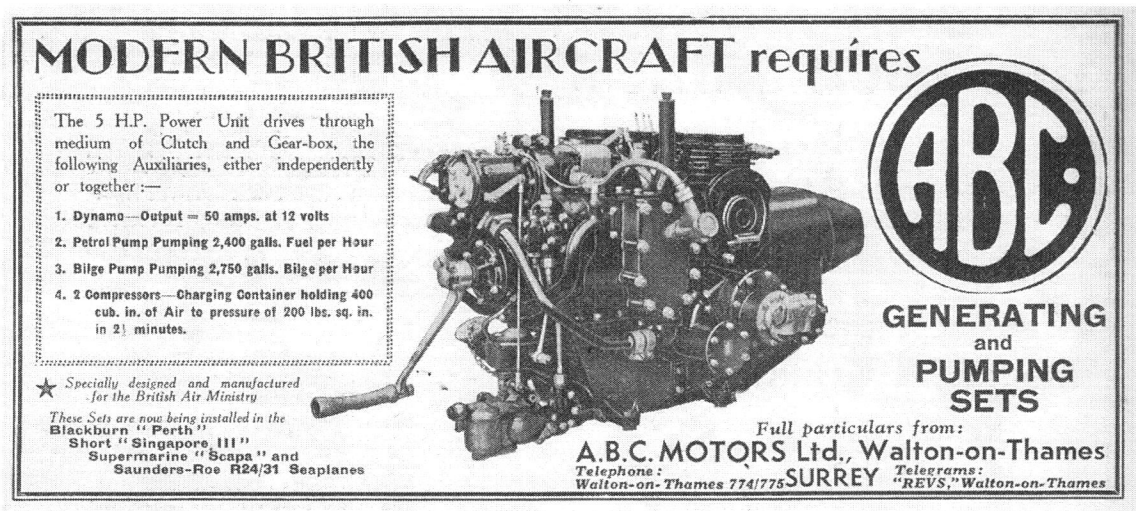

Mk.I auxilliary engine, 1933

Cpt. P G Robinson's Redwing I was designed by John Kenworthy of Austin Greyhound fame. Built at Croydon, the Redwing made its maiden flight in May 1930 with a 75hp ABC Hornet and achieved a 650 ft/min rate of climb and 92mph top speed, but as with the CAC Coupe, production Models II and III adopted the A-S Genet IIA engine.

A similar fate befell the Southern Martlet of 1929. This was developed by F G Miles who had designed and built his first aeroplane, the Gnat at the age of 19 years, before he had learnt to fly! Gaining his licence in an Avro 504 at Shoreham in Sussex in 1926, Miles then formed the Gnat Aero & Motor Company which was then promptly reformed as the Southern Aero Club with an offshoot, the Southern Aircraft company. His first successful design, the Southern Martlet (G-AAII), was a converted 'Avro Baby' whose original Green in-line 4-cyl engine was replaced by an 85hp ABC Hornet engine transforming it into a fast, agile single seater. First flown on 10th July 1929, it was soon replaced by an 80hp Genet II radial which, like the Hornet, gave a 112mph level flight and 1,100ft/min rate of climb. This was only bettered by the 120hp DH Gypsy II in-line engine used on later models. Three airframes followed, fitted with the deHavilland Gipsy II in-line in G-AAYZ, and a Genet Major radial to G-AAYX. Miles later developed the famous Miles Master and Magister monoplanes.

A Hornet also replaced the ABC motorcycle engine fitted to Flt.Lt. Crawford's privately built monoplane which he had at RAF Hinaidi, Iraq in 1925. Though flown in RAF colours this was most certainly not in RAF service!

As usual, the ABC's design criteria for the Hornet was years ahead of its time, and was fitted to only a handful of prototype aeroplanes which failed to make an impact on the light aeroplane market increasingly dominated by the conventional and ever popular DH Gypsy or the ADC Cirrus in-line four. Quite why the Hornet 'failed' is not known for in America in 1931, an entirely new 37hp flat-four appeared from Continental Motors. Their Model A-40, 1.9 litre (79mm x 95mm) engine was significantly smaller and less powerful than the 4.0 litre 75hp ABC Hornet, yet it quickly caught the imagination of aeroplane designers and the flat-four concept, quite literally, took off. The A-40 remained in production until 1938 against competition from the American 50hp Franklin of 1935 and the 50hp Lycoming of 1938. The flat four soon became the de facto light aero-engine, of which the Lycoming is now the most famous. Sadly, sales of the Scorpion Mk.II and Hornet barely kept the company alive. Production of the Scorpion ceased in 1936 soon after production of the Douglas flat-twin engine (which remained a serious competitor) had been taken over by a new company, Aero Engines Ltd, who developed it as their 'Sprite', and also planned to manufacture the Hispano-Suiza, G & J Weir 'Pixie' and other light aeroplane engines.

ABC auxiliary engines

In 1929 Lord Ridley (Matthew White Ridley CBE, 3rd Viscount and better known for his Brooklands Ridley-Special of 1931), joined ABC and with him now on board, the company reappraised the ABC auxiliary engines that Granville Bradshaw had developed during the Great War. Securing a new government contract, they set about designing new auxiliary engines. The first Mk.I appeared in 1932; it was a 115cc water cooled side valve, flat twin with alloy crankcase and machine turned steel barrels, designed for aircraft installation. Making much use of lightweight alloys, these Mk.I self-contained engine/self-priming pump units weighed only 146lbs and were soon being fitted to a new breed of flying boats: the Blackburn Perth, Short Singapore Mk.III, Supermarine Scapa and the Saunders-Roe A27 London.

During the last war, ABC was contracted to produce many hundreds of these all-in-one auxiliary motors for the RAF's flying boats, particularly the Short Sunderland and, like its WW I cousin, these drove a 50amp 12v 'Dynamotor' to start the engine remotely from the cockpit as well as charge the aeroplane's batteries; a 2,250gph fuel pump to replenish the tanks from fuel tender boats; a 300gph oil pump; a 2,500gph water pump to scavenge the aeroplane's floats and hull; and a 200psi air compressor with a 400cu.in. reservoir to charge reservoirs for starting the aero-engine. These were all driven separately or in combination.

In the later war years, a Mk.II appeared; this was a 174cc (54mm x 38mm) ohv flat twin which continued in production after the war, developing 5hp when centrifugally governed at a nominal 4,000rpm and which used external fans to cool the engine. The crankshaft was the self-same three throw motorcycle design of earlier models, which minimised the vibration associated with a normal twin. These engines were also supplied to HM submarines for a variety of tasks and for field use. The lightweight twin was also adopted by Auster in 1951 for their B3 experimental radio-controlled target-drone, of which little is known. Sadly very few of these auxiliary engines have survived.

ABC auxilliary engines of 1957 showing the V-4 Bee

These were joined by a 215cc Mk.III, driving a bilge pump and generator and a Mk.IV generator unit. A proposed Mk.10 flat-four was aborted at prototype stage but this led to a new airborne auxiliary engine developed in 1950, the 750cc, 15hp V4 'Bee' which primarily powered a 6KW 28v dc generator for the RAF's Blackburn Beverley transport aeroplanes of the 1950s. The last Beverley was withdrawn from RAF service in December 1967 while XH124, a Beverley C Mk.I, was the last aeroplane to land at RAF Hendon for display in the new RAF Museum. Days later this historic airfield, opened in 1910, was dug up for housing.

In addition to auxiliary engines, in 1943 ABC's William Wilson Bath patented a divided crankpin arrangement, retained by a through bolt for easier assembly of one piece big end-bearing con-rods. Later that year he patented a rotary vane pump whose rotor was mounted eccentrically to the pump body; the rotor had four sliding blades which, as the rotor turned, reduced the available space forcing the fluid through. ABC's other major war-time contract engineering work was for deHavilland, producing constant speed units for their variable pitch propellers and for the Admiralty, making high pressure air-valve units for torpedo propulsion as well as the manufacture of navigation gyroscopes. These contracts led to the factory being extended. Two of their engineering foremen, Joseph Brady and Stephen Ackland, received the British Empire Medal.

Vickers

Such precision engineering contracts continued after the war for deHavilland, Handley Page, Miles and Vickers and in 1947 George Urban Leonard Sartoris, who had developed transmission systems at APT Ltd in the 1920s, patented with ABC Motors an improved horizontally opposed engine design in which the balanced crankshaft turned by gear a pair of balanced countershafts to eliminate vibration.

In late 1950, with Thomas Dennis's son, Rene Maxime Dennis, now at the helm, Vickers Ltd bought ABC Motors Ltd and in 1952 a small experimental and electronic workshop was established within the Hersham works. ABC Motors Ltd remained an autonomous subsidiary and became the major subcontractor to Vickers Armstrongs (Aircraft) Ltd at their Brooklands (more correctly, Weybridge) works for the Vickers Viscount, Vanguard, BAC 1-11, VC-10, Concorde and the aborted TSR-2, mainly making VGS bolts, gears and, for the VC-10, throttle control boxes. They still however maintained a spares and service facility for their 'Bee' V-4 auxiliary engine.

By 1960, ABC Motors Ltd had become part of Vickers Armstrongs (Engineers) Ltd and though they lost their autonomy, they became a specialist, precision gear cutting facility for the group. Ironically, their parent Vickers Armstrongs (Engineers) was based at the old Armstrong-Whitworth Elswick ordnance works where they had built the ABC V8 aero-engine and where they now build tanks.

In 1960 Vickers, Bristol and English Electric merged their aircraft interests into the British Aircraft Corporation as a direct competitor to Hawker-Siddeley, but as part of Vickers Armstrongs (Engineers), ABC was outside BAC's control and began work producing valves for British Rail's new Sulzer diesel locomotive engines as well as precision inch/metric conversion drums on machine tools together with the production of 'Kaba' security locks. In 1969 ABC's works at Hersham came under control of Vickers Crayford and the Hersham works was gradually wound down with production work being transferred to Vickers' old Maxim armaments works at Crayford during 1970. The construction of the last Vickers aircraft, a Super VC-10, at Brooklands in February 1970 had no bearing on the closure of the Hersham factory in January 1971 and its 200 redundancies. ABC Motors Ltd remained a dormant company within Vickers until the late 1970s.

ABC's old Riverdene, Hersham site remains virtually intact and is currently used by Ian Allan Ltd, publishers of the famous *ABC* pocket books beloved by schoolboys of the 1950s and 60s. While their buildings are mainly from the 1930s/1950s, parts of the original ABC works and forge have survived.

Thomas Dennis died on 9th March 1965. His son Rene died on 27th July 1989.

De-icing

Granville Bradshaw returned to aviation work in the mid 1930s when temporarily living at Crawley Down, Sussex where he submitted a provisional patent (No. 26751 of 4th October 1937) for a 'de-icer' mat. Ice on a wing's leading edge seriously affects an aeroplane's ability to fly. This simple device, described in *Flight* of 3rd November 1937, was a thin metallic parallelogram lattice mat comprising many fine steel blades which could be distorted longitudinally and, by so doing, break up any ice on its surface - well, that was the theory. A sample was submitted to the Air Ministry but there is no record of their reaction and the patent was not granted. In the postwar years, 'Kilfrost' chemical anti-freeze pastes and fluids became the industry standard.

Man powered flight

Encouraged by the potential of rotating-wing 'Autogiros' developed by the Spaniard, Senor Don Juan de la Cierva in the 1920s, dreams of man-powered flight were once again aroused. Bradshaw developed his theories around the 'Katzmayr effect' of oscillating airflows over an aerofoil which gave it lift. This is demonstrated by vortices caused by a bird's oscillating and rotating wing action and is well demonstrated by the fluttering lift gained by blowing air over a fixed aerofoil, or for example, a piece of paper. Bradshaw built an improvised 'helicopter' rotor frame in his garden, to determine the energy and power required for vertical lift. The aerofoil rotor was attached to a vertical pole and turned by hand through a lever mechanism. It was also possible to alter the aerofoil's angle of attack for maximum lift. By means of a spring balance, he was able to calculate that a fit man can exert around 2hp of work; the most efficient output being from the arms through a rowing motion, supplemented by pedal power through cycling. Although this fell far short of the power necessary to gain vertical lift from stand-still, Bradshaw reasoned that, given the right design, 2hp would be sufficient to sustain level flight once airborne, using a rotor angled slightly forward to gain both horizontal and forward thrust - as indeed is the principle of helicopter flight.

Bradshaw testing his man-powered flight theory in his Sussex garden, 1937

To make pedal action more efficient and relieve the considerable strain on the muscles when going over top dead centre on full load, he devised a simple 'torque equaliser' which would, in effect, put effort into the crank as it approached top dead centre. A second chain, attached to the crank, turned a half sized pinion coupled to an adjustable 40lbs spring which was stretched easily as the crank gave its greatest torque approaching bottom dead centre, but as the pedal reached top dead centre, giving its least torque, the spring retracted and applied full torque to the smaller pinion, helping to ease the pedal over top dead centre. He experimented with his combined rowing, cycling and torque equaliser action in a small pedal car and reported good success, but his provisional patent (No. 13472 of 13th March, 1937) was not granted. Several cyclists tried out his torque equalisers for themselves and several reported back of its advantage, but clearly its benefits were considered insufficient to warrant the idea being taken up by cycle makers.

He also investigated 'reaction lift' used in the beating wings of an Ornithopter's vertical oscillation and experimented with a small 3 foot model glider with an upward beating wing. His experiments determined that as the power needed was twice that to lift its weight once airborne and to maintain flight, it would not have been possible to test the theory unless the glider, with a seemingly impossible maximum weight of 80lbs were manned, which brought him full circle!

Intrigued, the influential *Flight* magazine followed his progress throughout July 1937 and though doubtful of its benefits, praised and encouraged his efforts in this field. It also provoked much reaction with both supportive and dismissive comments from other experimenters. One stern critic was H C H Townend D.Sc FRAeS (of 'Townend ring' fame). Although he was doubtful that man could provide enough power for vertical lift, he too was intrigued by the 'Katzmayer effect' and believed it could possibly provide the answer as, like Bradshaw, he also reasoned that if the effect was reversed and the airfoil was oscillated to gain lift as well as the wings being designed to have a natural period of oscillation matching the natural speed of someone rowing and exerting sustained power, then it should, in theory, be possible to sustain manned flight without the large oscillations of the wings previously employed by

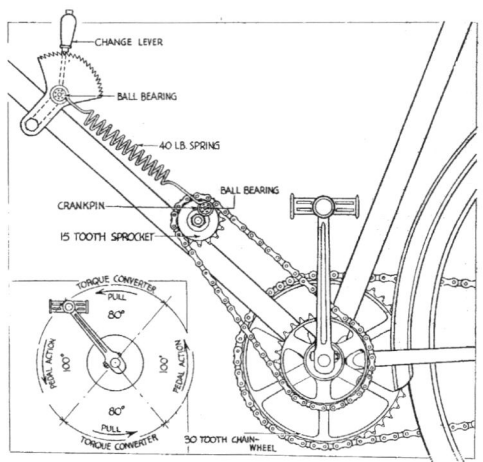

Torque converter

pioneers with their earlier, fantastic, Ornithopters. Townend was also keen for scientists to investigate this theory further and many ornithologists added to the debate by contrasting the energetic wing beat of ducks with that of the leisurely albatrosses whose tips alone 'powered' its flight through wing tip vortices. It was certainly one for further scientific research, but nothing more was heard of it for quite some time, then in the 1960s, researchers at Farnborough began new studies and in late 2007, research by Harvard School of Engineering mathematicians showed that such oscillations could indeed provide lift, but far short of that required for manned flight.

The industrialist, Henry Kremer shared Leonardo da Vinci's and Bradshaw's belief that man should be able to fly by muscle power alone. Kremer had developed resin-glues for the lightweight plywood panels used in the deHavilland Mosquito and, later, chipboard. He also pioneered lightweight glass fibre materials through his Microcell business. In November 1959, he offered a £5,000 prize for the first British man-powered and controlled flight over a one mile figure of eight circuit. In December 1960 he offered a £2,000 grant towards the prize, prompting Bradshaw to consider applying. He discussed it with Arthur W J G Ord-Hume who was showing interest in his toroidal engine. Ord-Hume referred him back to work in America in the 1920s and 1930s on 'cyclogyros' which used large paddle wheels either side of the fuselage, in lieu of wings, just like a lightweight Mississippi paddle steamer. In these, the paddles were small aerofoils which automatically remained horizontal but, by varying their angle of attack could, in theory, provide lift. Ord-Hume suggested a locking device which automatically set the aerofoils to a safe gliding angle of attack in the event of power failure, should it ever take off. This was not as far fetched an idea as it seems, for eminent German and American aeronautical engineers, including the US National Advisory Committee, were also experimenting with the idea and had high hopes of success, but the problem of making the machine light enough for sustained flight remained the major hurdle.

It was not until early 1960 that a 'cylcogyroplane' - a bicycle driven helicopter - was successfully built by a Briton. However, it would be a conventional bicycle powered, propeller driven glider which took the Kremer Prize in 1977. This was designed and built by an American, Paul MacCready, using exceptionally strong, lightweight man-made materials and the 'wing-warping' controls of pre-WW.I and which, on 23rd August 1977, allowed Bryan Allen to fly the 'Gossamer Condor' around a one-mile figure of eight circuit in 6 mins 22 secs, taking the, now £50,000 prize. On 12th June 1979, MacCready's 'Gossamer Albatross' crossed the English Channel taking the £100,000 cross-channel prize, 70 years after Louis Blériot.

7
ABC Motorcycles

Brooklands soon became the Mecca for exciting new technology with the famous 'Blue Bird' restaurant a favourite meeting place for pioneers to discuss the day's adventures and mishaps. Having the only fully equipped machine shop on site, Granville Bradshaw was often called upon by those testing their machines to make improvements to this or that. Following the rebuff he received from the War Office in 1912 over the military use of aeroplanes, and now facing a rapidly dwindling reserve of capital, he increasingly found his attention turning towards motorcycles and engines. He was still, after all, a keen motorcyclist himself, commuting daily from his home, Darby House, Sunbury-on-Thames on his trusty Triumph 500cc side-valve until, one day, he found himself in direct line of attack from a rapidly descending, side slipping, aeroplane! Throwing himself to the ground in self-preservation, his Triumph skidded from under him for some 20 feet before colliding with and being written off by the aeroplane.

He toyed with a new Douglas horizontal twin to replace the Triumph and soon became acquainted with S L (Les) Bailey, the Australian rider, driver and flyer who was the Douglas works rider. They soon became close friends and Bradshaw made several modifications for him in November 1912 creating, in effect, his first ABC horizontal-twin (60.9mm x 60mm) which developed 13hp at 5,000rpm. This improved engine included new machined steel barrels, an overhead valve gear and a cylinder head modelled on his aero-engine practice. It had steel conrods, one of which was in a Y-form off a three throw crankshaft. On December 17th, 1912 Bailey attempted the Class B (under 350cc) flying mile and kilometre records, but the compression ratio was too high for the fuel, so it was back to Bradshaw's

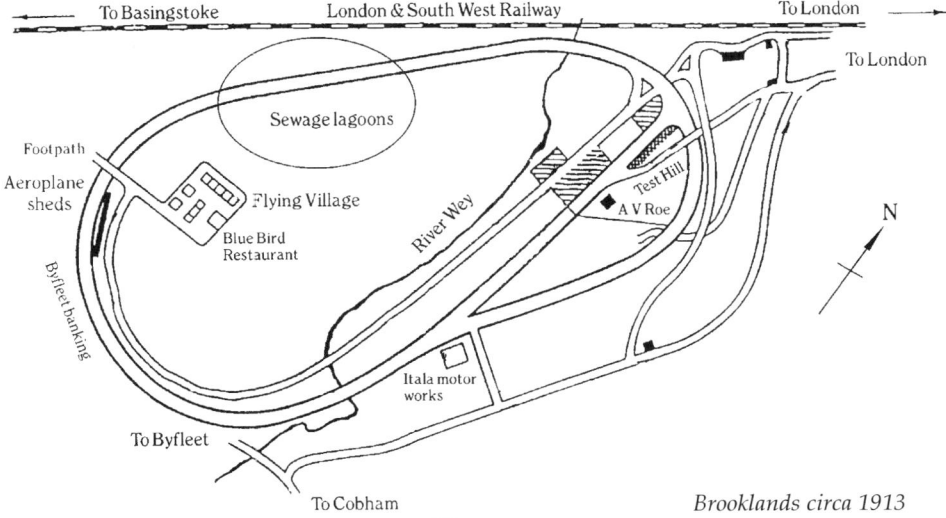

Brooklands circa 1913

1913 ABC engine. Note 3-throw connecting rods

works for compression plates to be inserted to reduce it to 6.0:1. He then took the kilometre record at 72.63mph beating that set on a Martin-JAP motorcycle. This was followed by the mile, reaching 70.04 mph but still with plenty of revs left before he ran out of track. On his second attempt, the carburettor worked loose and on the third and final attempt, a spark plug disintegrated ending his hopes. This unique Douglas-ABC hybrid was never raced again as Bailey had left for Australia, though he returned in 1914 and in 1920 became the Douglas Works Manager.

Another Brooklands pioneer Bradshaw helped with the odd machining job, was a former railway engineer, Walter Owen Bentley, who rode a 5hp Rex motorcycle with success in the One Hour race in August 1909 followed by a 5hp Indian in 1910. In 1912, 'W O' and his accountant brother Horace became directors of the Lecoq & Fernie motor-car dealership in Mayfair, London gaining the British concession for the French Buchet and the DFP (Doriot, Flandrin et Parant of Courbevois) motorcars. They soon became Bentley & Bentley. 'W O' was soon racing a DFP 12/15 2 litre model which was capable of reaching 55mph in its standard form, but a specially prepared 1913 example had reached 85mph. In visiting M. Doriot at DFP's works, 'W O' noticed a sample casting made from a revolutionary lightweight aluminium alloy cast in the form of an ashtray. 'W O' persuaded DFP to fit lightweight pistons, developed with the Corbin foundry, of 10% copper and 88% aluminium to increase its power to weight ratio. So modified, the new DFP 12/40 Speed Model of 1914 achieved 65mph as standard and in excess of 89mph on the track. Under the Bentley brothers' influence, Britain soon became DFP's best export market. This is believed to be the first use of aluminium pistons and it greatly influenced Bradshaw's later designs.

With the onset of war, W O Bentley was appointed to the Technical Board of the Admiralty Air Department and while acting as the Admiralty's Technical Liaison Officer at Gwynnes, 'W.O' proved the benefit of aluminium alloys on his new Admiralty Rotary (later Bentley Rotary BR.1) design with its steel-lined, cast aluminium cylinder barrels and alloy pistons. After the war, Bentley returned to motorcars and set up in Cricklewood designing his own superior car in 1919, using aluminium pistons - the rest is, as they say, history.

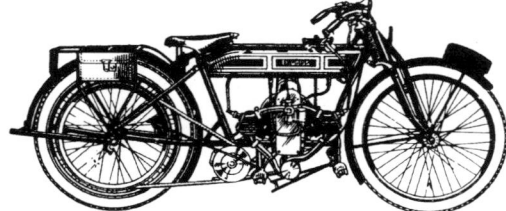

Left, Jack Falahee on a 1913 ABC engined motorcycle, and right, a contemporary $2^3/_4$hp Douglas

Inspired by Bailey's success with the his modified engine, Bradshaw designed and built his first ABC motorcycle engine in the winter of 1912/13, primarily as a replacement for those existing motorcycle engines fitted in the fore and aft, Douglas fashion, though this was itself developed from J F Barter's 'Fee' of 1905. Bradshaw's new engine was a 492cc, $3^1/_2$hp horizontally opposed 'square' twin (68 x 68 mm). Lubrication was by the conventional splash system; the engine weighed barely 40lbs.

Bradshaw preferred a square, or over-square, design as it produced higher revolutions for maximum power. He even chose 'Revs' as ABC's telegraphic address! Once again he chose the proven lightweight machined steel cylinder barrels commonly used in rotary aero-engines, turned and bored from a solid billet to relieve inherent stresses, with the finely turned, nickel-plated cooling fins of an eccentric pattern to provide greater surface area around the hotter exhaust valve. The crankshaft used a three-throw con-rod design to maintain balance: the central one connected to one cast-iron piston in the usual manner, while the two either side were of a lighter design and joined at the gudgeon pin of the opposing piston. Bradshaw's three-throw crankshaft design proved remarkably smooth running, free from the usual twin cylinder rocking motion usually experienced at low speeds. However, while it was well capable of running up to 6,000rpm, it ran roughly under load. Leslie Hounsfield of Trojan fame, who was almost as prolific an inventor as Bradshaw, used a similar forged steel 'V' con-rod arrangement for his famous twin cylinder two-stroke Trojan car engine of 1914. His fascinating story is told in *Trojan - Can you afford to walk?* by Rance and Williams.

The new ABC motorcycle engine was revealed to the public in April 1913 and was fitted into a Zenith frame which Freddie Barnes then raced on the Brooklands track. Improvements rapidly followed, starting with a pressure-feed lubrication system and in May 1913, a modified valve gear.

Due to complications in modifying other maker's frames to accept the fore and aft ABC engines, Bradshaw got both Ernest Humphries at OK-Supreme and the Collier brothers at H Collier & Son ('Matchless') to build special frames for him into which the new engine could be fitted. By June 1913 the ABC engine had reached production status and was soon selling to private owners of Douglas, Edmund, Matchless, Zenith and PV (Perry Vale) machines; one such PV was entered in the 1913 London-Exeter MCC trial and gained a Gold Award for completing the gruelling course.

George Brough also fitted an ABC engine into one of his father's (W E Brough) motorcycles with the intention of entering the 1913 Senior TT. Titch Allen, of the VMCC (Vintage Motorcycle Club), related the story told him by George that, having found his father's flat twin too slow, he fitted an ABC twin, which at the time was having great success in racing. However performance was not up to par. Investigating the problem, he discovered a poorly cast head had partly blocked the inlet port. That rectified, he entered for the TT, but during a practice run the engine 'blew up'. He quickly borrowed a JAP engine but was forced to retire in the 5th lap when the JAP's timing gear failed.

Jack Emerson

Jack Emerson was already a respected rider when he entered the 1912 500cc Senior race on a Norton, leading the way until his carburettor unscrewed! He joined ABC in early 1913 as their works tester, devoting much of his time to racing the new, modified ABC motorcycle engines. One of these engines was increased in bore and stroke (70.45mm x 64mm) with lighter rocker arms. Fitted into a belt driven, rigid motorcycle frame, built by Ernie Humphries, it was equipped with a vestigial, 30" long by 15" diameter, streamlined cone attached to the back of the saddle. It was on this machine that, on 13th January 1914, Emerson broke the 500cc flying kilometre record at 80.47mph, followed by the mile at 78.27mph. However he noticed considerable slip on the V-belt drive, so Bradshaw proposed a chain drive for a further attempt.

Jack Emerson on his rigid, 1914 ABC with streamlined tail on which he set a new speed record

Streamlining was still very much in its infancy. The cone had been made for them by the Martinsyde Aircraft company in the Brooklands flying village and everyone expected to see a noticeable drop in speed when the tail-cone was removed but, remarkably, Emerson recorded an even faster, unofficial, flying kilometre at 82.5mph! His secret lay not in the tail-cone but the fuel! Instead of using the standard Pratts' 'Pure Aviation Spirit' racing petrol, Emerson had chosen to use Benzole, without telling Bradshaw, and it was this which allowed him to keep on winning. When eventually let into his secret, Bradshaw experimented by machining $^1/_{16}$" off the cylinder to increase compression ratio and, finding no undue effect, got ever better results on the track. As Pratts were dependent on racing success to promote their petrol, they implored Bradshaw to use their standard "Pure Aviation Spirit", but when he experimented with their fuel in the modified engine, it pinked badly. So the story goes, this prompted Pratts, and others, to add Benzole.

Emerson's first attempt at the inaugural Brooklands TT of 1914 was thwarted by a faulty mechanical oil pump, but he came 2nd in the Senior event at 50.02mph over 26 laps, and 2nd at 57.55mph over 56 laps. At the June 1914 Oxford-Cambridge University race at Brooklands, E H Lees won two races at 64.71mph and 63.44mph on his ABC. When war broke out, all further civilian motorcycle racing came to an end, though in the Combined Services meeting in September 1915, Cpl. H G Hodgson won the 500c touring class at 44.78mph on his ABC.

Emerson's 1914 engine and belt drive

ABC Road Motors

Racing success persuaded Bradshaw to concentrate on motorcycles and he temporarily promoted the company, with its staff of six, as 'ABC Road Motors'.

With Zenith's works at nearby Baker Street, Weybridge (they moved to Lower Mill, East Moseley in 1914), Granville got to know Freddie Barnes quite well. Freddie had developed a unique variable speed belt drive and pulley system for his Zenith-Gradua, which he freely boasted was "the finest motorcycle in the world". "How so?", asked Bradshaw, to which Freddie responded that he "personally tested everyone made"! This prompted Bradshaw to road test and compare every make of motorcycle he could to ensure that it would be his ABC motorcycle which would become 'the best in the world'.

By late 1913, Bradshaw had designed a prototype motorcycle with a novel spring frame in which a lower rear swing arm pivoted off the splayed saddle down tubes, cushioned by laminated leaf springs fitted between the saddle lug and a vertical rear stay from the axle lug. He lodged four provisional patents in 1914 but these were later abandoned: (4503 01/02/14 Cycle frames; 4504 21/02/14 Cycle frames; 5400 03/03/14 sidecars; 11310 07/05/14 Cushion drive). It is not now possible to determine what these designs were, but it is interesting to note reference to 'cushion drive'. Was this perhaps a hub mounted rubber block cush-drive similar to the later Enfield?

By early 1914, ABC was offering a range of chain-driven motorcycles fitted with a 3-speed Armstrong gearbox and optional kickstart off the final drive chain. Production models began to trickle out in early spring; these boasted an improved lower mounted rear wheel leaf spring arrangement and conventional coil sprung front girder fork; they later adopted a leaf-spring front suspension. There was a 'Touring' model capable of 60mph; a 65mph 'TT' and a 'Brooklands' model, works-tuned to 70mph. Both the 'TT' and 'Brooklands' models retained the smoother three-throw crankshaft design with its scientifically designed eccentric cooling fins. The 'Tourer', however, had plain, machined steel barrels with an overhead exhaust and side inlet valve arrangement, but with conventional two-throw connecting-rod design. It also featured a Best & Lloyd semi-automatic lubrication system with a rider operated auxiliary pump. An under-engine tray protected the rider, engine and final drive from mud and dust thrown up from the road.

A restored 1914 ABC (AB Thomas)

Following much publicity over the ABC's potential on the track and at trials, the initial trickle of orders rapidly gained momentum and ABC raised production to half a dozen motorcycles a week. Bradshaw now earnestly sought larger premises and in the summer of 1914 earmarked a vacant, two acre site in nearby Hersham. With the outbreak of war in August 1914, Brooklands was commandeered by the Royal Flying Corps and those companies not directly involved in aeroplane production were unceremoniously evicted on 24 hours notice! A new factory was hastily constructed at Hersham and equipped with what portable machinery they could salvage from their Brooklands works.

Top, a 1916 250cc prototype with Albion 2-speed 'box. (GB)
Bottom, 1914 ABC chain driven combination (CMC)

While waiting for production to resume, Bradshaw coupled a Montgomery sidecar to his ABC, but the engine proved underpowered, so he developed an improved sidecar model and ran this prototype for several hundred miles. He also improved the rear spring arrangement with an upswept swinging arm in December 1914; this was granted as Pat No: 24189 in June 1915. Motorcycle production began again in early 1915 and these new models now featured a hollow head stock tube, through which the control cables passed, and adopted an adjustable head-stock cone bearing (Pat No: 191500090 granted 1916). Improved laminated leaf springs, separated by interleaves also featured (Pat No: 113345 laid 07/03/15), but more importantly, the frame was now fitted with a new, car-type, phosphor-bronze encased 4-speed gearbox with ratios of 16, 10, 6.25 and 4.5 to 1, controlled through a tank mounted gate-change gate. It also had a car type cone clutch. The gearbox was suspended below the frame. 1915 models now offered customers optional colour schemes to their standard battleship-grey and black.

The improved ABC was enthusiastically praised by' Ixion', a well known motorcycling clergyman-cum-columnist for *The Motor Cycle* magazine. It was ideal for sidecar work, at an appealing £75, but with the war now on, there was little market for the new ABC at home, however it was exported in limited numbers to Commonwealth countries where it was often raced.

In these early war years, many motorcycle riders volunteered as despatch riders with the British Expeditionary Forces in France, taking their own motorcycles with them. Though not adopted by the War Office, a small batch of ABC combinations was later ordered for the Mesopotamia campaign, under the auspices of the Indian army, but these were lost in the Mediterranean when the ship was sunk in passage to Alexandria.

By mid 1915, ABC was fully engaged in production of auxiliary stationary engines for army field pumps and generators, described earlier. Motorcycle production ceased at the end of the year. How many prewar ABC motorcycles were built is not known, possibly only a hundred or so: very few survive. Few rider reports exist, but one writer recalled how the long induction pipes required extensive lagging to prevent carburettor icing in the freezing winters! Another letter in *The Motor Cycle* reported how "extraordinarily comfortable and in the broad sense reliable" these ABCs were, though he too picked up on the induction problem.

Though motorcycle production had ended, one of the 250cc Firefly auxiliary engines was experimentally fitted in 1916 to a rigid, cradle framed lightweight motorcycle, coupled to a 2-speed Albion gearbox. It was also said that Granville was working on a 900cc version of the 500cc ABC engine with a vaned flywheel-cum-cooling fan together with an improved 4-speed gearbox with friction drive for a possible cyclecar, but little else is known of these two projects.

New technology

These were exciting times for pioneering engineers and it must be said that Granville Bradshaw was far less interested in manufacturing than he was in research and design. He became increasingly concerned with motorcycle frames and suspension systems to maintain stability at speed and improved upon his original rear suspension in late 1917 by having the engine pivoted off the same bracket as the extended lower rear wheel stays, thus permitting the fitting of shaft drive. It should be remembered that a flexible chain drive allows for geometric inaccuracies as the rear frame moved up and down; something which shaft drive does not tolerate. This design was granted under Pat No: 119480 in October 1918 and followed up by an improved double laminated leaf spring arrangement under Pat No: 132952 of 1919.

New racing motorcycles were invariably ridden to and from events straight from the works by their works rider, who had to sort out the tuning and numerous teething troubles en route. Racing certainly pushed engine, fuel, lubricant technology and knowledge of metallurgy to their very limits in the pursuit of new records. Record breaking, with its associated good publicity, was vitally important to manufacturers of petrol, oil, spark-plugs, magnetos, tyres and chains and they rapidly explored ever better designs to win more races and fame. Early petroleum fuels and oils were of quite variable quality with petrol spirit being prone to pre-ignition as the engines got hotter the faster they ran! Added to which, carburettors had to ensure the correct fuel:air ratio at all speeds, which demanded very careful attention to jetting and venturi design. Riders often sought help from the petroleum companies, who responded with anti-knock additives to control pre-ignition, leading to the introduction of Benzole and in the 1920s, tetra-ethyl-lead. Benzole, a coal tar distillate, permitted an increase in engine compression which improved power, performance and economy considerably, leading to demands for ever higher octane ratings to allow engineers to develop ever smaller, lighter and more efficient engines.

Engine longevity was improved by better lubricating oils which served a double purpose; oil both separated moving parts to prevent friction and seizure and helped dissipate heat. Early, heavy mineral oils were more suited to relatively slow moving, lightly stressed parts with sliding plain bearings, such as those found in steam locomotives: these oils had no cooling effect. High speed motorcycle, car and aero-engines however needed something quite different for their ball and roller bearings. Experiments by Professor Gough of the Royal Aeronautical Society had shown that in early aero-engine reduction gearboxes, the extreme pressure put on gear teeth resulted in pitting when using plain mineral oils, yet remained undamaged when using castor-oil. But while mineral based oils were superior for fast rotating and highly stressed bearings, it was also found that immiscible castor oil overcame the problem found in rotary aero-engines of the oil being washed off the cylinder surfaces with the inflow of petrol enriched air. Castor oil also remained popular in short races as it did not require a complex pumped lubrication system. The disadvantage of castor oil is its tendency to gum and form rubber like compounds at elevated temperatures.

Castor oil's immiscibility needs qualification, so I quote from Elliott A Evans MCIAE, CIME, Chief Chemist to C C Wakefield & Co in his *Lubricating & Allied Oils* of 1921: "Castor is the only common fatty oil which is soluble in alcohol and insoluble in petroleum. In common with petroleum it is soluble in benzene and aromatic hydrocarbons."

Its suitability to rotary engines was due to its chemical bonding by positive ion attraction to the iron surfaces. Its shear and lubricity was thus in the oil itself and, as castor oil was not readily dissolved by the petrols of the day, the chemically bonded oil was not washed off the cylinder walls by the petrol:air mix introduced into the Gnôme rotary engine's crankcase. Excess oil flung from the caged big end bearing was carried by the fuel:air mix into the combustion chamber and burnt in the exhaust gases.

Neither will castor oil dissolve in mineral oil except when dissolved in alcohol and refined by adding sulphuric acid. Indeed from late 1909, Gnôme successfully used a castor blended mineral lubricating oil for their Monosoupape radial, supplied by C C Wakefield and known as "Gnôme-Castrol". It was considered the finest aero-engine oil in the world and some 300 gallons a day was supplied to the RFC.

Its use in a 'Petroil' mix did not come about until the 1920s, long after the demise of the rotary engine, when it was found it would readily mix (but separate out, over time) in modern petrols with high levels of alcohols, napthas and aromatic hydrocarbons.

Castor oil has a characteristic, pungent smell when hot, and soon became associated with racing and all its wealth and glamour. Many a young rider added it to the most humble of lightweights in the hope of attracting the girls! As a consequence, sales of new mineral oils temporarily declined in favour of castor oil.

Castrol

This effect was noticed by one of the largest suppliers of mineral oils, Charles Cheers Wakefield, who had begun his business in 1899 after resigning from The Vacuum Oil Company (later known as Mobil oil). He offered three types of Wakefield oil: 'CW' an 'everyday' oil; 'C' motorcycle oil and from 1912, 'R', a castor based racing oil.

Bradshaw had become great friends with C C Wakefield and the story goes that he told him he should do two things: call his oil 'Castrol' to suggest it used castor oil, and to offer higher awards to racing drivers who promoted Wakefield's 'Castrol' oil! Wakefield took up the idea and registered the Castrol brand in late 1909. He also added, for good measure, castor oil essence to titillate the nostrils; he soon saw sales rise!

Castrol 'XL' was developed later for aluminium car and motorcycle engines while 'XXL' was a heavier oil for sports and high compression engines. In 1960 CC Wakefield Ltd was renamed Castrol Ltd. They produced their first multi-grade 'GTX' 20W/50 motor oil in 1968. Wakefield also gained control of W B Dick & Co (Holdings) Ltd and by the 1970s had become part of the huge Burmah Oil Company which had pioneered oil exploration in Persia in 1902, from which was formed British Petroleum.

8

The Sopwith-ABC Motorcycle

The Sopwith Aviation Company

In the inter-war years, Granville Bradshaw became better known as an independent consulting engineer, selling his designs to eager clients. His critics maintain his later, tainted reputation was self-inflicted through leaving his clients in a financially crippled mess as they tried in vain to make his fantastic ideas work. Yet today, we routinely hear of new ultra-safe, ultra-modern motorcars which have been designed entirely in house by faceless designers, having to be recalled urgently for inherent safety defects - defects which should have been picked-up in pre-production tests! Nothing has changed, yet there is rarely these days a public outcry or a denouncement of that (unnamed) designer.

Bradshaw's first post-war design job for ABC was the famous transverse-twin ABC motorcycle, designed for his friend Tom Sopwith to keep his factory in work. They had become good friends before the war but though both Sopwith and Harry Hawker spoke highly of Granville and his work, Tom was not impressed by the post-war stigma of the Dragonfly's reputation, though he did concede that Bradshaw was pushing contemporary technology to its limits at the Government's haste to win the war.

Born on 18th January 1888, Thomas Octave Murdoch Sopwith was the first son and, as his name suggests, the eighth child of Thomas Sopwith. In his youth 'Tom' became a keen yachtsman and in 1909 employed a young engineer, Fred Sigrist, to tend to his yacht. Unimpressed by the time it took to cross the Channel, compared to that of Bleriot's flight, he decided to learn to fly and on 22nd October 1910, bought a Howard-Wright biplane with an 40hp ENV engine. Taking to the air that same day, he almost killed himself in the process for having flown barely 300 yards, he hauled the stick back, rose into the air 40 feet and nose dived to the ground, breaking the propeller, undercarriage and a wing. Basing himself in Howard-Wright's shed (No. 21) to repair the machine, he then chose to buy another Howard-Wright instead! On 21st November, Sopwith took the precaution of some exploratory taxying before taking to the air and gaining his RAeC Certificate (No.31), then with barely 10 hours flying experience, he entered himself in the December *Daily Mail* contest competing against Cody and, incredibly, set a new all-British duration record of 107 miles in 3 hrs 12 mins.

Sopwith soon made the acquaintance of other pioneers at the Brooklands Flying Village, including Granville Bradshaw, then of the Star Company. On returning from a tour of America with Fred Sigrist in early spring 1911, where he had ordered a new Burgess-Wright biplane, he established a flying school at Brooklands, occupying sheds Nos. 29-31. He then replaced the Burgess-Wright's

50hp Gnôme rotary with one of Bradshaw's 40hp ABC 4-cylinder vertical water-cooled in-lines that had been developed for the Star monoplane.

His Sopwith School of Flying became hugely successful with its Burgess-Wright dual control biplane, Howard-Wright biplane and a pair of Sopwith-Farman and Blériot monoplanes. One of his pupils was a Major Hugh Trenchard DSO, of the Royal Scots Fusiliers, who gained his RAeC Certificate (No.270) on August 13th 1912 and was duly appointed Flying Instructor at the RFC's new Central Flying School, Upavon in September 1912. He is perhaps better known today as the 'father' of the Royal Air Force.

Another 'star' pupil was a young Australian, Harry George Hawker, who had come to England in 1912 to learn to fly. Gaining temporary employment with the Austro-Daimler engine company, Hawker finally landed his dream job under Fred Sigrist at Sopwiths and, once on the payroll, enrolled in Sopwith's flying school gaining his Certificate (No.297) on 17th September 1912 with the Sopwith-Farman. In October, Hawker entered for the British Empire Michelin Trophy No.1 Cup for endurance with the ABC powered Burgess-Wright spluttering into the British record books with a flight duration of 8 hrs 23 mins, winning the coveted Michelin Trophy on 24th October.

Now certain of his future in aviation, in December 1912 Tom Sopwith bought for £900 a spacious former roller-skating rink at 1, Canbury Park Road, Kingston-upon-Thames. Sited at the junction with Richmond Road, the brick building had a large, clear spanned hall which would allow the airframe's parts to be marked out on the floor - a practice copied from ship-wrights. Thus was formed in 1913 the Sopwith Aviation Company. Its directors included his sister, May Gertrude Sopwith, R O Cary (General Manager), Fred Sigrist (Engineering Manager) and H P Musgrave (Company Secretary). They initially employed six carpenters and metalworkers and by March 1914 had become a limited company.

Winning major contracts to build fighters for the Royal Flying Corps, Sopwiths were soon forced to buy shops and houses in nearby streets, taking up much of Canbury Park Road in their expansion, but by 1917, concerned at Sopwith's limited production capacity for aeroplanes, the Government asked him to sub-contract their manufacture; this Sopwith firmly opposed, but the Government went ahead placing hundreds of orders for Sopwith fighters elsewhere under the Munitions of War Act. With the launch of the National Factory Scheme, a 35 acre industrial estate was created on Ham Common at the junction of Richmond Road and Dukes Avenue, Richmond, alongside the River Thames, but rather than buy the factory, Sopwith leased it from the Government for aircraft assembly, at the same time keeping the engineering machine shops in Canbury Park Road.

By the Armistice of 11th November 1918, production at 'Aircraft Factory No.1', as the Ham Common plant became known, had reached 90 aircraft a week from a staff of 3,500, around 1,000 of whom were women. Some 16,237 examples of Sopwith's 32 aeroplane designs had been built, but with the Government immediately cancelling military contracts, Tom Sopwith realised that his factory now faced a very uncertain future; as indeed did the rest of the aircraft industry. Fortunately for him, there were no immediate signs that the recently placed RAF orders for the new Sopwith Snipe were facing major cancellations and he hoped Sopwiths would at least be considered as major contractors for the RAF's post-war requirements. Fortunately, having also been appointed to the Government's Civil Aerial Transport Committee in May 1918, Tom Sopwith saw great potential in a new civilian market and backed his hunches by building the Sopwith Dove two-seat sports biplane, derived from the Sopwith Pup fighter, as well as the Sopwith Atlantic with which he hoped to win the £10,000 trans-Atlantic prize. These were followed by the Sopwith Gnu civilian monoplane with its 3-seater cabin enclosed within a streamlined fuselage. However, with the post war depression and mass unemployment from the wholesale demobbing of servicemen, only 13 Sopwith Gnus and 10 Doves were built, most of these were exported to Australia.

Many in the industry faced almost certain bankruptcy as they fought for a share in the now severely limited aviation market. Some, such as Nieuport and BAT simply collapsed while others, such as Blackburn and AVRO, pinned their hopes on motor cars. Though Blackburn's venture was short lived, Alliott Verdon-Roe's designs for a prototype two-wheeler cycle-car and a motorcar succeeded with motorcar production being taken over in 1920 by Crossley when they became a major share-holder, with 60% of share capital in Avro, at which point Sir Alliott Verdon-Roe severed his links with Avro and in 1921 bought S E Saunders Ltd of Cowes, following its sale by Vickers; the company then became Saunders-Roe. As boat builders, Saunders had made the hull for Tom Sopwith's 'Bat-boat' spotter of 1913, literally a Wright pusher biplane on a racing boat hull, but in 1927, Verdon-Roe sold his major share holding in Saunders-Roe to Supermarine, the famous flying boat (and much later, Spitfire) company. By late 1927, Crossley were in financial trouble and in May 1928 they sold Avro to John Siddeley, formerly of Siddeley-Deasey.

Tom Sopwith however was very keen to survive as an aircraft maker, but to retain his highly skilled staff of woodworkers, he pinned his hopes on Bradshaw's ABC motorcycle and the odd engineering contract.

Designed in 11 days

To most people, the post war Sopwith built ABC, is the 'real' ABC, one still held in the highest regard for its innovative design and high standards of comfort. Many a legend surrounds this motorcycle: the most quoted being that BMW copied it. The other famous legend has it that Bradshaw conceived the design in his bath on Armistice night and built a prototype within 11 days against a wager with Sopwith that he could produce a new motorcycle in 21 days. The terms of the wager are reputed to have been that should he fail, he would pay Sopwith £100 per day for every day over the three weeks; likewise, he would receive such a sum for every day under. There is a large element of truth in both these claims!

Ten days after the Armistice of 11th November 1918, Bradshaw told *The Motor Cycle*:

"ABC have made all arrangements for the mass production of an entirely new design". This was followed a week later by the comment, "ABC are in a very enviable position. The war brought them close to some of the largest aircraft makers in the Kingdom. The (ABC) factory is now a 'laboratory' for research and design and priorities are a cycle-car and light motorcycle. As each design is perfected it will be passed on to the associated factories for mass production".

In December 1918 Tom Sopwith announced at a press conference:

"Nothing was decided until the Armistice was signed, then I approached Mr Bradshaw and asked him to turn out an experimental model. Mr Bradshaw said it could be done in three weeks. I told him I did not believe him but asked him to go ahead, with the result that the machine was placed on the road in eleven days and was a well finished job". Bradshaw then remarked, "On the twelfth day, Mr Sopwith's cheque arrived". The cheque was for £1,000.

Tom Sopwith later conceded there was 'some exaggeration' in these claims but an account recalled by his old friend 'Ixion' (the Canon Basil H Davies) in *The Motor Cycle* of 26th May 1932, supported by Granville's grand-children, paints a vivid picture of Bradshaw's determined, if unorthodox, approach to his design work:

"I remember seeing him start work one day. At that period of his life he regarded the working day as beginning in the evening, when the house was quiet. After a good dinner had been digested he stripped and got into a hot bath. The tap was left on and a small hole which had been drilled in

1918 ABC prototype (CMC)

the vent plug maintained the water at a steady level and the temperature too. A cross tray supported drawing board, instruments, smokes and a decanter. He stayed there until about 3am, by which time he had got through more designing than most conventional engineers would achieve in a week."

In a sense, Sopwith had provided the catalyst for putting into production Bradshaw's earlier ideas for a new ABC twin, postponed by the war, yet Bradshaw claimed that almost everything for the Sopwith ABC, from bearings, spindles, hubs and brakes, was freshly designed. His choice of a transverse engine layout came from his experience in aero-engine design in the belief that the fore and aft layout of his record breaking pre-war ABC merely caused an imbalance in the engine's cooling, limiting its racing potential. The laminated leaf spring suspension would have 'weakened centres', by virtue of thinner or waisted leaves, to improve flexing while interleaves would improve damping (Pat No. 113345 laid 07/03/17).

He followed his usual practice of beginning with a full size drawing of the engine and transmission and then adding the rear suspension, hubs, wheels and brakes, before designing the frame and adding the fuel tank. From these full scale drawings he could more easily assess stress factors and in this way was able to make the patterns and moulds for the integral gearbox well within the three weeks wager laid down by Sopwith.

The new transverse twin ABC was regarded by one and all as 'revolutionary' with a design and many features 20 years ahead of their time; he certainly lost no time in patenting them! He proposed a wide steel foot board with upturned toes to protect the engine from mud (Pat No: 139834 laid 12/06/18);

Left, the efficient mud shield, and right, a contented ABC rider (CMC)

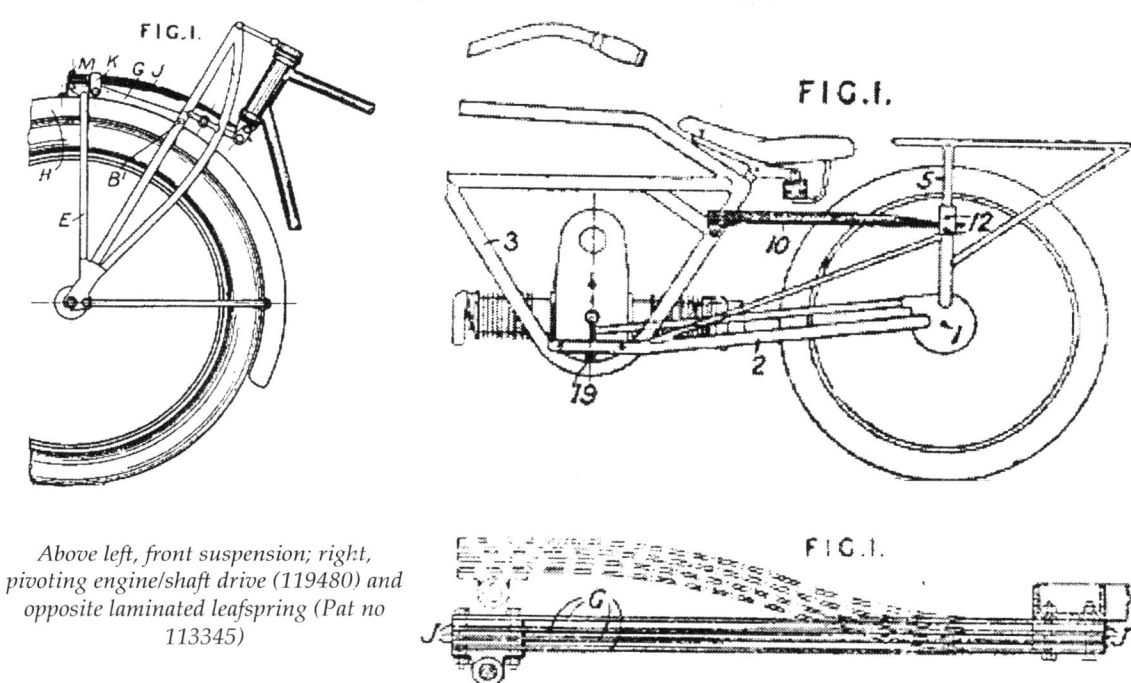

Above left, front suspension; right, pivoting engine/shaft drive (119480) and opposite laminated leafspring (Pat no 113345)

weather protection afforded by a large frontal, steel leg shield which incorporated adjustable louvres for improved cooling (Pat No: 140154 laid 22/01/19) and a new splayed sprung cradle frame (Pat No: 151032, laid 13/03/19) offering unparalleled engine protection.

The fully sprung, unit construction ABC, with its deeply valanced mudguards, weighed just under 240lbs and offered extensive weather protection, making it an ideal machine for the everyday rider. The 3hp, 398cc (68.6mm x 54mm) ohv engine was a modified 250cc (54 x 54mm) auxiliary unit (sometimes referred to as the 'Firefly') which retained Bradshaw's hallmark steel cylinders with turned cooling fins. The hemispherical, cast iron head had 46° inclined valves operated by forked rocker arms, closed by coiled springs. With its concave aluminium piston crown, the combustion chamber was almost spherical. A new three-throw crank with deep groove ball and roller bearings was lubricated under pressure from an internal, immersed pump, recirculating oil every three minutes at full revs. While the prototype had a three speed, car-type, unit construction gearbox, production models had four speeds. It used a car-type flywheel clutch, H-gate gear selector and roller bearings throughout.

1921 ABC twin

Bradshaw had considered shaft drive in 1917; his design was granted under Pat No: 119480 in October 1918, but leaf spring rear suspension presents major problems in frame geometry, alignment and movement which are not so critical with chain drive. Given its short development programme, the new ABC therefore had to be content with bevel gear final drive off the main shaft to a transverse shaft and thence by chain to the hub. The laminated rear springs had 'weakened centres' while the new spring front fork

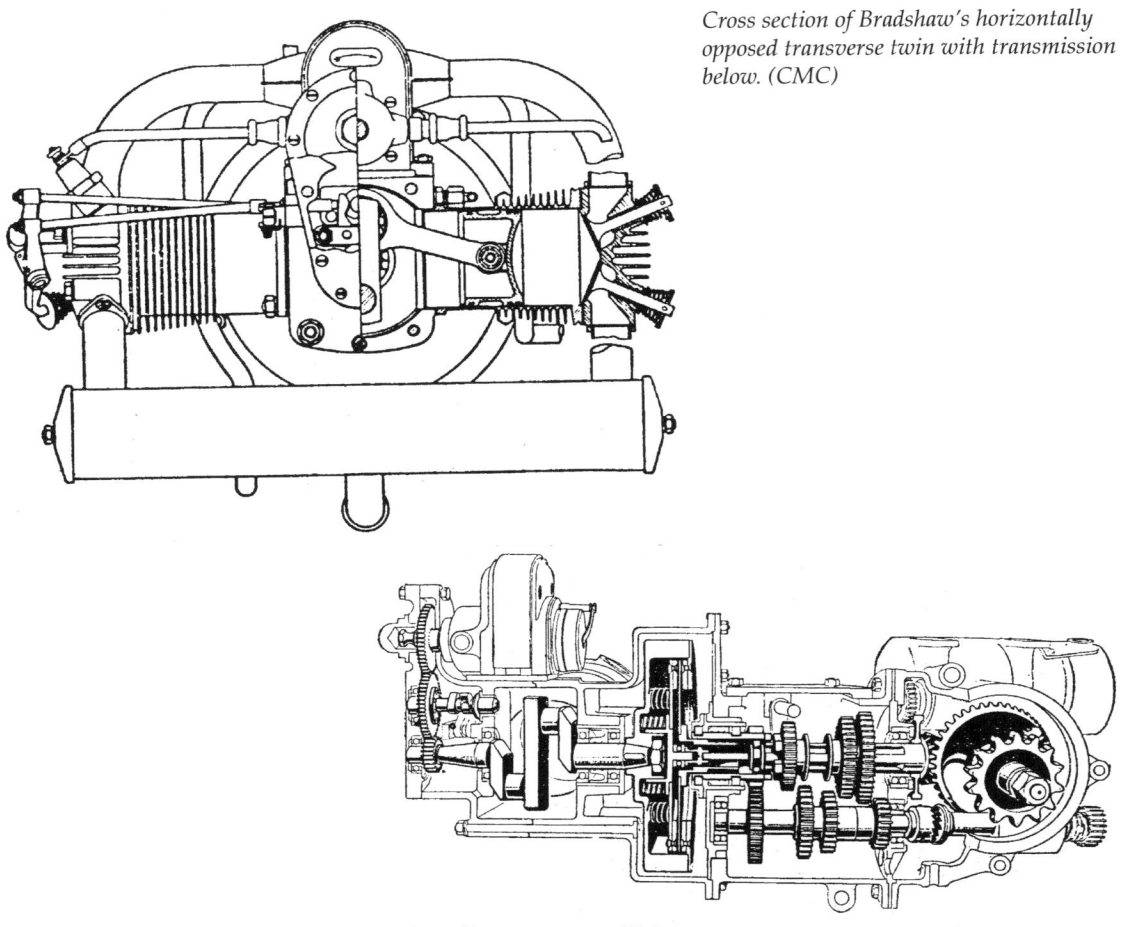

Cross section of Bradshaw's horizontally opposed transverse twin with transmission below. (CMC)

(Pat No: 139570 laid 01/02/19) had an additional forward bridged stay attached to the leaf spring mounting which acted upon the steering head by roller and link system. These gave it an extremely comfortable ride, even by today's standards! Consider also the ABC had internal expanding drum brakes - front (4") and rear (6") - quite an advance on the more usual and quite ineffective bicycle or dummy belt rim brakes.

Production was planned at 100 per week and in common with other motorcycle makers, the sub-assemblies would each be completed by one man before moving to another for final assembly into a complete motorcycle. The finished machine was then coupled to a Froude dynamometer to check its power output at the chain sprocket; it had to record at least 8bhp to pass muster. Advertisements pushed the ABC's ability to romp up the famous Brooklands test hill in second gear complete with a laden side car, prompting the question, "Why have a clumsy 8hp drunkard to do the work which the 3hp ABC can do so much better?" (Note that hp is a taxation class not to be confused with power output, bhp.)

Confident of the new motorcycle, Sopwith's directors resolved at an extraordinary general meeting in March 1919, to change their company's memorandum to carry on in business as "manufacturers of motor cycles, motor cars and conveyances of all kinds" and in addition, "coach and carriage makers, cabinet makers and all kinds of furniture hardware....". Much to the amusement of Harry Hawker, their contracts included making kitchen utensils.

Full details of the new ABC were released in March 1919 with the announcement that 'production was imminent'. Its revolutionary, advanced design was met with rapturous applause. Offered with magneto at £70 or with a gearbox driven Lucas dynamo electric lighting at £98, it now weighed only 175lbs, well within the lower tax limit. While pre-production models were in ABC's pre-war grey and black, riders were promised a new gunmetal-blued finish.

Unfortunately, in Sopwith's eagerness to get production on stream to pay his employees, large numbers of costly components were ordered before the ABC had been fully tested, with the inevitable result that many minor modifications were needed - these merely added to costs and delays. Yet despite further promises in July 1919 that production was still 'imminent', no deliveries against forward orders had been made by the time of the autumn Olympia Motorcycle show.

Now enhanced by a new, if rather inadequate kick-start, the ABC was a show sensation. The motorcycle world was agog at its brilliant design. As 'Torrens' of *The Motor Cycle* later described, the ABC stand was 'besieged' with visitors more so than at any other time he could recall! Orders flooded in to Sopwith's temporary showrooms at 1, Albemarle Street, London, from where he had earlier run, in partnership with Phil Paddon, a motor-car dealership in 1903. The building housed many offices including the Society of British Aircraft Constructors. Throwing caution to the wind, Sopwith opened another showroom and depot at 65/67 South Molton Street, W1 and, keen to emphasise their proud aeronautical heritage to, hopefully, gain aircraft orders he displayed their new Sopwith Dove sports biplane alongside the new ABC motorcycle.

Racing and trials

Jack Emerson and Ronald Charteris both enjoyed success in MCC trials attracting further orders, which, unfortunately merely exasperated those still awaiting delivery and with further racing success, demand soon far exceeded the planned production rates. Granville claimed some 40-45,000 orders were received - a figure suspiciously overstated - but nevertheless, such was the potential that 'half a million pounds worth' of plant and machinery was ambitiously installed at the works to meet the demand.

The ABC got off to a good start with Jack Emerson, as their rider, winning the first post-war race at Brooklands at 66.7mph in the April 1920 Victory Handicap. In those early days the ultimate goal was very much the hour record and in 1920 Jack Emerson took the record twice at 67.93mph and 70.44mph. His preferred approach was to rev the engine to its maximum of around 6,000rpm, and then promptly drop the clutch for a frighteningly fast standing start; the ABC seemed quite happy with this abuse! Bradshaw claimed that Emerson won nearly half his races at Brooklands in just one season. Eric Porter also gained success at Brooklands and other national and international events often against more powerful competitors. Tuned examples,

Jack Emerson won the 1920 Victory Handicap on an ABC works racer at 66.7mph

Jack Emerson testing an ABC racer

such as those by Stephen Bassett, reached 80mph but for the man in the street, they were more than content to cruise all day at a 45-48mph and enjoy the ABC's surprisingly brisk performance.

Disappointment

Although the ABC had managed to achieve the magical "60mph and 60mpg" goal, the love affair was not to last for by the time deliveries had began to trickle through in late spring 1920, post-war inflation had forced the price up from £70 to £150. Worse, high unemployment rose further as a result of industrial unrest across the country for better conditions and wages than those imposed under the Munitions of War Act which had forbidden strikes and lockouts. Much worse was to come following the devastating Moulder's Strike which plunged the country into ever deeper economic depression.

And then the technical faults came to haunt them. As a result of inadequate pre-production testing, the valve gear was found to be unreliable due to the flimsy rocker gear bending or throwing the push rods free of the engine at high revs. One of the main problems with early ohv engines was the correct setting of the tappet clearance on a cold engine, for as the engine got hot, the barrels expanded longitudinally, but faster than the exposed, air cooled push rods; if the tappet cups were not sufficiently deep, the push rods tended to pop out, but if the rider were lucky, these just might have dropped onto the ABC's foot board and be easily found lying under the engine, trapped in an oily gunge. Fortunately, as riders' reports indicated, a cylinder head and valve gear could be stripped and reassembled within 45 minutes at the roadside! Bradshaw also concluded that the valve springs were too short, for at full revs, when subjected to the greatest heat, the springs tended to harden and lose their ability to snap the valves closed quickly enough, causing further problems. This he rightly blamed on the riders for taking advantage of the engine's ready ability to freely rev. But then, being over-square, the engine had to be revved hard to get full power. A temporary remedy was effected by an extended heat shield and twin concentric springs, which he covered under Pat No: 132827, lodged in 1919, but the exposed valve gear remained prone to rapid wear from road grit and a rather haphazard oil mist supply from the suction pump lubrication system which proved quite inadequate. This was soon solved by an external Pilgrim pump, with emergency hand-pump, to the main bearings.

The kick-start was prone to skipping at the ratchet quadrant, but fortunately it was quite easy, with its low set saddle, to 'paddle off' in second gear to start the engine. The axial crankshaft also lacked adequate thrust bearings and tended to float, causing much noise. Following a chance meeting with Tom Sopwith, Walter Moore briefly joined Sopwiths in 1920 to help solve problems with the flat twin. Walter had worked with Joseph Barter in developing the original Fairy Fée engines, becoming Douglas' design engineer in 1913; his position was taken by Les Bailey. Walter then joined Nortons, before rejoining Douglas in 1945.

As more faults emerged, overall reliability become unacceptable to the usually tolerant motorcyclists, to the extent that Sopwiths was soon flooded with warranty claims, mostly over the valve gear but fortunately, independent companies had found cures to the ABC's inherent flaws, such as fitting race proven 'Celerity' valves and, later, improved rocker gear from R S Inglis Ltd of Upper

Sopwith's Canbury Road headquarters and works (Sopwith)

Marylebone Street, London. These modifications and others from Taylor-Young and BEW (Brookland Engineering Works) finally transformed the ill prepared ABC into a much loved and reliable mount. The engine was even fitted into the GSD motorcycle between 1921 and 1923. But these improvements came too late to save the ABC for other external events had overtaken this casual on-going development process, and had put Sopwith's very future into jeopardy.

The end of Sopwiths

The oft quoted accusation by motorcycling journalists that it was Bradshaw and his ABC that had brought about Sopwith's collapse is mischievous and ignores more pressing external pressures.

The lucrative refurbishment of war weary Sopwith fighters for the RAF had come to an end and the anticipated demand for Sopwith's civilian aeroplanes had failed - only 23 aeroplanes of all types were ever built. To add to their financial woes, under the Government's 'ten year peace plan' they saw little need in rebuilding the RAF's war-time might. Far worse was to come, for under the Government's war-time contracts, companies so engaged were governed by both the Munitions of War Act, which restricted the profits made, and the Finance No.2 Act of 1915 which severely penalised profits in excess of £100 over those of the pre-war years. The Government's urge to restore their much depleted treasury reserves caused them to call in these dues, presenting their many contractors with huge tax demands for immediate payment. Had the ABC been yielding a healthy profit, all may have been well but despite attempts to cure the ABC of it's many faults, the fundamental problem remained that Sopwith Aviation and their skilled employees, were not adept at large scale motorcycle production.

To make matters worse, the scheduled start of production in September 1919 was hit by industrial strikes at the works, followed by a nationwide Moulder's Strike which halted deliveries of castings delaying production of the first handful of motorcycles until January 1920. With further strikes at Sopwiths in August 1920, the directors saw little merit in continuing with the motorcycle, especially in the light of a noticeable slump in the motor trade through worsening national unemployment. In these very uncertain economic times it also became impossible to secure further business loans. Sopwith's directors saw little hope of recovery in the near future and wisely decided to safeguard their assets from creditors and the Government's excess profit taxes by going into voluntary liquidation, just as had the Bristol & Colonial Aeroplane Company in March 1920. Thus, on 10th September 1920 with the company suffering from yet more strikes, production at Sopwith's Richmond Road works stopped.

On 13th September an emergency general meeting was held to resolve "that the Company cannot by reason of its liabilities continue in business" and so, on 15th September 1920, Tom Sopwith announced, in despair, that the works would not be reopening after these latest strikes and on 5th October 1920, the company went into voluntary liquidation.

Richmond Road works

Their lease on the Richmond Road works duly reverted to the Government who quickly passed it to Leyland Motors Ltd for rebuilding ex-War Department Leyland G type lorries. The post-war slump and cancellation of Government orders had also hit Leyland very hard, for they had just started production of the luxurious 7.3 litre 'Straight Eight' car designed by their chief engineer, J Parry-Thomas. At £2,500 for the bare chassis, it is not surprising that only 13 were built!

When Leslie Haywood Hounsfield of Trojan Ltd sought a manufacturer to build his prototype light utility truck, based on his new under-floor engined two-stroke car, Leyland jumped at the chance and accepted the £5 royalty payment terms on each £175 van produced. In November 1920, they formed a new company, Leyland-Trojan Motors with production at the Richmond Road works getting underway in 1921. Leslie Hounsfield resigned as Managing Director of Trojan in August 1923 to become Leyland's Chief Engineer for the Trojan project, and production soon reached 85 per week with many going to the RAF as their first utility truck. But with Leyland making a massive £750,000 loss in 1923, they pressed Trojans in 1925 to reduce royalties to only £1 per vehicle to reflect a drop in its price, but Trojan resisted which ultimately caused Leyland to terminate the agreement in late 1928, whereupon production was transferred to Trojan's Vicarage Road works near Wandle Park, Croydon; some 17,000 or so Trojans were eventually built. In the late 1950s and early 1960s, Trojan produced their famous post-war van at Croydon alongside the Heinkel bubble car and Elva sports car.

In 1930, the Richmond Road works became an assembly plant for the new Leyland Cub lorry and bus chassis but when the lease expired in 1949, it was taken over by Hawkers for their experimental workshop. The factory was demolished in 1993 for housing, but the administrative block with its glorious marble facade remained as part of British Aerospace's Military division.

H G Hawker Engineering Company

As the Sopwith company's assets far exceeded their debts, they were able to repay these and their shareholders in full. Indeed, Sopwiths had come out of the liquidation with a healthy cash reserve, much of it from compensation for cancelled Government orders for around 2,000 Sopwith Snipe front line fighters and around 400 Salamander infantry attack fighters. Wisely Tom Sopwith had maintained remortgage payments on the Canbury Park works and retained the old Sopwith hanger at Brooklands.

On paper then, Tom Sopwith and his fellow directors held an unblemished record and were able to restart in business without the constraints imposed on bankrupts. Thus, on 15th November 1920, the H G Hawker Engineering Company was formed at Canbury Park Road. The directors comprised Harry Hawker, Fred Sigrist (Managing Director), V W Eyre and F I Bennett (Company Secretary). The 'Hawker' name was chosen to legally distance themselves from the Sopwith company, which was still being wound-up and which prevented Tom Sopwith from joining Hawkers as a director until much later. Meanwhile, Hawkers had acquired from the liquidators all of Sopwith Aviation's rights to aircraft (together with the supply of spares to the RAF's surviving Sopwith Camels), engines, motorcars and the ABC motorcycle, of which Bradshaw claimed there were parts sufficient for 3,000 motorcycles still held at Sopwiths at the time of their closure.

Hawker two-stroke motorcycle

Despite the drop in demand for the ABC, serious attempts were made to resurrect production, even Henry Ford was suggested as a saviour, but nothing came of it and only a few more ABCs were assembled at Hawkers up to 1922. Jarvis of Wimbledon also built a few until 1923, using stocks of new parts.

Enthusiasts entered five ABCs in the 1922 London - Lands End MCC trial. All five completed the course taking two Silver Medals and two Golds; the President's Cup was won by A G Wall who, in a last minute switch, abandoned his Velocette for an ABC - a machine he had never before ridden! G Maund also won the Duke of York Junior Handicap at Brooklands in May 1922 with a rigid framed ABC.

Hawkers even produced a basic motorcycle powered by a 292cc two-stoke engine; who designed it and whether it was already on the cards at Sopwiths is not known, but there was much logic behind this for cash flow during Hawker's formative years was crucial and, thankfully, there was still a demand for an inexpensive motorcycle. Alas very few were made. Harry Hawker raced one with some success, usually rolling up to the circuit in his Rolls-Royce car before clambering aboard his race prepared two-stroke. Harry was by now an accomplished racer on the Brooklands track and at Whitsun 1920, piloted one of Louis Coatalen's 5-ltr Sunbeams. Lapping at 108.98mph, he won the Lightning Handicap despite engaging 2nd gear by mistake, having slipped out of top at high speed! The following year he entered a Weller designed 1,495cc ohc engined, streamline bodied AC competing against Count Zborowski's mighty Sunbeam V-12 inspired Mercedes, "Chitty", but to little avail. On June 3rd, 1921 his AC reached 105.14mph in the flying half-mile; this car went on to set new standards after his death.

End of a legend

At the 1919 Aerial Derby, Hawker had elected to fly one of the ABC Dragonfly powered Sopwith Snappers (F7032/K-149) which the company held in store, but as the Dragonfly engine was still on the Official Secrets list, its participation as a private entry was banned by the Air Ministry. Sopwiths then built the Sopwith Schneider float plane (G-EAKI) with a Cosmos Jupiter radial for the first post-war Schneider Trophy race in late 1919, taking it to Hythe, on Southampton Water, for its trials, but its floats had been damaged and filled with water! After the race the airframe was converted to a land plane for the 1920 Aerial Derby at Hendon, and fitted with the now de-restricted Dragonfly engine. Christened 'The Rainbow', Hawker came second, but was disqualified for failing to correctly observe the recently modified course. The fastest contestant was a Dragonfly powered Nieuport 'Goshawk' but it too failed to complete the course and second place finally went to J H James in his Dragonfly powered Nieuport Nieuhawk. 'The Rainbow' was later refitted with a 500hp Bristol Jupiter which took it to 175mph; it came second in the 1923 Derby.

The fateful Nieuport Goshawk

Immediately after the war, Harry Folland built three Nieuport 'Goshawk' demonstrators as potential sports racers; one of them, G-EASK, had already reached 166mph, taking the British record, and was lent to Harry Hawker for the 1921 Derby. On 12th July 1921, four days before the event, Hawker rode to Hendon on his two-stroke Hawker motorcycle to carry out a test flight but following a carburettor fire, the Goshawk nose dived into the ground. Eye-witness accounts confirmed that the crash was preceeded by a carburettor fire. Hawker was actually thrown clear on impact and, although badly burned, managed to ask if the aircraft was all right. He died at the scene, ten minutes later. His death was a terrible personal blow to both Tom Sopwith and Granville Bradshaw.

Bradshaw always laid blame for this fire on negligent flight mechanics. The coroner's report revealed that Hawker suffered from a tubercular spine and concluded that the uncharacteristic nose dive could have been corrected, allowing the aircraft to be brought to land relatively safely. His failure to do so, was most likely brought about by a haemorrhage causing a temporary paralysis of the lower limbs.

The official Air Ministry Accident Investigation report was not made public for 50 years. It concluded that the fire was caused by a carburettor float chamber cover which had become completely unscrewed, but not lost. Both the Nieuport and ABC engineers who had prepared the aircraft confirmed it had been correctly fitted but the cause of only one out of three chamber covers working loose was never investigated. However, having forward facing carburettors and lacking a full exhaust system (for which Bradshaw could not possibly be held responsible), the leaking fuel had been blown onto the hot exhaust causing the fire. The report recognised that Hawker had managed to subdue or put out the fire, but due to his medical condition was unable to regain control. No blame was placed on the engine.

Several significant changes were made at Hawkers which secured the company's future in these uncertain economic times and included Fred Raynham, a former Avro test pilot, becoming their Chief Test Pilot along with Sydney Camm, a young draughtsman at the Martinsyde aircraft company who joined following a personal recommendation by Tom Sopwith. Hawkers were soon awarded a new RAF contract for the 380hp Bristol Jupiter radial engined Woodcock fighter and with this, Hawker Engineering ceased motorcycle production. By 1925, Sydney Camm had become Hawker's Chief Designer; he went on to design the famous Hawker Hurricane. In 1933 Hawker Engineering became the Hawker Aircraft Company and in 1934, they bought Gloster followed by Armstrong-Siddeley in 1935. In 1936 they established a satellite assembly plant at Langley, near Slough and when in 1949 the

Leyland lease expired on the Ham Common, Richmond factory, they regained it for their experimental work. In 1950 they gained tenancy of the former RAF airfield at Dunsfold, Surrey to where they moved aircraft production from Langley and then, their experimental workshop.

The transverse twin lives

At a very early stage both Tom Sopwith and Granville Bradshaw hoped to sell the ABC motorcycle in France. Having fitted both Gnôme and le Rhône rotary aero-engines in Sopwith's fighters, they approached the newly merged Gnôme and Rhône Engine Company at their Victoria Street, London offices in the hope that they would build the ABC under licence. And so it was that in April 1919, Graham Fenton took delivery of a pre-production ABC motorcycle for evaluation and began construction of an assembly line for the new Société Francaise des Moteurs ABC at 118, Rue la Boétie (also recorded as Rue de Boétic), not far from the main Gnôme et Rhône factory in Boulevard Kellerman, Paris. After a few teething troubles, production got under way, overseen by Cpt. Kenneth Bartlett, who later became a director of Bristol aircraft.

In stark contrast to Sopwiths, Gnôme et Rhône were well acquainted with engines and precision engineering methods; they soon became adept at motorcycle production and gradually ironed out many design problems and, by 1920, had adopted a cranked gear change lever. They even exported 'ABC' motorcycles to Britain following Sopwith's liquidation, but these differed slightly in their centre stand and suspension which were designed to better suit French roads. By 1923, Gnôme et Rhône were designing their own range of motorcycles and even produced a 498cc version of the ABC engine which, it is said, was also offered as a light aero-engine. Although Gnôme et Rhône produced motorcycles up to 1959, they remained better known as an aero-engine company, building the Bristol Jupiter radial in the 1930s under licence; they were nationalised in August 1945 as SNECMA (Société Natonale d'Etudes et de Construction de Moteurs d'Aviation).

It is estimated that around 2,200 ABC transverse twins were built. Their serial numbers appear to commence with '1,000' with suffix letters (eg: 1010A), presumably indicating pre-production models. Around 200 ABCs have survived; one was presented to the Science Museum in 1953.

The 'failure' of the ABC was a deeply felt personal disaster for Granville Bradshaw as everyone accepted that the concept was right. He later conceded that it was he alone who had made the fatal mistake of not allowing a motorcycle manufacturer, such as Phelon & Moore, to develop it properly before going into production and concurred with Tom Sopwith and his fellow directors, that they had probably done the right thing by liquidating the company as they were simply not geared up for motorcycle work.

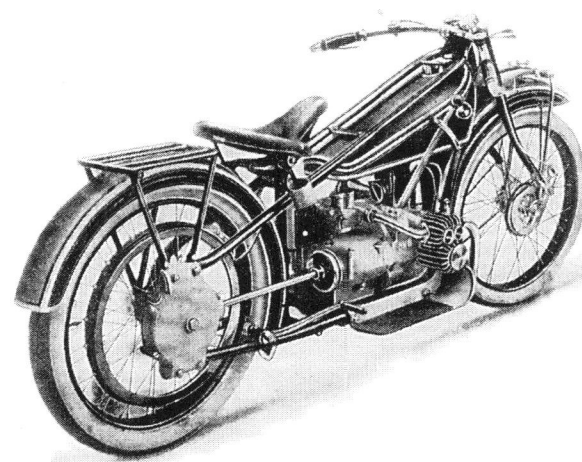

The 1923 BMW transverse twin

Without a shadow of doubt, had the time-pressed Sopwith Aircraft company allowed Bradshaw time to thoroughly test his prototype and, had they not have gone into voluntary liquidation denying Bradshaw the opportunity to develop out those faults, it could have been an entirely different story with the ABC rather than BMW (launched at the Paris Show of 1923), holding the production record of over 80 years for a transverse twin design. Bradshaw made a controversial claim that the BMW was copied from his ABC when BMW's engineers approached him at the 1920 Olympia Show to negotiate a license for its manufacture. His claim was hotly denied by Rudolf Schleicher, BMW's chief engineer, though clearly the ABC must have had quite some influence!

It is worthy of note that in 1940 the Harley-Davidson company in America announced that it was to manufacture an American version of the BMW R71 developed to meet the potential needs of the US Army operations in North Africa. This was a side-valve flat-twin, shaft drive motorcycle. Production commenced in 1942 but only around 1,000 of these Model XA 740cc twins were produced before the project was abandoned. The Indian Company also briefly produced in 1941 the Model 841, 90° 744cc transverse V-twin which however remained experimental. It would be nice to think this evolved from the Bradshaw designed P&M Panthette, but though the Indian was developed in both shaft and chain drive, predating the more famous Moto-Guzzi V-twin, it was Guzzi who perfected the V-twin design.

Motorcycle magazine correspondents continued to praise the ABC's advanced design, handling and comfort well into the 1940s; very few indeed derided it or its inventor. Even Edward Turner, of Triumph motorcycles, praised the ABC at a lecture given by Bradshaw in 1944, confessing that he too had bought one after Sopwith's collapse as he was intrigued by its design. Turner had not the slightest doubt that the concept was perfect, if years ahead of its time, and concluded that it was only the quality of construction and manufacture which had let it down. He maintained that had it been given the necessary, extended pre-production engineering (or the Joe Craig treatment, as he put it!) to iron out unreliability, then the ABC would still have been in production in the 1940s.

Joe Craig was the development engineer and racing manager at Norton Motors, and later AMC. He praised the ABC in *MotorCycling* of January 1944 as:

"...bristling with unconventional features, the majority of them brilliant and sound in conception and apparently heralding the approach of a much more advanced type of motorcycle. Had it been given the benefit of sound design at the right time it would have stayed the course", to which he added "now it has been fully developed by a firm in a foreign country and it has built up a very enviable international reputation".

Yet to Craig, the mere mention of the name 'Granville Bradshaw' was like a red rag to bull! He became one of Bradshaw's sternest and most vociferous critics. Craig was a firm advocate of big singles and most dismissive of Bradshaw's obsession with the side valve transverse twin ABC and leaf suspension. Indeed, at one public lecture in 1944, he described Bradshaw as "that gentleman who designed weird things in his shelter during air-raids". Craig was also passionate about ohv valve gear, yet Bradshaw pointed out that the BMW's ohv engine was very prone to accident damage in a fall, whereas the narrower, side valve ABC could easily be picked up, dusted down and driven off! Bradshaw also reckoned that a low compression side valve 500cc engine could achieve the performance of an ohv engine by fitting a low pressure supercharger, but Laurence Pomeroy soon threw a spanner into the works by dismissing Bradshaw's proposed fan type 'blower' as having insufficient peripheral velocity at the impeller's tips on such a small, relatively slow running side valve engine; however Pomeroy conceded that a blown side-valve engine was feasible, in theory, if fitted with a positive displacement blower.

Alas, Bradshaw's many dreams of a post-war ABC never materialised, though several of his ideas were put into practice in designs for Phelon & Moore.

9
Skootamota

Pioneering motorcycles often had their simple engine clamped to a bicycle frame, driving the rear wheel by leather belt around a large rear wheel pulley. More often than not, the engine barely afforded 'pedal assistance' to the cyclist on steep hills but as their designers yearned for ever more speed and performance, motorcycles became ever more powerful and sophisticated. Yet there remained a considerable demand for the most basic of motorised bicycles for the masses by which to get from A to B with the minimum of fuss, bother, expense and need for protective attire.

Popularly known as 'cyclemotors', most designs came from Europe, but as early as 1899, when today's conventional 'safety bicycle' was still in its infancy, Edwin Perks and Frank Birch secured a patent (Pat No. 7928) for a self contained, motorised wheel which replaced the rear bicycle wheel (or front if preferred). The design was taken up Singer & Co, cycle makers. George Singer had worked at the Coventry Sewing Machine Co's bicycle division, under James Starley, before forming Singer in 1900. First offered in 1902 in both a bicycle and motor-tricycle form, the cyclemotor proved very popular. Perks' other inventions were as varied as Bradshaw's and included a wire cutter for military bayonets but more bizarrely, he proposed a fanciful vertical lift "float-aeroplane" (Pat No. 22809 of 1908). In its simplest description, it comprised a massive canvas panelled 'in-line engine' whose huge reciprocating fabric pistons were fitted with flap valves. These were operated by a crankshaft driven by chain from a conventional petrol engine mounted within a suspended boat shaped gondola. As the pistons moved up and down, they would in theory have 'sucked it up into the air'. The crankshaft was also fitted with a propeller for forward motion while the gondola's engine was fitted with a propeller for an optimistic take off from water! Quite.

To counter the Singer model, Arthur William Wall, a prolific motorcycle patentee, designed in 1909 a 2-cylinder, horizontally opposed two-stroke engine complete with road wheel which bolted to the rear bicycle frame and propelled the bicycle wheel by a coupling. This was improved upon in 1914 with a single cylinder 'Auto-wheel' assembly which offered 100mpg and a top speed of around 16 mph, slightly faster than an average cyclist! Wall proclaimed his design as "the long awaited link between cyclists and motorcyclists" and in 1915 it started to be built under license in America. Financed by Sir Arthur Conan Doyle, A W Wall produced several other novel motorcycle designs at his 'Roc' works, originally in Guildford but from 1907 at Aston, Birmingham. Though a doctor, Conan Doyle was fascinated by new fangled machinery, but his greater fame came from writing his Sherlock Holmes stories. Holmes was named after his favourite American writer, Oliver

The Skootamota (right) beside an Excelsior Welbike in 1944

One for the ladies, May 1919

Wendell Holmes, while, it is claimed, he based his Doctor Watson on Arthur Wood who worked with Wall. While Sir Arthur made his literary fortune, Wall died in poverty.

The popularity of these cheap clip-on, or auto-wheel, devices was revived after the second world war with the French 'Velosolex' friction roller drive bicycle of 1946, the similar British Trojan 'Mini-motor' of 1954, the 'Cyclemaster' auto-wheel of 1951 and the clip-on rear wheel drive BSA 'Winged wheel'. Even Phil Vincent, famous for his expensive 1,000cc twins, was forced to produced a 48cc 'Firefly' cyclemotor attachment to stay in business. Granville Bradshaw also offered a miniature 'Bumblebee' V-twin cyclemotor in 1946 but it is for his 'Skootamota' of 1919 that he is better known in this field.

Skootamota

The driving force behind the Skootamota appears to have been Gilbert Campling (b. 20/6/1889) who established Selsdon Engineering Ltd, a motorcar garage, at No 1, Brighton Road, Croydon before the war. At outbreak of war he and his engineering partner (by now Selsdon's chairman), R Brown MIME, quickly realised the need for engineered components and gained Government contracts for the manufacture of fuses for munitions, and work on submarine engines. Their objective turned to precision aero-engine work with components for the Gnôme rotary aero-engine, now built in Britain, and other engines. The Brighton Road works soon proved too small and by 1915 they had taken on larger premises, the 'Camco Works', within a triangle formed by Brighton Road, Sanderstead Road and Purley Road, surrounded by terraced houses. They now also assembled production machinery, including lathes. After a major extension in 1917, they had in their employ 624 staff at the Armistice.

As Air Ministry contractors, they now traded as Selsdon Aero & Engineering Ltd with their old Brighton Road works constructing the 'Imber' rubberised self-sealing aeroplane fuel tanks, while the Camco Works concentrated on components for ABC (they were soon assembling the Gnat, Wasp and

A restored 1920 Skootamota (BMJ)

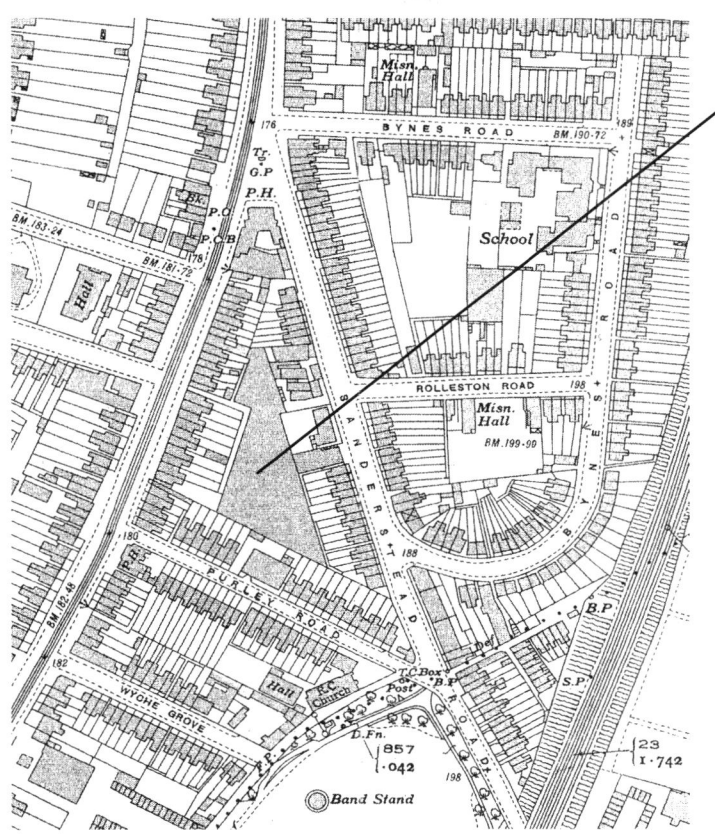

The site of Selsdon Engineering's Camco works

ABC auxiliary engines) as well as intricate components for the Gnôme and, later, the Kauper-Sopwith and Constantinescue interrupter gear for firing machine guns through spinning propellers.

At war's end, Gilbert Campling secured from Walton Motors sole manufacture and sales agency of the ABC aero-engines, hoping to continue manufacture of a now civilian Gnat and Wasp, along with a new W-24 1,000hp aero-engine under development by W Hooper. Campling also reformed his No.1 Brighton Road works as Southern Counties Garage Ltd, 'Motor vehicle and aeroplane dealers and suppliers'.

But despite their best efforts, aero-engine orders were not forthcoming so, on 1st January 1919, Gilbert Campling, as Gilbert Campling Ltd, secured the exclusive 21 year rights to the 'Skootamota' (except in France, where it was presumably held by the Societe Francais de Moteurs ABC), paying a 15/- royalty to ABC per machine. In September 1919, he amalgamated Selsdon Aero and Engineering Ltd with Gilbert Campling Ltd to now specifically manufacture and market the Skootamota. The new company had a share capital of £200,000; their directors included R Brown, Charles Jermyn Ford (Chairman of Edison-Swan Electric Co), A Warne Brown (engineer and former Manager of James Russell & Sons) and D H H Temple (engineer).

When talking about the Skootamota in *The Motor Cycle* of 26th August 1937, Bradshaw stated that, "it was no concern of mine except that it used half of a miniature flat twin that had been turned out by me for some years for the Air Ministry. I was, in fact strongly opposed to the project". Regardless of his supposed 'opposition', Bradshaw enjoyed the Skootamota's success and fame and happily claimed to have 'invented' the scooter. Designed to carry an 80 lb load, it was ideally suited

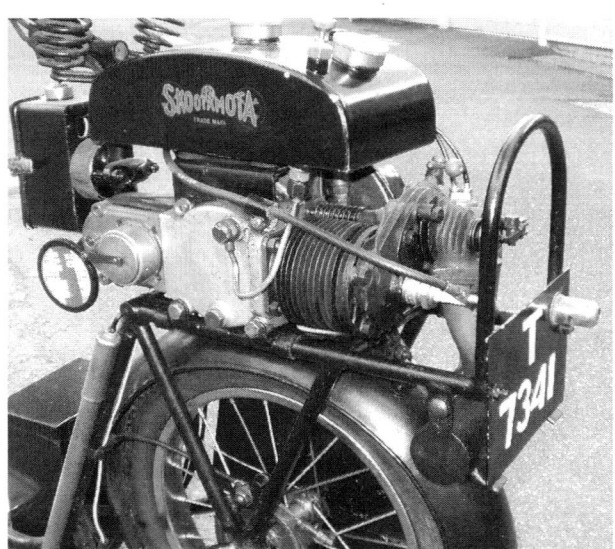 

Skootamota engine and transmission (BMJ)

to doctors for carrying their medicines or AA patrolmen; they were temporarily used for telegram deliveries.

Advertised as 'a runabout for the lady, doctor or businessman', then as now, the scooter appealed more to the Bohemian set than the dyed in the wool motorcyclist. Described as 'Everywoman's joy-car' its advantages were even promoted as health giving and full page advertisements in magazines such as *Cycling*, of May 1919, were at pains to emphasise how easy it was for a lady to jump on and ride 'off to the village, shopping; then to the Post Office afterwards, those long delayed calls; back for lunch. Afternoon?, Golf or bathe; or perhaps just a pleasant little ride out somewhere to tea!'.

One dealer, Marshall & Milnes of Manningtree Lane, Bradford emphasised that 'this is not a toy' and proudly boasted that a Skootamota had carried a 17 stone man from London to York! Flat out at 15 mph, that must have been a very long and most uncomfortable journey for it had no suspension, relying instead on low tyre pressures to absorb road shocks with the, then miniature, 16" x $2^{1}/_{4}$" tyres on 14" rims. The same simple 'suspension' system was adopted by Lawrie Bond for his post-war 'Minicar'. Unfortunately the low pressure tyres, small wheels, long steering tube and short headstock make for a most disconcerting progress!

The Skootamota was, in effect, a large motorised child's scooter, employing a lightweight but very strong, duplex tubular steel frame with a step-on platform which did not require the rider to straddle the engine, making it ideal for daring lady riders. By the time it had reached production, the Skootamota had received a saddle mounted atop the boxed-in single-cylinder, 125cc ohv engine fitted above the rear wheel, driving it by chain. Curiously there was no clutch or kick start, for the rider merely 'scooted' along in gear to start - it truly was more of a child's scooter than the buyer had imagined. Such simplicity was pointedly described in advertisements: 'The most untechnical mind can grasp ABC driving: A push - step on - you're away! Switch off - brake - step off.'

As the road wheel turned all the while the engine ran, albeit slowly, progress was controlled entirely by a valve lifter operated by the left handle-bar lever. To stop, the rider literally killed the engine! There was, thankfully, a single cable operated front brake but fragile young damsels were assured 'Too high speed is impossible.'

Skootamota advertisement from The Aeroplane, July 1919

The engine was, in essence, one half of Bradshaw's war-time ABC Firefly 250cc auxiliary engine with the usual over-square 60mm x 44mm format and machined, finned steel cylinder; and it looked quite dated. The CAV magneto was bolted to a crankcase flange where once would have been the second cylinder.

Weighing only 98lbs, the Skootamota was offered at its launch in early spring 1919 for only 45gns, fully equipped. There was also a 'commercial' model fitted with a box-type cubby hole seat and a 'Colonial' model aimed at Commonwealth markets, complete with folding parasol!

Marketing and distribution was handled by Gilbert Campling, also as Gilbert Campling Ltd, now based at 1B, Albemarle Street, Piccadilly, from where he briefly ran Selsdon Aero and Engineering Co Ltd. The building also housed Sopwith Aviation's London offices and those of a small engineering firm, Ernest Theodore White & Co.

The first Skootamotas were made at the ABC works in Hersham during early 1919. Production models, with 16" rims, were produced from July at Selsdon's Sanderstead Road works. With "firm orders for 11,000" Skootamotas by September 1919 (and "orders in hand for a further 11,000"), optimistic production rates of 100 a week were raised to 500, necessitating the purchase for £27,500 (in £5,000 annual instalments) of the recently vacated Somerton Works, Cowes on the Isle of Wight. These had been built in 1916 at a cost of £42,387 (inclusive of plant and machinery) by John Samuel White & Co Ltd for assembling aeroplanes for the Great War.

Liquidation

For reasons not clear, possibly to avoid the war-time excess profits tax or because of a petition to nullify his marriage that year, Gilbert Campling put his business affairs (Gilbert Campling Ltd) into voluntary liquidation in November 1920. Skootamota production appears to have ceased in late 1920 or early 1921, soon after which Samuel White & Co sold the Somerton airfield and works to S E Saunders Ltd for aircraft production. In the 1960s, with their aircraft work having ended, part of the Somerton Works became home to Enfield Industrial Engines (a subsidiary of Enfield Cycle Co) who in conjunction with Tube Investments and The Electricity Council took over development of the Enfield 8000 electric city car, built for them in CKD form in Greece.

Campling took up a new London home and offices at 90, Jermyn Street, London in 1921, which he later shared with Bradshaw, and orders to wind up his Selsdon Aero and Engineering Company Ltd were put into place in November 1922. The Sanderstead Road works were vacated by late 1923 when the Griggs motorcycle company transferred from Twickenham where they had just started assembling the Wooler motorcycle, alongside their own lightweights; they also undertook general engineering work, but had closed by 1925. His family home after the war was at Oak Lodge, Felbridge, East Grinstead, but his marriage to Audrey Strickland Pierce of Nutfield in 1914, soon came to an end. Gilbert failed in his petition in 1920 to nullify the marriage and thereafter stayed at 90 Jermyn Street, but Audrey succeeded in her pursuit for divorce in 1925 when, for reasons unknown, Gilbert had moved to Sedgley Park, Stockport where he then married Muriel Cooper. This too was a short lived and unhappy marriage, leading her to petition for restitution of conjugal rights, leading to divorce.

Gilbert's former business affairs were dissolved in January 1927 and that of Selsdon Engineering in late 1928. Having retreated to Mill Hill, London in 1927 he then married Margaret Mackay Robertson in 1931; they soon moved to Bickley, Bromley, Kent, living around the corner from his parents', Rosa and John Joseph Bexfield Campling's home at 7 Park Hill, into which he moved after his father's death. He became involved with Herbert Louis Read, an inspirational automobile, aeronautical and electrical engineer and patentee and in 1931 they co-patented light sensitive switch systems for automatically illuminating advertising displays etc (Pat No. 383287) in the presence of, or by, prospective clients. He then became associated in 1936 with Radio Furniture & Fittings Ltd of London, loudspeaker and radio cabinet makers, developing a versatile, interchangeable, divisible wooden housing system, built into house walls or partitions, for accommodating radio-gramophones, speakers and other electrical appliances, storage or service hatches for which he received his second patent, Pat No. 507285 in 1938. Around this time the company was installing wired radio and television systems in offices and flats; they were taken over by Redifussion in 1945 in readiness for the revived television services. Gilbert died at Bromley in 1970.

Scooters

The Skootamota was the first of its type to win an open race for 'motor scooters' and, at the North Birmingham Automobile Club Reliability trials, a Skootamota won both a Gold Medal and the Bayliss Trophy. It is believed around 3,200 Skootamotas had been sold by 1922. Its main competitor was the British made 'Unibus', also produced between 1919 and 1922, by Gloster in order for them to stay in business following cessation of aircraft orders. Others included the American 'Auto-ped' and the simple 'Autoglider', built between 1919 to 1922, which had a 269cc Union two-stroke engine mounted above the front wheel, driving it by chain. The Autoglider came in two versions; the Standard, with the rider standing and the de-Luxe fitted with a seat above the valanced rear wheel.

In stark contrast to the Skootamota, the Unibus used a curvaceous pressed steel and aluminium alloy unitary construction body of quite modern design offering excellent weather protection. Its construction reflected Gloster's origins as coach and railway carriage fitters as well as their recently acquired skills assembling aircraft during the war. With full front and rear suspension, the rider sat above the rear wheel in comfort. The single cylinder two-stroke engine, with 2-speed gearbox, was mounted behind the front bulkhead with shaft drive to an underslung worm drive on the rear wheel hub. The rider was provided with a foot pedal clutch and brake. It proved to be a very satisfactory model but, as the 'Rolls-Royce' of scooters at 95gns, it was far in excess of the purchasing power of its intended market during these depressed post-war years. Production trickled on into 1922 with only a handful being sold. However the Unibus undoubtedly influenced Bradshaw's thoughts on post-war motor-cycle designs, as discussed in the later chapters.

In October 1942, 'Carbon', of *MotorCycling* fame, enthusiastically announced that he had recently discovered a Skootamota still in regular use. In 1956, Bradshaw received news from the owner of a Skootamota offering it for free for presentation to a museum. Granville pleaded with his brother Ewart to get hold of it and present it to a local museum; one had already been accepted by the Science Museum in 1948 to join his three aero-engines.

Despite its gawky looks, the Skootamota was without doubt the most successful of these early scooter attempts - or as *Classic Motor Cycle* later put it: 'The best of a bad lot'!

The scooter dream remained unfulfilled until 1946 following the genius of the famous aircraft and helicopter designer, Corrandio d'Ascanio's 'Vespa' (Wasp). It seems uncanny that Bradshaw's proposed cycle-motor V-twin engine, also of 1946, was called the 'Bumblebee' or *Bombo* in Italian!

Like the Sopwith-ABC Skootamota and Gloster Unibus, the Vespa was designed to keep the Genoa aircraft works of S A Piaggio & Cie in business following Italy's surrender in April 1945. The Vespa scooter had a 125cc two-stroke engine and was first exhibited at the 1946 Turin Show. It was an instant success and prompted a swarm of imitators. It was even licence built in Britain during the 1960s by Douglas motorcycles. Other axis aircraft makers, such as Heinkel, offered scooters and, like Messerschmitt, inexpensive three-wheelers, but contrary to popular legend, the Fend designed invalid tricar of 1948, which was redeveloped in 1953 by Messerschmitt as a tandem tricar, did not use surplus Me.109 fighter cockpits!

Bradshaw's next involvement in scooters was in April 1959 when he redesigned the steering and geometry on the Bond P3/P4 to improve its handling.

Tracked motorcycle

An intriguing tracked motorcycle was developed in the 1920s. It is not known if Granville was involved, but it did use an ABC flat twin engine which, like the Skootamota was at the back of the frame mounted above the driven rear wheel. A long, reinforced rubber track connected this to the front wheel. As with the Skootamota, the tubular steel frame was low slung; the rider sat amidships. Steering was accomplished by a tiller which distorted the forward frame to the left or right with the flexible track 'bending' accordingly, though some pretty strong guides would have been required to prevent it de-tracking.

No other details are known; it was probably intended for military use.

10
Oil Cooled Motorcycles

Granville Bradshaw had spent considerable time during the Great War on aero-engine cooling and the means of improving thermal efficiency. He patented copper plated steel cylinders, modified the shapes of the barrels to increase the exposed surface area on radial engines and discovered, through his experiments, that much of the engine's heat from combustion was dissipated by the crankcase and lubrication system.

This he proved by removing the cooling fins from a cylinder barrel which caused the temperature to rise some 12%, but by increasing the quantity of oil circulated, and the surface area over which it flowed even by a small amount, and then subjecting the oil-sump to an air flow, the temperature at the cylinders dropped by up to 40%. In further experiments, he found that about 25% of the total heat dissipated by the engine was collected from the cylinder head and distributed to the cooler crankcase by the circulating oil. More importantly, the increased oil supply evenly spread the heat around the entire engine, minimising hot-spots and reducing the risk of head distortion. He reasoned then that, if a flow of oil was directed at the valves and piston skirt, in excess of that needed for lubrication, it not only transferred heat to the air cooled sump but also dampened noise made by moving parts, especially that of the valve gear. His oil cooling theories were supported by the influential Professor Gibson in a paper to the Institute of Automobile Engineers in 1920.

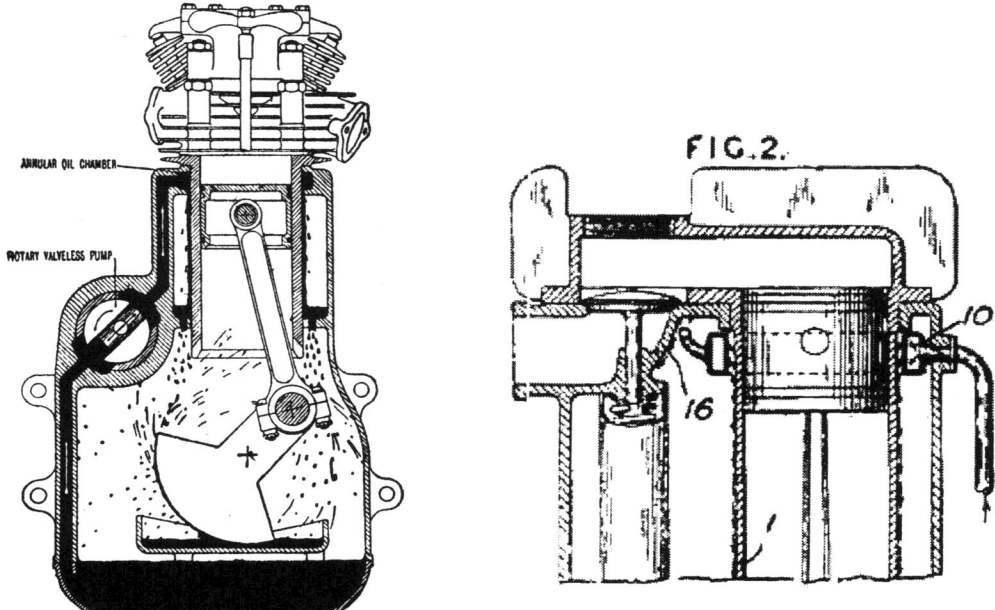

Left, schematic form for the oil cooled engine, and on the right, the oil cooled valve chest, cylinder and annular well (Patent No 194907)

On 21st June 1920, Granville lodged patents for the principle of cooling cylinder barrels by enclosing them within the crankcase and spraying oil over their surface; Pat No: 169255 was granted in September 1921 and he later secured a US patent. He went on to patent oil spraying of the outer wall of the piston and valve-gear chambers (Pat No: 194907 laid 22nd February 1922) and, on a vertical engine, forming a well around the top of the barrel which, when flooded, allowed oil to overflow around the whole of its outer surface transferring heat to the sump.

He further considered the question of engine noise. Early air cooled aero-engines were notoriously noisy through a combination of clatter from the exposed valve gear, the high frequency sound from detonating gases and the lack of acoustic damping through their, by necessity, lightweight construction. Valve clatter could easily be reduced by fully enclosing the valve gear and increasing the lubrication of moving parts. The latter he readily accomplished in his famous P&M Panther cast-iron ohv motorcycle engine of late 1923.

The density of materials and their sound absorbing capacity began to fascinate him. In the pioneering days of Brooklands, an aeroplane had been fitted with a toughened glass windscreen to protect the pilot in flight and, in an endeavour to prove its safety, the pilot, standing well clear, encouraged bystanders to shoot at it with a pistol. Apart from minor splintering, the screen remained intact. In the 1930s, Granville was watching the production of moulded glass pressure vessels and noted how heat moulded and toughened glass could be used for wrap around windscreens improving a car, or aeroplanes' aerodynamics. Knowing that glass was a good sound insulator, he also realised that toughened glass could withstand very high pressures and variations of heat making it possible to produce a 'quieter', glass cylinder block and head!

Many believed that water cooled engines dampened sound, but as Bradshaw pointed out, water was a very effective transmitter of sound, hence the ability of whales and dolphins to communicate over many miles underwater and hence the efficiency by which sound detectors easily picked up noise from submarines or ship's propellers many miles away. Sound travels four times faster through water than air; in dry air it is 763 mph, yet in water it is 3,154mph. In later experiments he found sound levels from a water cooled engine were significantly lower when drained of water, relying on the air in the water jackets to damp the sound.

Oil-cooled engines

His combined experiments convinced Bradshaw that the ideal engine should be air cooled externally and cooled internally by copious amounts of oil from an air-cooled sump. The oil would be circulated by pump through the crankshaft to plain main and big end bearings, thence by spray jets to the barrel and valves at the rate of half a gallon per minute.

And so it was that in 1920 he developed a 349cc (68mm x 96mm) vertical single cylinder engine for motorcycles and industrial use; it developed 7bhp at 3,500 rpm. There was also both a 496cc (68mm x 68mm) and a prototype 700cc (76mm x 76mm) horizontally opposed twin for the aborted Lea-Francis cycle-car. These were followed by a 1,094cc opposed twin and a 1,300cc V-twin for motor cars.

It will be noted that the 349cc single was a long stroke design rather than his usual square or over-square format as he considered the longer barrel was necessary for better oil cooling. He also used a cast-iron barrel and detachable ohv cast iron head for better sound absorbency. Adjustable roller tappets and push rods operated the rocker gear, which were automatically lubricated by a new rotary, valveless plunger pump in the timing chest, producing jets of oil drawn from the air cooled sump. The usual aluminium alloy piston was fitted with a floating gudgeon pin retained by brass buttons, but as the oil cooled engine had its barrel within the crankcase, withdrawal of the gudgeon pin was

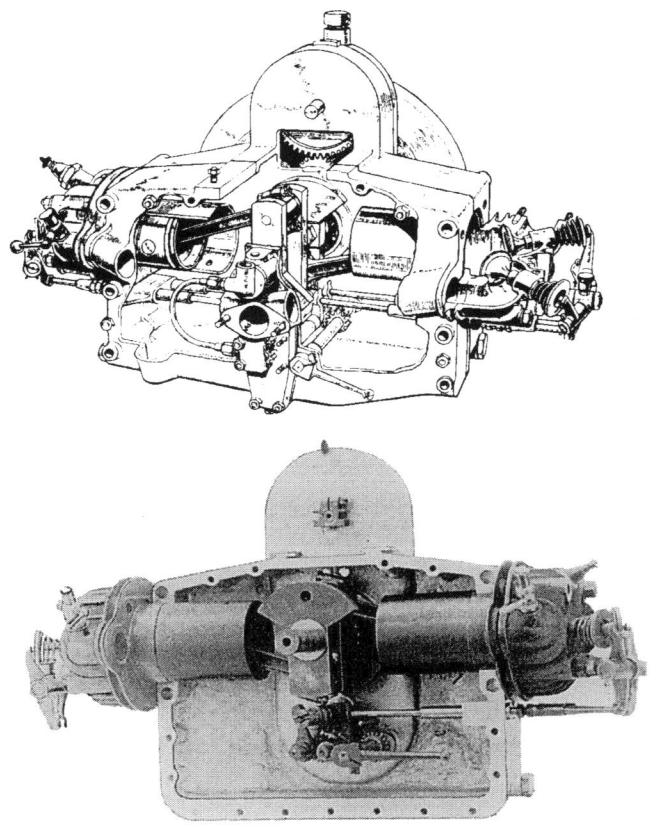

500cc horizontal oil-cooled Zenith twin engine

achieved by fitting removable plugs in the casting which allowed the pin to be drifted out. For compactness, these engines had an external 'bacon slicer' flywheel. The engine weighed 40lbs and measured $18\frac{1}{2}$" tall, $13\frac{1}{4}$" long and 10" wide.

Bradshaw, says in a letter to *Motor Sport* in 1960 that he was asked 'to design a motorcycle engine for a firm in Preston' - perhaps it was Bert Houlding of Matador who had asked for a single cylinder version? Working prototypes were built in late 1920 to Bradshaw's drawings by Ernest Theodore White & Co who shared offices at 1, Albemarle Street, London W.1 with Gilbert Campling. Gilbert had employed George Herbert Jones, a specialist in carburettor design, at his Selsdon Engineering works and it appears it was George who, in consultation with Bradshaw, resolved many of the minor production engineering problems in the manufacture of the pre-war ABC flat twins. With Campling having gone into voluntary liquidation, it appears George joined Freddie Barnes at Zenith motorcycles. Freddie was most impressed with Bradshaw's oil cooled flat twin's potential and it appears it was he who instigated prototype production of both engines by James Walmsley & Co. of Marathon Works, Frank Street, Preston, Lancashire in 1921.

Quite why Walmsleys were selected to build the engines is not clear, but nevertheless they were to be distributed to the motorcycle trade by Gilbert Campbell from 90 Jermyn Street, London W.1, who was acting as Bradshaw's marketing agent. Bradshaw had also been approached by W H Dorman Ltd (who had taken over the Adams' ABC V8 aero-engines) to also build his oil-cooled engines. Dorman's Board Room minutes record on 22nd November 1923: "It was agreed that due to the lack of finance the company could not undertake both the manufacture and selling of the Bradshaw motor cycle engine but that they were prepared to undertake manufacture for a separate company". As they currently had no spare capacity at their Foregate Street works, Stafford, Dormans set about building a new factory.

Dorman Board Room minutes record on 21st February, 1924: "Correspondence dated 7, 19 and 20 February 1924 with Granville Bradshaw and Chairman's of 18 and 20 February were submitted and approved". Bradshaw's first salary appears to be in May 1924. This most likely relates to the preparation for production of the oil-cooled engine, then at Walmsley's. There is also an obscure reference elsewhere to a 'Hornet' aero-engine in late 1924 (which had been presumed to be early work on the ABC air-cooled flat-four, which was not in fact fully developed by ABC until 1927) yet drawings in Dorman's archives include some by James Walmsley & Co, prefixed 'X', such as X35 of 23rd March, 1925 for an inlet tappet

Top left, oil circulation (on Dorman engine). Top centre and right, early Walmsley oil-cooled engine

guide in a 'Dorman-Bradshaw aero-engine, type Hornet'. This suggests an aero-engine version of the Bradshaw oil-cooled twin, as first promoted by Gilbert Campling in early 1923. Were these drawings then for a prototype oil-cooled Dorman 'Hornet' aero-engine, which never entered production?

Bradshaw's aero-engine designs had passed from ABC in late 1917 to Walton Motors, an ABC subsidiary under Charteris, but when Harper-Bean removed Bradshaw from ABC in 1920, they retained him only as a consultant engineer. Although based on his 1919 ABC cycle-car engine, Bradshaw is not generally credited for either the 1924 ABC Scorpion or the later 'double Scorpion' Hornet air-cooled aero-engines; indeed there is no mention of any work on these in his archives. However, when ABC Motors Ltd regained its independence in December 1923, and doubtless secured the Walton Motors' aero-engine designs, did Bradshaw become a consulting engineer to them, and had he approached Dormans in 1924 to become manufacturing sub-contractors; and is that to what the Dorman Board Room minutes of 1924 refer? Nevertheless, Walton Motors was wound up in January 1925 and, in the event, the 'Hornet' name was adopted for the new ABC flat-four 'double Scorpion'.

Meanwhile, things were not going to plan at Walmsleys. This was a new venture for them and they soon found problems working to Bradshaw's general arrangement drawings and meeting the tolerances required of them. With Bradshaw committed to work with Belsize on their new oil-cooled motorcar, it was agreed between him and Freddie Barnes that George Jones should be despatched to Walmsleys to help them 'interpret and improve' his engineering drawings. What should have been a temporary secondment, turned out to be a 4-year stint as Walmsley's Project Manager! Within a year, production of the oil engines, particularly the 349cc single, was rising fast. Bert Houlding's production records reveal the first oil engined 'Matador' appeared in November 1922 and that the first two Walmsley-built engines had a 'square' crankcase/block, thereafter rounded, as continued by Dorman.

Dorman oil-cooled engine (s/n B6041) (GB)

Bradshaw had shown a preference for sporty short strokes, but these nominally 8bhp long stroke oil-cooled engines had superb pulling power and proved ideal for sidecar work. Walmsley published dozens of testimonials from satisfied owners in their sales brochure of 1924, all of which, naturally, were written in the most glowing of terms. Many reported excellent performance without a hint of overheating, of success in trials and races and economy of fuel at 120mpg and 3,000mpg for oil. But it proved difficult to sell in a market entranced by Triumph's new budget 500cc side valve. To make matters worse for Walmsley, the trouble-free 500cc horizontally opposed twin, though much loved at Zeniths for their unique Gradua, found little interest elsewhere primarily due to the expense of having to design a new frame to accommodate the much longer engine. Only a handful of twins were sold to Andrees of Germany and Mineur of Belgium. In 1923 an improved 500cc twin, now with enclosed rockers, was also promoted by Gilbert Campling for light aeroplanes. As with the ABC Scorpion, the engine was reversed, presenting the exhaust valves and ports to the air-flow to help cool the engine. Although the twin now developed 12bhp at 3,600rpm (a specially tuned version developed 18bhp), few sales were forthcoming and as a result, Walmsley abandoned the twin and then, in May 1925 following a petition by creditors to wind-up the company, production of the single cylinder engines also came to an end.

W H Dorman & Co Ltd

In July 1925 production rights to the single cylinder oil-cooled engine alone was taken over by W H Dorman & Co at Tixall Road, Stafford. Along with the rights came George Jones who oversaw Dorman's initial production. George later joined OK-Supreme.

By the end of 1925, Dormans had paid £2,600 to James Walmsley for oil-engine stocks and parts. Bradshaw duly received a salary (£96/month in February 1926 - equal to a Senior Civil Servant), expenses and royalties split 45% to Bradshaw and 55% to Walter Haddon, Dorman's owner and Chairman in joint ownership. This undoubtedly applies to Bradhaw's oil-cooled patents, Pat Nos: 194907 (oil cooled valve chest) and Pat No: 169255 (oil cooled cylinder) as Dorman engineering drawings of 14th September 1925, Nos.9787 and 9787A (for name plates on the Dorman-built Bradshaw oil-cooled Sports (at $1^{3}/_{8}$" high) and standard (at $1^{7}/_{8}$" high), respectively), state "Sole manufacturer W H Dorman Co Ltd Type 350cc

Left, Dorman Standard. Centre, rear view and right, Dorman Sports engine

Patents 169255 and 194907". Dorman engines are generally recognised by a round, rather than square cylinder block and a different timing chest design; their engine numbers appear to start around B6000. As with the Walmsley engine, they carried a bold 'The Bradshaw' legend cast in the crankcase side.

Although some were supplied for industrial use, Dorman's $3\frac{1}{2}$hp, 4.9:1 CR engine was primarily intended for fast motorcycle touring, but an easily fitted high compression piston, offering 5.9:1 CR, made it suitable for occasional racing. In January 1926, Dormans enclosed the push rods and in May 1926 offered a 6.75:1 CR Sports version for competition work, with polished heads and porting, and roller bearing big-ends which added to the complication of assembly and repair. The Sports engine also differed slightly by using a $\frac{1}{4}$" BSF stud screwed into the gudgeon pin end to allow it to be speedily extracted. It later adopted a cylindrical extension to the forward sump. By reworking the engine. Dormans eventually extracted 17bhp when tested on the Heenan and Froude dynamometer, quite a remarkable output for a 1920s 350cc engine. They then enthusiastically entered a team of three oil-cooled OK motorcycles in the 1926 TT, under the watchful eye of Eric McTie.

Toreador Engineering Company

Bert Houlding adopted the name 'Moveo' for his first motorcycle. Loosely translated from the Latin as 'I go', this was the telegraphic address of Loxham's garage, his sponsors in the 1910 TT, but his Houlding & Co production motorcycles adopted the 'Matador' name after his Matador Works, Balfour Road, Preston. His later Matador Engineering Co of Deepdale, Preston had a successful racing history with the famous sidecar rider, A Tinkler. Houlding fitted the standard 350cc oil-cooled Bradshaw single in his 'Matador' motorcycle and in 1923, Matador-Bradshaws came first in the French Grand Prix de Marseilles and 4th and 12th out of 72 entries in the 1923 Junior TT. A disagreement with his business partners caused Bert to leave Matador Engineering in August 1924. Returning to Balfour Road, he set up a new company with Chris Shorrock in June 1925, called Toreador Engineering Company at Ribble Bank Mills, Bow Lane, Preston, which produced an improved, but cheaper, range of motorcycles until 1927.

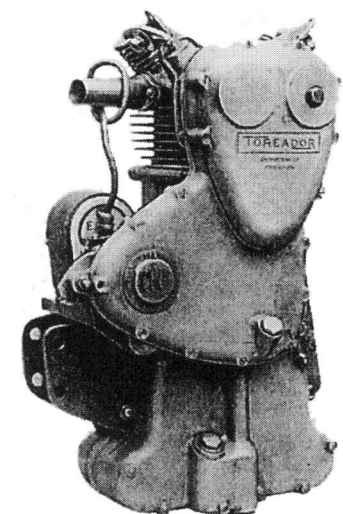

Toreador ohc engine

In 1925, Houlding, George H Jones and, it appears Bradshaw, developed a 350cc (68mm x 96mm) twin overhead cam-shaft oil cooled racing engine which was a great improvement over the standard Walmsley ohv model; whether Walmsley had been involved is not known. The chain driven cams were contained within a large bronze housing, bolted to the crankcase and cylinder head. The crankshaft ran on roller bearings with a large double row caged big end bearing. It had limited success and in 1927 Houlding left to pursue an interest in motorcars, setting up the Moveo Car & Engineering Company in Preston, developing a 6-cylinder Meadows engined 'Moveo' luxury GT car chassis of 1931/32.

Oil engine users

The ohv oil cooled singles were used by Alldays & Allon, Andrees (Germany), Banshee, British Standard, Cedos, Comfort (Italy), Connaught, Coventry-Mascot, DFR (France), Diamond, Dot Motors, Excelsior, FM Molteni (Italy), Henley, Imperia (Germany), Liver, Mars, Martinshaw, Massey-Arran,

Left, DOT Bradshaw Super Sports from 1928 and right, a Connaught-Bradshaw sports

Matador, Mineur (Belgium), Montgomery, New Coulson, New Scale, OEC, OK-Supreme, Omega, Orbit, PV, Quick (Switzerland), Reiter (Italy), Rovin (France), Royal Ruby, SAR (Italy), Saxel, Sheffield-Henderson, Simplex (Holland), Sirrah, Sparkbrook, Toreador and Zenith. Others are still being discovered by VMCC Bradshaw specialist, Sid Wilkinson! Many of these companies fitted Blackburne engines as an option to the Bradshaw oil engine.

Freddie Barnes's Zenith company of Hampton Court, were one of the first to fit both the 349cc single and the 500cc horizontal twin oil cooled Bradshaw engines. Bradshaw had done work for Freddie at Brooklands in the pre-war years with the earlier ABC 500cc air cooled flat twin with which Zenith had some racing success. These were followed in the 1920s by several wins on 990cc ohv JAP V-twin models. Among Zenith's riders was now Jack Emerson (formerly of ABC), Bert le Vack, C T 'Count' Ashby and Tommy Allchin; the last three also rode for Phelon & Moore at various times. The friendship between Bradshaw and Barnes lasted all their lives and in the early twenties having left ABC, Bradshaw acted as a design consultant to Barnes.

Dot Motors Ltd. of Manchester was founded by Harry Reed, a blacksmith and cycle-maker who in 1906 won the Flying Kilometer Handicap at Blackpool on a Peugeot twin. In 1907 he exhibited his first 'Dot' ('Devoid of Trouble') at the influential Stanley Show. Reed went on to win the twin cylinder class in the 1908 TT. In September 1922, he formed Dot Motors Ltd and adopted the new Bradshaw oil cooled engine as on option to the JAP or ohv Blackburne. Four Dot-Bradshaws were entered in the 1923 TT but they all retired during the event, leaving only the Dot-Bradshaw VPT sidecar combination, ridden by Reed himself, to finish in the sidecar TT coming in 6th - his was the only 350cc to complete the course. For the 1924 TT, Reed chose to fit only Bradshaw or Blackburne engines to four Dots, together with Reed's combination, but they all suffered overheating or seizure in practice and three were then refitted with JAP engines, but Jack Emerson, then riding for Dot, stuck to his Bradshaw engine and came a respectable 6th in the Junior class, but was forced to retire the same machine in the Senior event. This was to be their last TT entry with Bradshaw's engine.

Left, the German Imperia Model B and right, the PV spring frame TT model

Left, OK-Supreme. Right, Zenith twin, see below right for engine details (BMJ)

In their advertisements Dot boasted of their "all-Lancs" Dot-Bradshaw with its Preston built engine. They also claimed the company to be the largest motorcycle manufacturer in the world using an oil-cooled engine. They offered a range of six models including a Sports, Touring and Tradesman's Combination, selling in 1924 at between £57 and £82; they continued to list a Dot-Bradshaw Sports or Super Sports with the improved Dorman built Bradshaw engine as late as 1928 when Dorman ceased their manufacture. These Dot-Bradshaws were highly regarded by their riders 'for pep and stamina', 'glutton for work ... pulling power of a 500cc sv....'

OK-Supreme produced a range of motorcycles and also had an enviable racing record with the likes of 'Count' Ashby, Frank Longman and Joe Sarkis all of whom later rode for Phelon & Moore. OK produced just the one Bradshaw model in 1925. Whilst Reed lost control of the financially insecure Dot company in 1926, the underlying economic climate also saw Matador and OK-Supreme cease trading in the late 1920s.

In France, Roaul de Rovin won the 350cc class on his Rovin-Bradshaw in the Bol d'Or Sidecar race, setting a new course record. The adventurous DFR company of France even raced a supercharged 248cc oil cooled motorcycle with great success between 1925 and 1927. It is believed this was a sleeved down 349cc Walmsley single coupled to a simple supercharger, fitted beneath the saddle.

Leonard Henderson was an aeronautical engineer who had established a sidecar business in Sheffield in 1919; his first sidecar used a floating axle. He produced the odd Blackburne engined single between 1919 and 1923 and also established a factory in Liege, Belgium. Sheffield-Henderson-Bradshaws won the 350cc class in the Liege-Paris-Liege 500 mile road race and, ridden by M. Michaux, the Grand Prix de Charleroi held at Laneffe.

Walmsley engine in a restored Matador (Sid Wilkinson)

Even Kenneth Bartlett of Gnôme et Rhône, who produced the Sopwith ABC motorcycle, won the 350cc class in the Grand Prix de Marseilles on a Walmsley oil engined motorcycle; broke the hill climb record at Calais and won the Drap d'Or race. In South Australia, a Beart-Bradshaw won the Five Mile Championship, establishing a track record for machines of up to 600cc, while another came second in the 100 mile road race. Max Beart (no relation to Francis Beart of Norton fame) built about five of these motorcycles, modifying standard frame designs into which both single and twin cylinder Bradshaw oil-cooled engines were fitted. One single (s/n TT20B), believed to be from a DOT, was being raced in the 1960s and is now under restoration.

Over 150 awards, at home and abroad, were claimed by Bradshaw oil-cooled motorcycles in general competition, reliability trials and racing, far more than had so far been achieved with his air-cooled engines.

Oil Boiler

Though popularly known as the 'Oil-boiler', there was no evidence that Bradshaw's oil-cooled engines ever did!

It proved to be a pleasant and reliable engine though, unfortunately, as the oil pump pick up pipe was at the back of the shallow sump, it could fail to draw oil on a steep downhill descent. It was also reported that the rather small big end bearing bolts tended to shear, but as Bradshaw wryly observed at the Manchester lectures in 1944: "The designer can only test his design to destruction; he cannot guarantee there will not be some exceptionally ham-fisted rider who will prove even more destructive than that test. A seized piston, for example, can break any big end bolt". Other reported problems related to the magneto which, being set too close to the barrel, could overheat affecting its performance. The keyed-on external flywheel also had a tendency to come off, despite it being retained by a large nut.

A former Dorman engineer recalled two general arrangement drawings in Dorman's archives for a 500cc 4-cyl Bradshaw designed motorcycle engine (no further details are known). It appears a prototype was built as a photograph was known to exist. It is not certain what caused the oil engine's early demise. Some felt that it was due to the external flywheel being deemed unfashionable, a stumbling block since there was no means of fitting it inside the crankcase without a major redesign. Dot were certainly the largest users, but with their failure to support the engine in the TT, orders fell and there was no longer sufficient demand to justify Dorman's continuing production after 1928, so the company concentrated on their water cooled road, industrial, marine and from 1932 diesel engines, competing against L Gardner & Sons and others.

Dorman's Board Room minutes record the last royalty payment to Bradshaw in 1932. This most likely relates to the oil cooled engines, but Col Reg Grey recalled that Bradshaw was known to have undertaken 'major development work' for Dormans on their early diesels - but to what is not known, for their diesel designer was William Mitchell (later Dorman's Chief Engineer, after W Dalton). The first Dorman 2RH diesel of 1930 (110 x 180mm) used the German, Franz Lang 'ACRO' patented air-cell pre-combustion chamber system, but under a restricted license to non-road use. The British designed Ricardo (swirl-chamber) head first appeared on the Dorman 2RB diesel of 1932.While Gardner diesels are synonymous with ERF, Foden and Atkinson lorries, it was a Dorman diesel which was first fitted to the latter two famous marques. Dorman's strength however, was in plant, marine and locomotive engines - though they did produce the odd air cooled engine in the 1940s followed by a wider range of 4 and 6 cylinder in-lines and V8 engines in the 1970s. These reintroduced a pumped oil jet to cool the piston skirt as it descended towards BDC. In 1961 W H Dorman & Co Ltd merged with English Electric and their diesel engine business eventually passed to Perkins, who had also brought Gardner diesels under their wing.

Suzuki

In 1985, the Japanese motorcycle maker, Suzuki, launched a 'revolutionary' oil cooled, lightweight, all alloy 750cc engine. They boasted of its 'unique' oil cooled cylinders which eliminated heat related problems encountered with the engine's high performance. In a compact engine, largely hidden away within the motorcycle's aluminium frame, air cooling was no longer adequate but water cooling added considerable weight, which reduced performance. Their engineers hit upon the 'novel' idea of using copious amounts of cooled engine oil, sluiced over the hot cylinders as well as jets of oil directed up, under the piston crown to draw heat into the sump. The oil was circulated by two large capacity pumps through an air-cooled radiator. Following a letter from Geoffrey Bradshaw, Suzuki were more than happy to acknowledge Granville's pioneering work on oil cooled engines, half a century earlier! Their 'unique' oil cooled engine remains in production today.

The Bumblebee

In the summer of 1945, Bradshaw designed, as a private venture, a tiny aluminium alloy, oil-cooled 'Bumblebee' V-twin side-valve, multi-purpose, auxiliary engine which could be cradled in one's hand. Whereas he had boasted his ABC motorcycle had been designed in 11 days, he conceded the Bumblebee had taken him 11 months.

Designed for vibration free running and minimal mechanical noise by using fully enclosed valve gear, it weighing only $8^1/_2$ lbs. Bradshaw was at pains to point out that his new engine was a refined alternative to the cheap and cheerful, noisy and smelly two-strokes widely available for lightweight motorcycles, motor mowers and portable power plant for generating electricity in homes, caravans or industrial sites. For these various applications it could be fitted with either a kick-start or rope pull-start.

The 100cc (39 x 41mm), 90 degree V-twin, had a 6.4:1 CR and developed $3^1/_2$ bhp between 4,500 to 5,000rpm. Its squish-head combustion chambers were fitted with the then new, miniature 10mm spark plugs, fired from an advanced combined centrifugal ignition and timing device of Bradshaw's own design. A novel austenitic cast steel cylinder liner with valve plate (b[1)] in the patent drawing) was

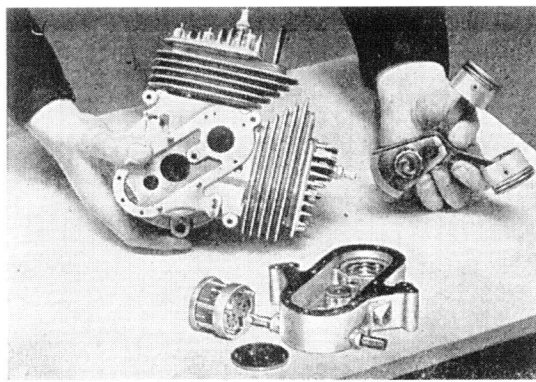

Above, the Bumblebee was a small engine! Below, Bumblebee engine with hydraulic pump

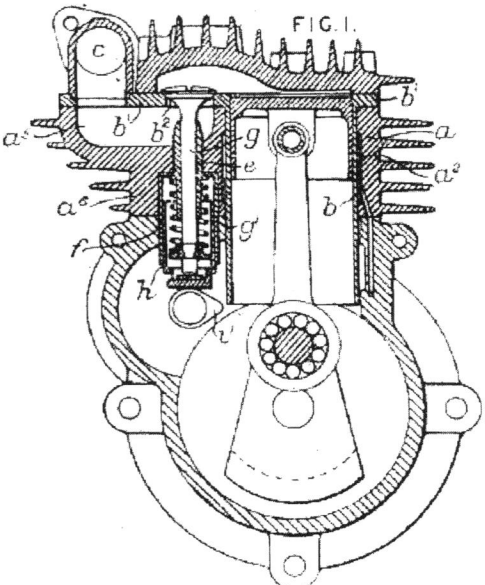

Above, cross section of Bumblebee engine. Pat No 614040, below, exploded view

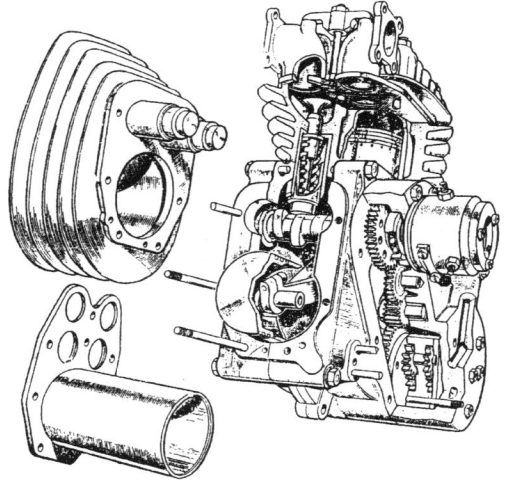

machined at its top to provide a small reservoir of oil fed from a pair of pumps to act as both coolant and noise damper; this formed part of his patent (Pat. No: 614040) laid in July 1946 and granted in December 1948.

Its pre-production development was taken up in early spring 1946 by Messers Gillett Stephen & Co at the Atlas Works, Great Bookham, Surrey (they were more famous for the Blackburne engines) who then fitted it into a tradesman's bicycle, so modified to accommodate the engine, clutch and two speed gear in the bottom bracket, following the principle established by Joah Phelon in 1901 with his Phelon & Rayner motorcycle.

With an estimated 250mpg, Bradshaw had high hopes for the engine and intended that it be available as a kit at bicycle dealers, complete with bicycle bottom bracket, clutch and sprocket set. He suggested they could also fit the Bumblebee to new tradesman type frames and sell at one-third cheaper than a new two-stroke auto cycle.

Announced in July 1946, *MotorCycling* was most impressed by the one they had inspected at Gillet Stephen's works. They were particularly struck by its compactness, the vibration free running from a finely balanced crankshaft, its enclosed design, copious lubrication and the fact that it was designed for long periods of running without overheating. Bradshaw promoted the Bumblebee with the slogan, 'Throw away those pedals!' but when asked by journalists "what happens when the engine fails and you have no bicycle pedals?", Bradshaw was quick to dodge the question by pointing out that it was rare indeed to see riders of pedal assisted auto cycles cycling when they had broken down!

Agricultural use

With post-war British industry under tremendous pressure to 'export or die' and supplies of raw materials restricted to exporters in a desperate bid to reduce the war-induced national debt, many sound projects were shelved or abandoned altogether.

Little more was heard of the Bumblebee until 1951 when Bradshaw revived the engine, promoting it for 24 hour industrial use with a large cooling fan, protected by a cowl. The output shaft was fitted with a combined flywheel and clutch housing, a power take off spline for alternator or generator and a chain sprocket, allowing ready use for many different applications. The Bumblebee still faced stiff competition from established two-stoke engine makers from home and abroad, but in 1953 he gained a design contract with J E Shay Ltd for a new Rotoscythe motor which he developed from the Bumblebee; this is discussed in a later chapter.

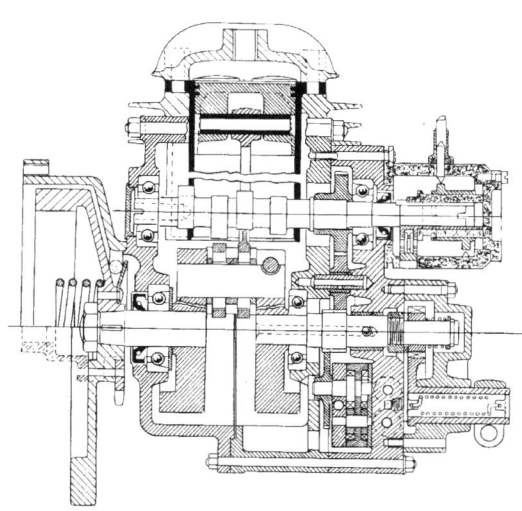

Cross-section of Bumblebee industrial engine showing camshaft driven magneto, power take off, flywheel and/ or gearbox bell housing

In 1953/54, Bradshaw was approached by a well respected Canadian engineer, Douglas James Lemery, to develop a small chain saw engine to allow one-man operation to de-limb trees. It should be borne in mind that the standard chain saws of the early post war years, such as the German 'Dolmar' or the British 'Danarm', were heavy two man felling saws powered by a two-stroke motorcycle engine. The 'Tornado' chain saw even used the mighty 250cc Villiers 2T! On 23rd January 1956, an agreement was signed between Lemery and Bradshaw in which Bradshaw agreed to help establish a Canadian company to produce chain saws with his new (unidentified) engine in return for a 33% (other letters say 25%) interest in the business, but the project floundered. Bradshaw freely admitted that this was 'through neglect', but his solicitor's letters more helpfully point out that the terms neglected to give Lemery the rights to free use of the engine for which, having invested heavily in the project, he was not prepared to pay more! Douglas Lemery went on to form the Lemery-Kolve Chain Company.

Blackburne engine

Granville Bradshaw knew Thomas Gillett from their Brooklands days when Gillett maintained Graham Gilmour's Anzani engined Blériot monoplane. Blériot later established a factory at nearby Addlestone. Gilmour was a much loved dare-devil pilot. On one occasion, when summoned to the Magistrate's Court to answer a charge of reckless driving, he flew there to face the court which so impressed the magistrate, that Gilmour was let off with a nominal fine!

During the first war, Gillett Stephens made radial aero-engine components many of which were for the Royal Aircraft Factory. They also became the British 'foster parent' to the Liberty 12-cyl watercooled aero-engine which the Americans had rapidly developed in 1917 with their entry into the war. It was redeveloped by Gillett-Stephens with air-cooled cylinders, which showed great promise, but other developments overtook this interesting hybrid, and it was abandoned. Like Bradshaw he became a member of the Royal Aeronautical Society and also received the OBE for his war work. In 1921, Tom Gillett became a director of AC Cars while Gillett Stephens & Co took over production of the famous Blackburne motorcycle engine.

The origins of the Burney-Blackburne engine lie in Geoffrey deHavilland's 499c motorcycle of 1902, to which Cecil and Alick Burney bought the rights in 1904. At the time the Burney brothers were working at Willans & Robinson Ltd at Rugby. By coincidence, AC Cars had taken over Willans & Robinson's original works at Thames Ditton. In 1912, the brothers left to form, with Henry Blackburn, a pioneer aviator and financier, Burney-Blackburn Ltd to produce motorcycles; he had no family connection with the Blackburn Aircraft company. Henry left the business soon after, at which point the brothers added an 'e' to their name to become Burney-Blackburne Ltd. The brothers enlisted as despatch riders at the outbreak of war leaving the company in the hands of their new business partner's father, G Q Roberts, but on their demobilisation in 1919, they sold their motorcycle interests to OEC of Gosport but in 1921 Blackburne engine production moved to Great Bookham, Surrey along with the chief designer, Harry Hatch, formerly of JAP. Now as part of Gillett Stephens & Co, Blackburne engine production continued until 1937 at which point Gillett Stephens turned to hydraulic aircraft landing gear in response to Britain's rearmament programme.

11
Bradshaw's Prophecies

A regular contributor, by letter and article, to *The Motor Cycle* and *MotorCycling* in the inter-war years, Bradshaw also wrote the 'Technical problems discussed' column for *The Autocar* as well as occasional articles in *Flight* magazine. Many were written under a nom de plume, usually 'Designer'.

These articles were generally written up from inspired, spontaneous jottings and then presented to the editor in bulk every few months from which were selected appropriate contemporary or provocative articles on which he was often asked to expand. Only by reading his original articles in their correct chronological order, as opposed to published order, can it be appreciated how much of a visionary he was. It was from these many articles, and his war-time lectures at Institute and club meetings, that caused Granville Bradshaw to be looked upon, and consulted as an 'elder statesman' in the vetting of new inventions and ideas.

While many of his articles and comments were controversial or profound, they generally provoked more response in agreement than against - both could be quite heated! Such letters were sent to Bradshaw for him to respond personally, or through the magazine. Then, as now, editors exercised their discretion over what they published in order to add fuel to the fire and gain extra sales or act as page fillers; one wonders how many ridiculous arguments presented by unknowing readers were published just to provoke further debate!

'Technical Topics'

In one of his July 1927 'Technical Topics' articles for *The Motor Cycle*, he implored future designers and manufacturers to concentrate on machines for the day-to-day rider, having both the capacity for a pillion passenger and the power and comfort to keep the rider content with the marque as he progressed through his profession, just as a car owner progressed to a better, more comfortable and powerful model.

Despite his long held views on flat or V-twins, he increasingly prophesied the adoption of motorcar type in-line four cylinder engines, with integral gearbox, giving an ideal weight distribution and handling - such as the Reliant 750cc powered 'Quasars' and 'Phasars' of the 1970s. His ideal design would be a high revving, short stroke 300cc vertical in-line four for the most efficient output, 65mph performance and good economy. This would employ a cast iron head on a cast aluminium, unit construction block. The frame, with its long wheelbase and soft springing would cosset the rider in total comfort. He later proposed a hydraulic, step less transmission.

By the end of 1927 he had produced detailed drawings for his proposed 4-cylinder in-line with a forged steel, I-beam backbone frame, as per his P&M Panthette of 1926. He had reportedly secured a provisional contract from a major, but unidentified, motorcycle manufacturer in 1929, but as the market was still mainly demanding singles, the project was dropped - fortunately so perhaps, for soon after came the Wall Street crash and three years of deep economic depression. It is tempting to think this

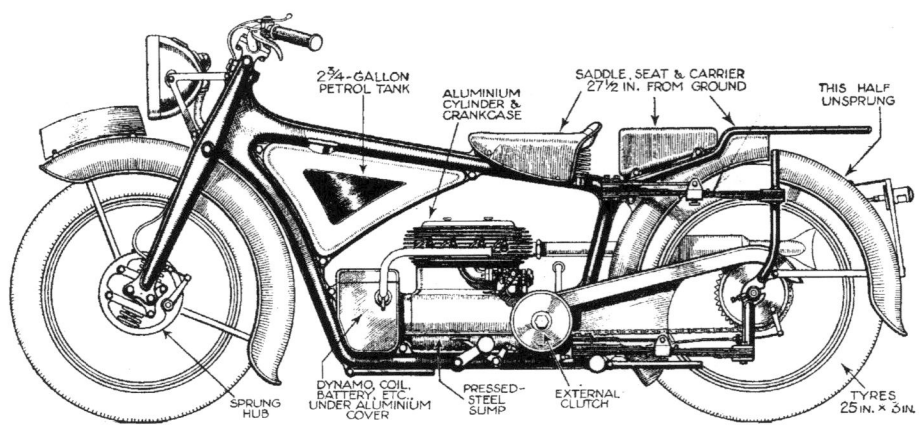
Artist's impression of the £20 in-line four-cylinder of 1937 (CMC)

company may have been Phelon & Moore, but a 4-cylinder would have been a major risk for them, given the phenomenal success of their 600cc Panther single.

In the autumn of 1937, Bradshaw prophesied that it should be possible to mass produce a 300cc 4-cylinder in-line motorcycle at a seemingly impossible target of £20 - far lower than the bargain basement P&M Red Panther 250cc ohv single, then selling at £29/17/6d. His £20 target was ridiculed by experts, including his friend 'Ixion' who claimed Bradshaw was "talking out of his hat", but once again many came to his defence by pointing out how Henry Ford and William Morris had both slashed the price of their cars without an appreciable drop in quality. Ultimately, though, it would be the retailer who, by setting his own profit margin, determined the success, or failure, of a budget machine. And then again, a canny retailer could expand the model's appeal (and raise his profit margin) by offering luxury add-ons for a few pounds more, such as, Bradshaw suggested, a supercharger which could be easily fitted to boost performance and undercut the opposition. This is still of course how today's manufacturers and retailers make their profit.

Bradshaw maintained that the secret to low production costs came from weight saving designs and he repeatedly pointed out that steel and aluminium alloy were bought by the pound weight; halve the weight - halve the cost! He therefore proposed a properly engineered and immensely strong 'tubular' frame formed from pressed-steel half-tubes, complete with lugs, seam welded, to which would be attached lightweight tubular steel sub-frames for the engine and rear wheel. It would have a 58" wheelbase and a centre of gravity barely 1" above the wheel spindle; a novel leading link front hub, with integral counterbalanced coil spring suspension and damping (for which he submitted Provisional Patent No. 18955 on 9th July 1937), together with his Bi-flex leaf spring rear suspension (Prov Pat No. 22630, 18th August 1937) adapted from his ABC and later developed for the Panther in-line twin of 1939.

In the December 1939 issues of *MotorCycling*, he proposed a new, 350cc supercharged flat-four ohv ABC with a streamlined Unibus type enveloping body. It would have an integral gear box with gear change by floorboard heel-and-toe gear selection. Surprisingly, he proposed torsion bar suspension, fore and aft just as he now proposed for motorcar designs. This dream machine caught the readers' imagination and he happily returned to the subject in the October 1942 editions of *MotorCycling* magazine, proposing a thoroughly modern, post-war, flat-twin ABC motorcycle, discussed below, which brought together his experience of 300,000 miles riding motorcycles and 500,000 miles in cars; he revised these figures in November 1943 to 1,000,000 miles in over 35 years of accident free driving, which prompted a doubting reader to calculate that, allowing for holidays,

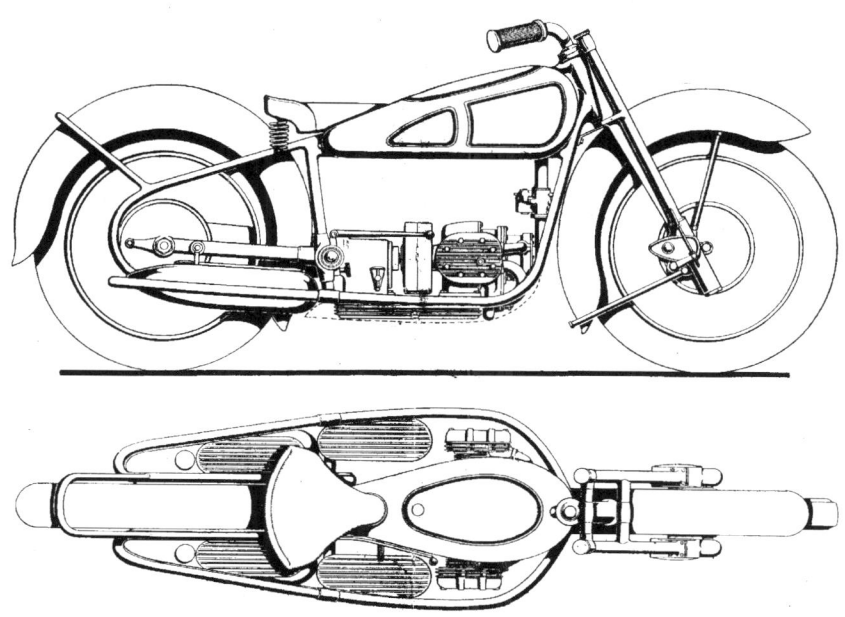

Artist's impression of the post-war ABC of 1942. Note the floor boards and teardrop fuel tank (CMC)

sickness, war-time rationing, overhauls etc, Bradshaw would have had to have driven 167 miles each day for 35 years, non stop!

The future

Bradshaw regularly received criticism and derision from readers and contributors, such as 'Slide Rule' and 'Parvisimus', as well as respected engineers such as Laurence Pomeroy of Vauxhall car fame; yet it was usually Bradshaw's, not their ideas which became standard practice in motorcycle and motor-car design such as telescopic forks, swinging arm rear suspension, smaller wheels and even of tyres ready fitted to rims for quickly bolting onto the wheel - a common practice on post-war scooters! Bradshaw also predicted oil-cooled engines, leading link forks and interconnected brakes - a well proven benefit in the post-war Italian Moto-Guzzi. Today's trend for stylish motorcycle and scooter design also closely match those of his pre-war vision.

He also foretold the collapse of the British motorcycle industry through its inherent conservatism and complacency towards foreign competition and boldly criticised them for being dependent upon fickle marketing ideals - for far too long, they had concentrated on the volatile, but 'certain' market, of the youthful rider, and had overlooked the more faithful utility rider, who required a basic, reliable and trusted workhorse - a market long since dominated by foreign manufacturers. He also drew attention to the lucrative 'mature rider' market and what we now call 'born again riders', who sought greater refinement over performance. To prevent the inevitable collapse, Bradshaw was keen to espouse the need for an Institute of Motorcycle Engineers to develop the potential of motorcycles and enhance the industry; he was prepared to personally fund it. His fears for the industry brought him considerable criticism at the time! He also painted a dismal pictures of 3mph traffic 'queues' stretching for miles and miles and of the need to remove cyclists off the main highway for their own safety. He also reported that many cyclists refused to use those cycle paths already provided for them - nothing changes!

Many motorcyclists agreed wholeheartedly with him and saw Bradshaw as the saviour of the post-war British motorcycle industry, but it was not to be and yet, thirty years on, most of his prophesies have proven true. The reader is drawn to Bert Hopwood's *Whatever happened to the British Motorcycle industry?*, published in 1981.

GB 3-cylinder radial

Incredibly, before the rotary aero-engine had been developed and flown, Charles Redrup was building his 'Barry' 2-cylinder rotary engined motorcycle. The engine lay within a special frame in line with the wheels below the tank and drove it by belt or chain. Redrup also developed the first practical radial engined motorcycle designs in 1920, with the 304cc 'Redrup' motorcycle followed by the 396cc, and later 994cc, British Radial motorcycle designed by J E Manes. Its 120° ohv engine was built by C B Redrup and the frame by Chater-Lea. Though a smooth running engine, it encountered major air cooling problems and was particularly prone to damage due to its poor ground clearance. Charles Redrup also produced a particularly smooth running, 2-row 6-cylinder radial motorcycle engine in 1922, but it had little success. In 1927, working with Crossley, he developed a 5-cylinder engine for an experimental motor-car chassis; in 1946 he developed an experimental axial-radial motorcycle engine.

During the war years, with dreams of a perfect peace ahead, Bradshaw's visionary designs received much praise, even if they were controversial or bizarre! In 1941, he proposed in the pages

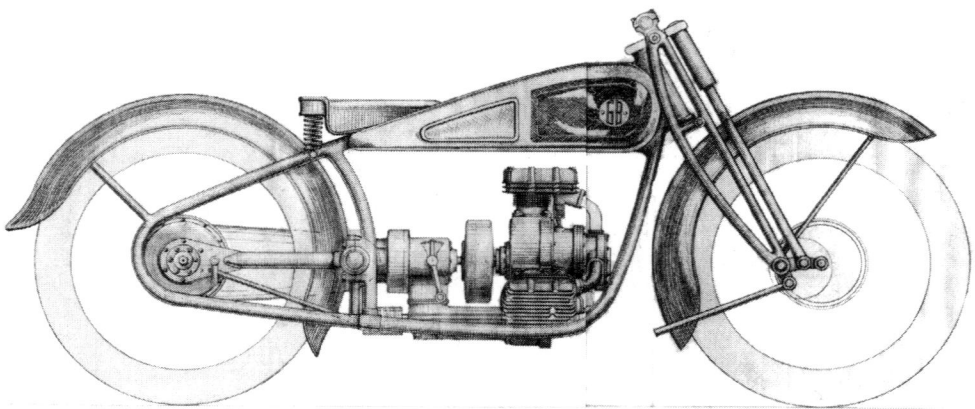

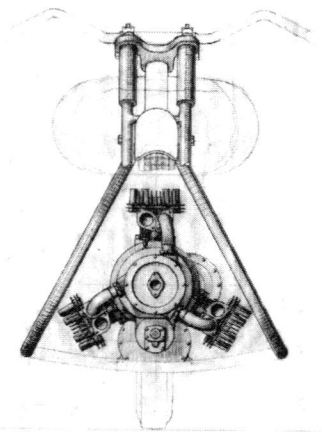

Bradshaw's pencil sketches of the lovely GB 3-cylinder radial showing the widely splayed protective frame (GB)

of *The Autocar*, a 5-cylinder radial, air cooled rear engined motor-car. Though this was not built, a 3-cylinder Redrup radial, rear engined, two door 'Peoples' Car' prototype was designed and built in 1945 by Monty Beaumont of the short lived Beaumont motorcycle fame (1921-1922); his Kendall car was equally short lived.

Bradshaw followed his radial engined car design with a proposal in *MotorCycling* of autumn 1942 for a post-war ABC flat twin motorcycle with a loop frame, similar to that which he proposed in 1939, for a motorcycle with scooter type floor boards for rider and passenger. It would have shaft drive and a swinging arm rear suspension using a torsion bar with an additional compression spring damper when carrying a pillion rider. Further comfort came from telescopic front forks, a short leading link front hub and interchangeable, smaller diameter wheels. Handling was to be further improved by lowering the centre of gravity by having twin fuel tanks mounted on the lower rear suspension swinging arms - perilously close to cast aluminium silencers and exhaust pipes - with the fuel being pumped to the carburettor. The dummy 'fuel tank' would be became a convenient cubby hole for gloves and documents - a feature common in motorcycles of today. He proposed a lightly supercharged side-valve engine to derive the power of an ohv engine and, surprisingly, a preference for an easy to use hand lever starting device working through the gear-box similar to that of pioneering motorcycles.

This same frame was also proposed for a radical 350cc, 3-cylinder radial engined 'GB' motorcycle which he drew up in September 1942. His pencil sketches (as opposed to those bland contemporary motorcycle magazine artist's impressions), reflect his skills and talents as a draughtsman in a beautifully styled motorcycle. It would have a widely splayed tubular steel protective outer frame, much like that of his earlier ABC but with a conventional fuel tank, a girder front fork with short leading link suspension and telescopic shock absorbers. The rear hub was the same swinging arm design; the rod which acts upon a hefty coiled compression spring in the bottom bracket is seen in the sketch.

It increasingly looks like the 'GB' motorcycle may have been developed for George Brough as Brough-Superior had let slip that they were proposing "two, three-cylinder (Bradshaw designed) rotary valve and side valve" engines. These were probably to be built by a new company, GB (Nottingham) Ltd, which had been setup with Leslie Mark Ballamy and Major Sheepshank. Ballamy had worked with Bradshaw on a 'horizontal cylinder' engine which suffered 'poor vane sealing' (was this the war-time Turner air-pump described later?) and had worked in 1937 on a shaft drive, worm gear rear hub Brough-Superior Dream 996cc flat four motorcycle. The association with Bradshaw had finally led to Bellamy meeting George Brough, and the 3-cyl radial engine being tested in a Dream frame.

According to Titch Allen, founder of the VMCC who knew George Brough well, Bradshaw's 3-cyl radial engine was originally developed for a small, remote controlled torpedo to attack German U-boats which had been attacking with great vengeance Britain's crucial Atlantic shipping. His first design used bronze disc (or rotary) valves, but after the project was cancelled, he offered a second development engine, with conventional side valves, to his friend George Brough. Both designs had a primitive, lightly supercharged (but quite ineffective) induction system, gear driven off the crank at $1^{1}/_{2}$ times engine speed. While the disc-valve engine was not a success, the side-valve engine did work well and an example survives, now in running order.

The side-valve radial engine drawing shows only induction pipes to a forward carburettor chamber and lightly charged 'blower' induction system. The deeply finned sump below the engine, extends back under the gearbox, serving both. An external flywheel and enclosed clutch is fitted. The gear-selection is by a swinging foot pedal rather than the conventional 'up and down' movement. The 'GB' really is a most attractive design; what a shame it was never built! The plans lay dormant until 1960/61 when, convalescing at Cowes, he resurrected the idea, but there appears to be no further reference to it in his archives, so one must presume the idea was abandoned.

Lectures

At the outbreak of war, Granville Bradshaw was contracted by the Admiralty on £2/10/0d a day, to work at a 'secret location somewhere in Surrey'. It was suggested he should go into uniform and be Commissioned at a rank to reflect his technical standing, but Granville preferred to remain a civilian which would allow him to follow his own profession as a consulting engineer. By preserving his independence, he would also be free to write and give lectures, the most important of which were held in 1944 as the prospect of peace seemed ever closer. The full texts are recorded in *The Motor Cycle* and are well worth reading; the following is a brief appraisal.

On 23rd January 1944, Bradshaw presented a lecture to the Sunbeam Motorcycle Club at the RAC Club in London under the topic, 'Motorcycles which might be a pleasure to ride' in which he outlined his ideal machine. Among the audience were Bertram Marians (P&M), Edward Turner (Triumph), A G Wall and eminent delegates from BSA, JAP and Royal Enfield, among others. The chairman, F W Pinhard, remarked that, "It is a tribute to his (Bradshaw's) standing in the motorcycling world that we have such a sparkling representation of the leading firms in the industry" at this meeting.

The theme revolved around achieving cleanliness of design, maintenance, rider comfort and pleasure, and the reduction of excessive noise. Bradshaw competently explained how each could be overcome by proper design and engineering; in their turn, the audience and speakers praised Bradshaw's ABC (Edward Turner especially so), his Panthette and ohv Panther engine designs. But one member of the audience, D S Heather, responded: "Listening to Mr Bradshaw, I at first felt it necessary to rush home and take my place in the workhouse. Then, as always, I became a little disillusioned. He is so eloquent that he carries me away completely and it is only when I think things over afterwards that I can get away from his thoughts"!

Bradshaw's 'eloquence' was evident at a well attended series of lectures on the future of motorcycle design at Chorlton Town Hall, Manchester, between May 1944 and late summer 1944, which again drew eminent speakers and officials, mainly from the motorcycle industry. At the first meeting, the chairman, Donald H Smith (who in the 1950s wrote excellent books on diesel engines) announced that there was little need to introduce Granville Bradshaw:

"... as he is famed for having designed the ABC, one of the best British motorcycles ever produced. This machine made a complete breakaway from orthodox practice and it is to be regretted that the German BMW is the sole surviving example of a type introduced by an Englishman, Mr. Bradshaw". He went on to say, "Few designers have been more talked about in the past 20 years than he. Moreover he himself has never hesitated to jolt his fellow designers out of the ruts in which many of them appear to travel. Mr Bradshaw designed flat-twin engines before 1914 and then he burst upon the world in 1919 with the transverse flat-twin engined, unit construction and spring framed ABC. It was a complete break from standard practice. Whatever small faults it had, it was motor cycle design as it should be, the complete homogeneous machine rather than a development of the motorised bicycle. It was a completely designed single track motor vehicle. It shared with the original Scott the distinction of being a motor cycle that never could have been anything else but a motor cycle".

Thanking the Chairman, Bradshaw began his speech by emphasising the need for the rider to feel part and parcel of his machine; this applied as equally to the day-to-day utility rider as it did the sportsman. His ideal was now a 1,100cc, 250-lbs lightweight with overdrive top which "growled like an angry lion" to which 'Ixion' responded that many years earlier, as motorcycle technology developed and they became better and stronger, so did their weight. L J Metcalfe offered that he was of the belief that extra stiffness and strength came from extra metal, but as Bradshaw observed, so too did their cost as steel was bought by the pound!

Bradshaw didn't see the answer in more costly lightweight alloys, but rather in more thoughtful engineering design through improved metals, comparing the strength of lightweight steel tubed motorcycle frames with the stiffness of light alloy strutted aeroplane wings over that of unitary construction, steel panelled designs. However, he conceded the inevitability of unitary construction in future designs. Metcalfe further questioned Bradshaw's demand for less weight, which he saw as less strength and lower safety, to which others concurred referring to the regular shearing of big end bearing bolts on the Bradshaw oil engines through a perceived obsession with lightness. Metcalfe then went on to express the need for stronger and lighter chromium based alloys to reduce mass and weight, but Bradshaw rightly observed that many alloys age harden and are not so easily repaired by owner riders.

Bradshaw also foretold increased power and performance from ever smaller, multi-cylinder, high-compression engines with lighter flywheels, in which the engine generates the kinetic energy for momentum like the radial engine. Its increased power would come from greatly improved higher octane fuels, simultaneously yielding better fuel economy; that factor remains as true today as it ever did.

He also felt it was essential a motorcycle should have 'walkability', with a low centre of gravity, so that Mr Utility Rider could easily restore the machine to the vertical when its tendency was to fall over. This is self-evident when pushing a BMW flat twin compared to pushing a V-twin Moto-Guzzi! Furthermore a low centre of gravity was necessary for comfort and directional stability. He also felt the fuel tanks should be set low down with the rider sat in a feet forward position, directing the machine through hub steering for the ultimate steering control as was well demonstrated on the Ner-a-car. Such a design would allow a semi-enclosed bodywork for all-weather riding without the need for special clothing. He also felt the scooter was the design of the future, provided visionary engineers and manufacturers would allow it! It was indeed this lack of clear vision and understanding of the scooter - compared to a lightweight motorcycle - which failed those British motorcycle companies who dabbled in the scooter market, finding their traditional retailers were hostile to them. Only Douglas, who license built the Vespa, made a commercial success of them in Britain.

Of sidecars, which represented a quarter of all motorcyclists, Bradshaw felt that as these added considerable weight, the motorcycles had to be built heavier and as this combined weight approached that of a motorcar, he saw sidecars disappearing from the utility market - which indeed they did following the Suez crisis of 1956 which encouraged ever lighter, cheaper and more economical cars.

"Leave room for criticism"

It was through his visionary and pioneering work that Bradshaw was often regarded as being far too opinionated. At one meeting, the chairman introduced Bradshaw by saying that such were his often controversial remarks, he could usually be relied upon to "put the cat amongst the chickens". In one witty editorial, *MotorCycling* sought to compare Bradshaw's opinions with those of George Bernard Shaw and coined the phrase 'Bradshavian'. There was many a sparring match between him and fellow motorcycle design engineer-cum-journalist, Joe Craig, former Racing Manager at Norton Motors. These sparring 'debates' had the full encouragement from the editors! Yet the response from their readers more often that not supported Bradshaw's controversial views; many pointed out the undeniable success of other 'unconventional designs' such as the Scott.

Further examples are to be found of his controversial opinions on engine and suspension technology. A firm advocate of four-stroke ohv engines for their higher efficiency, he emphasised the fact that side-valve engines were smaller, cheaper to produce, quieter and, as they were designed for smooth, low speed pulling power, he could easily improve performance with low pressure superchargers. But though he disliked two-strokes for their inefficiency, he willingly chose the valve-free, two-stroke cycle in his later toroidal rotary engines. In the 1920s, diesels were 45% efficient, compared to only 25% for petrol engines but they had lost their initial appeal due to vast improvements in petrol engine technology, efficiency and economy. He later saw the simpler two-stroke diesel (more correctly compression ignition), as the only viable solution to the future of his revolutionary toroidal rotary engine. He was also a firm advocate of leaf springs, but he later proposed torsion bar suspension in motor-cars.

While many questioned or dismissed his theories, very few charged him with hypocrisy - that only came in the 1980s. As he rightly pointed out, invariably the dissent expressed by his critics was simply due to semantics, but if Bradshaw was proven wrong, he would readily accept it. Almost without exception, pre-war magazine editors publicly supported Bradshaw's views and theories but while he readily accepted praise, he much preferred constructive criticism and recalled the personal advice from the master of the art, Sir Winston Churchill:

"If one's writings mildly please the reader but there is no response, the writer is not much good. If they bring in a few letters of appreciation the writer is still not much good, but if they bring in storms of abuse, then and only then, can the writer begin to feel that he really is achieving something!"

Of course, such controversy kept the editors happy as readers' letters filled their pages and wallets! But was Bradshaw wilfully playing to the audience, or had had he been simply following the further advice of Churchill: "Leave room for criticism. Create a controversy. By that you will be much better known".

Bradshaw's competence as a design engineer was openly debated over many years, particularly as a result of his contentious theories expounded above. As Montague Tombs (who often test drove Bradshaw's Wind-waggon observed), "Bradshaw has the knack of making us take stock of what we know, either to prove or disprove his usually startling theories and suggestions". But his reputation was recognised and honoured in the 1980s when, prompted by Geoffrey Bradshaw, the German car maker Audi, stated in an advertisement for their new five-cylinder Audi 100: 'We're not the first people to be attracted to the advantages of a five cylinder engine. The legendary British engineer Granville Bradshaw waxed lyrical in *The Autocar* of April 1940 over its virtues and eulogised over the Gardner 5-cylinder diesel. "A five" he said "is probably a better engine than either a four or a six"'.

In fact, L Gardner of Patricroft had been offering 5-cylinder diesels since their first automotive diesel of 1930, the Gardner LW. Though defunct, the marque is still highly respected and widely regarded as the 'Rolls-Royce' of commercial diesels.

Air cooled Peoples' Car

Throughout the 1930s and 1940s, Bradshaw prophesied about the post-war motorcar in regular 'Technical problems discussed' articles for *The Autocar*, who proudly introduced them as: "A famous engineer deals with technical matters in a way that can be readily understood by the ordinary motorist".

In *The Autocar* of 1st April 1938, he firmly predicted the demise of leaf spring suspension in favour of independent coil springs and radius arm suspension. Later that year, he predicted bonnet-less cars as the ideal design giving the driver unimpeded forward vision. A rather sceptical correspondent recalled 'a scientific observation' that if a driver 'were able to see the road for less than 15 feet in front of him, he would suffer giddiness'. This point was not supported by motorcyclists, nor by Sir John Cobb who, when sat ahead of the front axle on his four wheel drive, Napier Lion W12-cylinder aero-engine powered Napier-Railton, took the land speed record at 350.20mph in 1938, then to 369.70mph in 1939, raising it again to 394.20mph in 1947, where it remained until 1963 when challenged by Craig

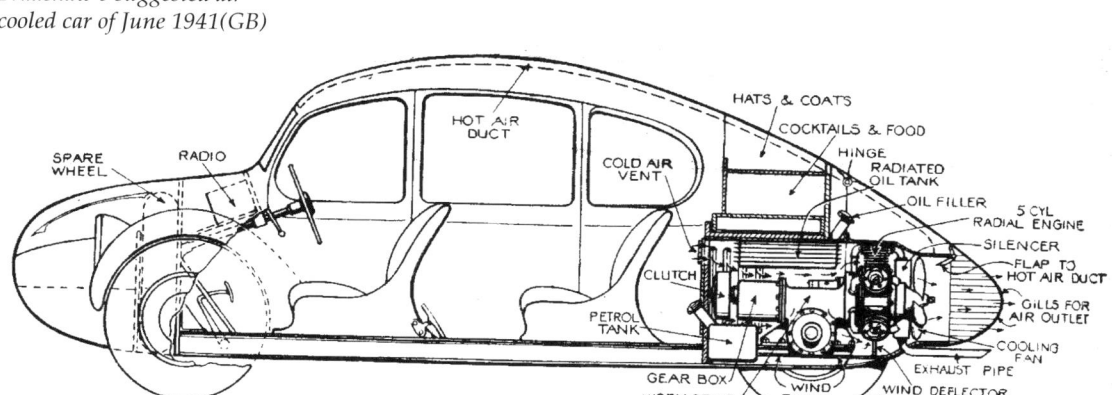

Bradshaw's suggested air-cooled car of June 1941(GB)

Breedlove and, in 1964, by Sir Malcolm Campbell, who also sat well ahead of the wheels in his Proteus jet-engined 'Bluebird'.

Without inventors and 'dare-devils' tackling life's problems, and ignoring the sagacious predictions by 19th Century scientists that passengers in a railway carriage would suffer curvature of the spine from velocities in excess of 20mph, there would be no progress! And, as Bradshaw observed in 1938, "an aeroplane and bird both make the same approach to landing; but the bird gives three or four reverse thrusts before landing and walks away - it knows something that we have not yet discovered". He further commented that he was "a great believer in Nature's experimental department. No scientist has yet found anything that is mathematically unsound in Nature's structures".

By 1941 his vision of 'a car for the future' had developed into an aerodynamic, tear drop-shaped, 4-seat, rear engined, fan cooled, high-efficiency, 5-cylinder radial with rear wheel drive. This concept had been pioneered in Czechoslovakia with the revolutionary, aerodynamic, air-cooled V-8, 'Tatra 77' of 1934; a design which inspired, or was copied by, Ferdinand Porsche for his 1938 'Volkswagen'.

While the short 'bonnet' on Bradshaw's car would barely have room for a hold-all and spare wheel, the passengers at least had the luxury of a small cocktail cabinet in the rear window-shelf! His radial engine was developed from a 3-cylinder job already 'under development for military use'. Engine cooling was by fan, drawing air through louvres in the rear body, over the inboard rear brakes and thence past the radial cylinders. His proposals received much praise in the letters pages and everyone seemed eager for the car to enter production after the war, though many decried the use of an air-cooled engine preferring a pressurised water-cooled engine to achieve maximum thermal efficiency. Yet, it is a matter of fact that the rear-engined, air cooled Volkswagen Beetle, rejected by both Britain's Ford and Rootes under war reparations, holds the production record of over 23 million cars in over 50 years.

Indeed in 1945 Roy Fedden actively worked on a similar streamlined 'British Volkswagen' with a 3-cylinder 1,100c radial with torque converter transmission developed in 1943. Also in 1945, W O Bentley revived the idea of an air-cooled 5-cylinder radial engined car. His was to have the engine mounted horizontally, above the front, driven wheels. A prototype was duly built but, to quote Bentley, "the vibrations and harmonic balance... were something of a headache", although this must have been an improvement over the original 5-cyl rotary powered Adams-Farwell of 1906. Perhaps Bradshaw's earlier, conventionally mounted proposal might have provided 'W O' with the answer? Given Bentley's venomous distaste for Bradshaw's ABC Dragonfly radial over his own BR.2 rotary, this was doubtless a very satisfying minor psychological win for Granville!

The 'Wind-waggon'

Getting their ABC aero-engines from Redbridge to Brooklands in 1911 posed an interesting challenge to Bradshaw which was resolved by 'road testing' the aero-engine en-route. They bought a scrapped 16hp Minerva, wicker-work bodied, chain-driven, wire wheeled car (it is suggested this was Charteris' old car) and then stripped the rear section down to its chassis. A wooden tower, which held petrol and water tanks, was erected at the back together with a slidable frame on which was bolted the aero-engine, complete with its 2-bladed propeller and protective outer guard. The slidable frame was coupled to a massive spring balance by which they were able to measure the propeller's thrust. By setting the engine at a slight incline, the propeller's thrust barely disturbed the finely ground road dust of those pre-tarmacadamed days. Thus propelled, the now artillery wheeled Wind-waggon proved an uncommonly successful vehicle, reportedly giving a most smooth and comfortable, if noisy, ride to the driver and passenger sat in the luxuriously padded leather bench seat. The car returned to Redbridge on the Minerva's 2.3ltr engine ready for the next delivery.

Affectionately known as the 'Wind-waggon', these deliveries were mainly carried out at night on clear open roads, but they had their exciting moments. The rear brakes were hopelessly inadequate for the speeds at which the car was propelled and on several occasions frightened the occasional passing horse, pedestrian and motorist. On one such, the propeller's thrust blew over a milk cart, spilling milk across the road; but with swift recompense to the milkman, all left in good spirits.

When Bradshaw transferred the ABC engine operations to Brooklands, production at Redbridge came to an end and the Wind-waggon's body, Minerva engine and transmission were removed. A rudimentary three quarter inch thick, wooden panelled sloping bonnet, to create 'drag', and large wheel arches were now fitted along with a pair of rearward-facing seats at the side for observer/engineers to check the engine under test on the Brooklands circuit. It was also briefly used as a mobile test bed by a 'local propeller maker' Lang, Garnett & Co of Riverside Works, Weybridge on an 80hp ABC V8 aero engine. Formed in 1913, Mr Lang had a long association with Sopwith Aviation. Although Armstrong-Whitworth were manufacturing the ABC aero-engines at their Elswick plant and had also begun to design and manufacture a fabricated, welded hollow steel propeller blade, it is unlikely they would have used the Wind-waggon.

The Wind-waggon was said to have been capable of 65mph on the Brooklands circuit with 'four men and a dog' and was claimed to have far out-performed any other vehicle, reaching 40 mph up the famous Brooklands test hill. But as the wheels were not directly driven, its record could not stand. Staff at *The Autocar* got to test drive it in August 1912, and enthused over its standing start when, unleashing maximum thrust it provided effortless delivery of smooth power. They also commented on its potential in normal motor cars but conceded it would be almost impossible to design an acceptable saloon car which didn't stir up the road dust, though this, they felt, was no more of a problem than the dust drawn up from a car's tyres at speed on the pre-tarred roads.

Bradshaw's Wind-waggon was not the first propeller driven vehicle, for there is a report of an air-screw propelled motorcycle in France around 1906. This may have prompted its further

The Wind-waggon in its original road-going form outside the ABC works, Brooklands (GB)

The Wind-waggon at Brooklands in its modified form with an ABC in-line 4-cylinder aero-engine

development by Marcel Leyat of Meursault, Cote d'Or around 1913, of a skeletal monocoque construction, enclosed cabin 'Heliocycle' tricycle which led to several four-wheel versions being built and sold. Known as a *mobil á hélica* (or *Hélica*, French for 'propeller'), the Leyat employed a forward mounted tractor propeller powered by an ABC (and other) flat twin motorcycle engine. It had leaf sprung axles and rear wheel steering, which doubtless made it quite perilous to drive! In this form he secured a French patent followed by a British patent (Pat No. 137037 laid 21st December 1918; granted 3rd June 1920) which quaintly describes it as a 'Ship adapted to run on land'. A 3-cylinder 35hp Type Y Anzani radial aero-engine was reportedly fitted later, although production ceased in 1921. There is a report of an ABC engined Leyat being bought by M. André Jacquemin who later removed the engine and restored it to its proper place, into an unidentified hydroplane. An ABC 1,200c Scorpion engined Leyat 2H exits at the National Motor Museum, Beaulieu and an original Leyat in the LeMans museum.

Just before the Great War, Angus Maitland built a handful of 'Beacon' wicker-work bodied light-cars at his Beacon Hill Motor Works, powered by a JAP V-twin (85mm x 110mm). He then experimented with a belt drive from the JAP V-twin to a rear mounted propeller and reported excellent acceleration, but the propeller's thrust created havoc among other road users. However, he believed this could be overcome 'by suitable means' but his experiments came to an end on the outbreak of war, but a Sizaire-Berwick propeller driven armoured car was developed by the admiralty in 1915. Success only really came with the 'Traction Aérienne ' powered variously by a flat twin 1,500cc aero-engine or Anzani V-twin and built by La Traction Aérienne of Neuilly, Seine from 1921 to Layet's patents. It had a four wheel tandem body with beam axle suspension, four wheel braking and front steering. They were also sold as the 'Eolia' and were offered as an open two seater, enclosed saloon, or delivery van and were capable of 50mph. Although production ended in 1926, a further streamlined three wheeler prototype appeared at the Montlhéry circuit in 1927.

In 1943, Bradshaw once again advocated propeller driven motorcars but critics were quick to doubt the efficiency of propeller thrust over a driven rear wheel. National Physical Laboratory tests had shown that a propeller was 70% efficient at transmitting its power, compared to chain drive at 98% efficiency; that a car's rear axle was 95% efficient and that a motor car's engine delivers only 90% of its power to the rear wheels. He also pointed out that his Wind-waggon had attained 65 mph at Brooklands and predicted that a Rolls-Royce Merlin engine, from the Spitfire fighter, would allow a streamlined racing car to take the land-speed record.

But doubts were still raised by experienced pilots who observed the way in which wooden wheel chocks held back a powerful 4-engined bomber. To this Bradshaw countered that their propellers's thrust was inclined to the ground and not directly to the rear and in fact on Brooklands' grass field, full throttle would easily push the Wind-waggon along even with its brakes fully applied! Bradshaw was more than confident that the modern variable pitch propeller blades could instantly apply reverse thrust requiring only normal car brakes for slow manoeuvres and for holding the vehicle stationary when at idle. But despite its potential, the over riding concern remained that of the powerful thrust of air directed at anyone behind. Once again, Bradshaw reported his own experiments which showed this was not as great as people feared and with a reduced diameter from contra-rotating propellers, turbulence and noise could be reduced without loss of power. Bradshaw's experience with his Wind-waggon at Brooklands confirmed that it handled just like a normal car, dispelling another concern over its control.

Although the project was not progressed, correspondence continued in *The Autocar* throughout 1943 and 1944 and much was made of the successful use of air-screw propelled, polar expedition sledges over snow and water since Mr Durston's patented design of 1914. Indeed, propeller driven sleighs were common in Russia during the 1930s for covering vast expanses of snow. Cynics suggested that if air-propulsion was superior to 'friction drive' (car tyres, ship's propellers) then naval engineers would have adopted it in boats, yet since the 1930s, propeller driven punts have been, and continue to be used in swampy areas of America and of course with the later hovercraft.

Consultant

After Bradshaw left ABC in 1920, he followed his own path as a consulting engineer with the Belsize-Bradshaw motorcar. He also found success with his designs for Phelon & Moore and continued to submit related patents. Among them was an interesting device to meet ever increasing road traffic legislation.

His design for an electric parking light was submitted in March 1920 (Pat No: 164458). The case was of a tough material with painted interior so that should the protective, coloured glass lens be broken, it would still display a coloured light. The protective face could be cast in any design and thus ideally lent itself as a brake operated 'STOP' indicator.

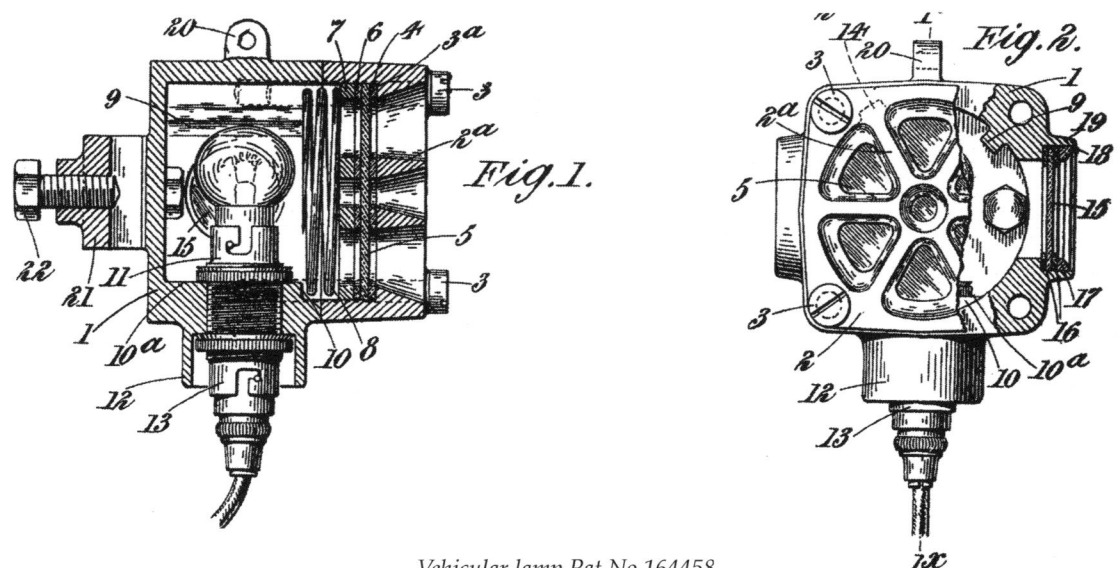

Vehicular lamp Pat No 164458

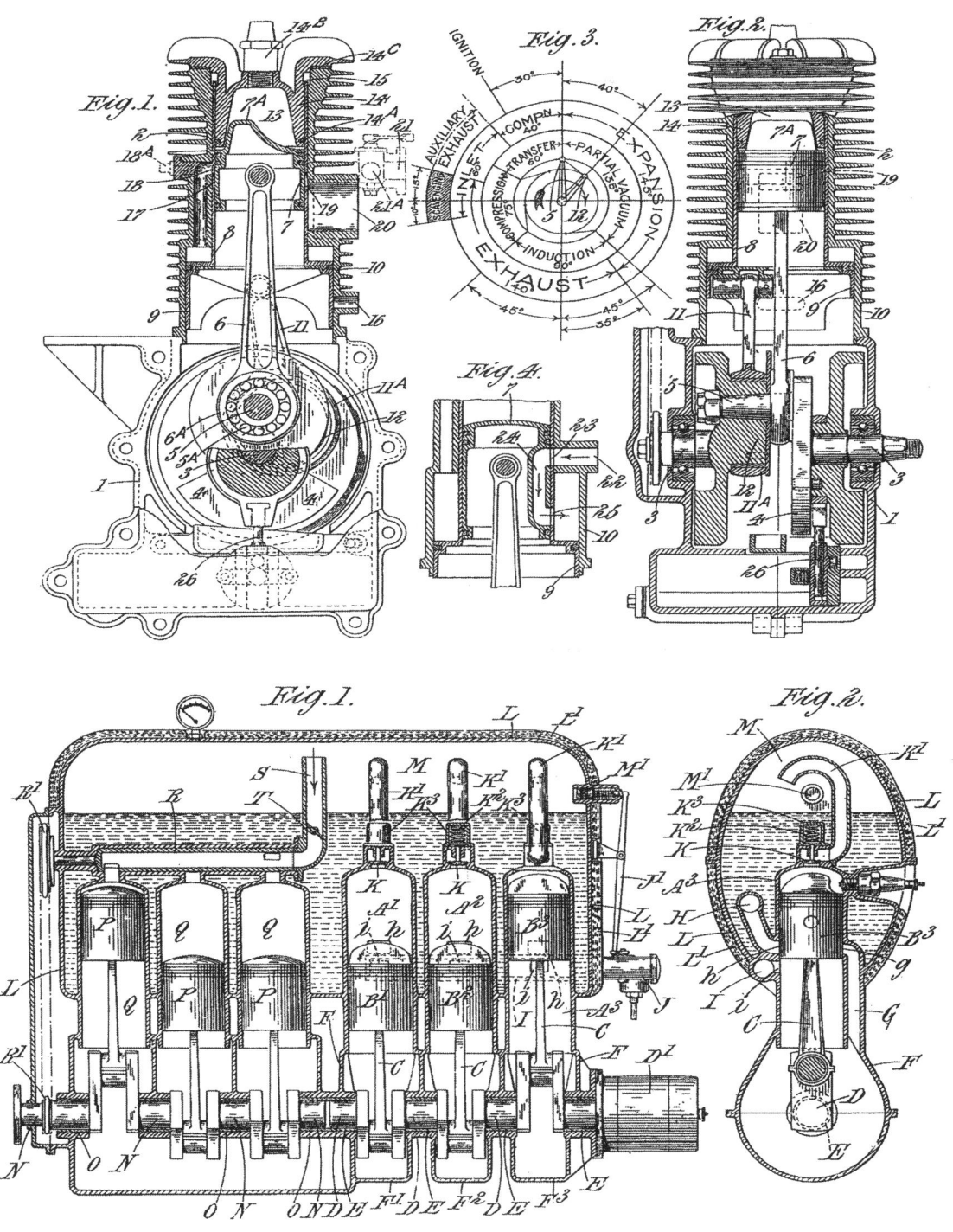

Top, two-stroke oil system Pat No 225275. Bottom, Bradshaw's intriguing hybrid engine showing the 3-cylinder two-stroke 'generator' engine on the right and the steam engine on the left. Note the massive water jacket.

In 1923, he patented a self-pumping lubricating system for two-stoke engines (Pat No: 225275) using a stepped piston. The larger, lower section operated a sleeve valve which not only drew in the vaporised fuel as the piston descended but also progressively opened exhaust ports to fully scavenge the combustion chamber. As the piston rose, the larger section compressed the petrol/air mix before entry to the chamber to, hopefully, increase power. It could also be used to draw in oil to form a 'petroil' mix and equally be developed as a fuel injection system for two-stroke diesel engines. Well, that was the theory! It is not known if the engine was ever built, or for whom, but some of his ideas were later tried in his Pulsation Motor in the 1950s.

Perhaps his most curious invention was a hybrid two-stroke petrol and steam engine. For whom and why this was designed and patented is not known; it is unlikely to have been for the Belsize-Bradshaw. However it was lodged in November 1923 and duly granted Pat. No: 230145. The 'engine' had two independent crankshafts on a common axis. One was a two-stroke 3-cylinder petrol 'generator' engine, whose cylinder barrels and heads were entirely within a massive water tank which thus worked like a kettle from which the steam was transferred to a matching 3-cylinder steam driven engine forming the other half of the engine block. The steam also operated an automatic throttle control for the two-stroke engine. By these means, the instant reaction of a petrol engine to generate the steam, coupled with the high torque of a steam engine to generate motive power with instant forward or reverse motion (dispensing with the need for conventional car gearbox or clutch), would make its application widespread and include motor boats, plant equipment and road locomotives. That again was the theory!

It is unlikely the engine was ever built; besides which, by 1930 legislation and taxation had sounded the death knell for steam with its greater torque and economy over the new and lighter automotive diesel engines. Accordingly, proponents of steam road locomotives, such as Fodens, abandoned steam altogether, leaving Sentinel to soldier on alone until their last steam lorry of 1950. Those interested in the transition between steam and diesel lorries should read L T C Rolt's superb autobiography, *Landscape by Machines*.

Bradshaw's final motorcar patent came in 1938 when he submitted a provisional patent (No. 3061 laid 1st February 1938) for a simple, semi-automatic 'Hill-holder' transmission band-brake; this was granted as Pat No: 512514. By using a ratchet (which acted only in reverse) on the first gear pinion and mating it with a similar splined ratchet, a transmission brake on the output shaft automatically came into operation as a car or lorry started rolling backwards on attempting a hill start. In reverse or any other forward gear, the brake was overridden. Although *The Autocar* praised the Hill-holder in February 1938, it does not appear to have entered production, yet similar devices were fitted to American cars of the 1930s, such as the Studebaker Commander and President.

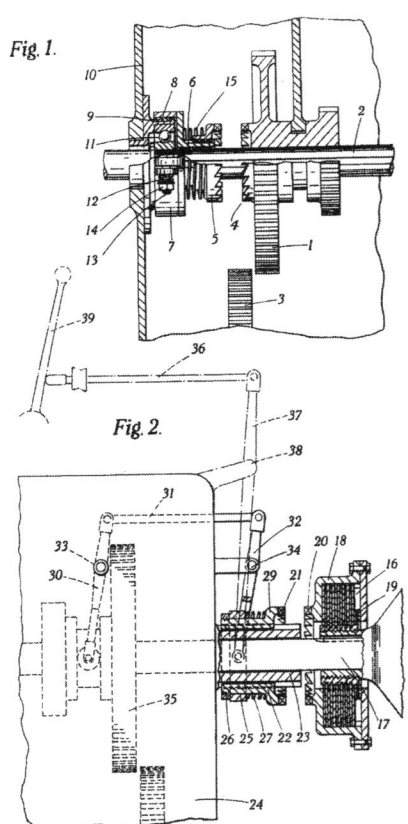

The 'Hill Holder' of 1938, Pat. No 512514

12
Phelon & Moore

Having fully parted company from ABC in 1921, Granville sought a new home and base in Central London. Being one of their earliest members, he stayed temporarily at the Royal Automobile Club in Pall Mall, until he could take up residence in late 1922 at 30, Hill Street, Kensington with easy access to his London offices at 90, Jermyn Street, shared by Gilbert Campling. The Club was a regular meeting place for pioneers of the automotive industry, one of whom was Bertram Marians, Phelon & Moore's London based Sales Director. And so it was that one day in the spring of 1923, Marians approached Bradshaw to design a sports version of their successful new P&M 'Panther' $4\frac{1}{2}$hp, 500cc side valve motorcycle.

The P&M engine and frame was unique in that, by replacing the conventional motorcycle front down tube with the engine, itself encaged within four long through bolts, the engine became a structural member and in so doing, presented the barrel and head in direct contact with the oncoming flow of cooling air. For this principle, Joah Carver Phelon was granted Patent No. 3516 in 1901. His Phelon & Rayner was Britain's first all-chain driven motorcycle and was also licence built by Humbers. By incorporating Richard Moore's patented two speed gear and wedge clutch (Pat No. 6191) into the bicycle bottom bracket, it soon gained for P&M a solid reputation as a reliable sidecar machine, fast tourer and an excellent 'Colonial' motorcycle for the rougher terrain of Britain's wilder empire. After extensive trials at Brooklands in 1912/13, of which Bradshaw was undoubtedly aware, the well proven P&M $3\frac{1}{2}$hp motorcycle was selected as the standard machine for the new Royal Flying Corps.

The P&M $3\frac{1}{2}$hp side-valve model developed into the $4\frac{1}{2}$hp of 1920, still with its sloping engine, but now with a new pattern of 'aerodynamic' cooling fins lying horizontal to the ground for maximum airflow. On whose theory this was based is not known, for Bradshaw's own experiments on aero-engine cooling had shown that, due to eddy currents between the fins, what mattered more was the surface area exposed to the air flow, not the angle. P&M reverted to conventional radial fins for 1921 though Bradshaw did use horizontal fins in later designs.

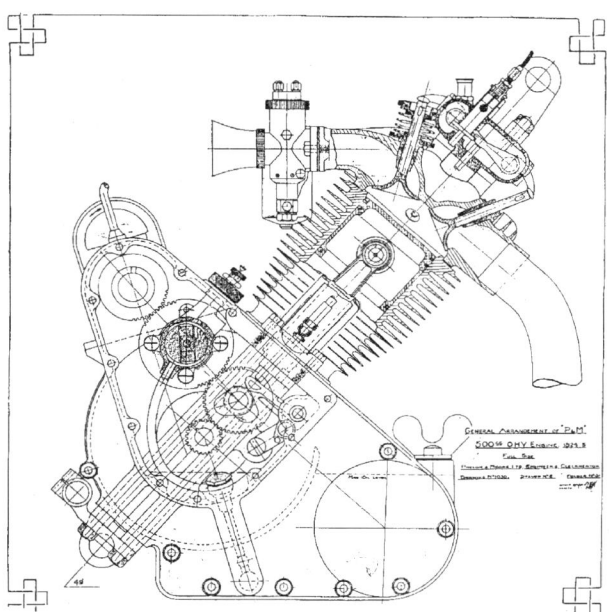

Bradshaw's ohv version of P&M's famous 500cc 'sloper' engine

1925 P&M Panther ohv motorcycle (CMC)

Bertram Marians hoped that with Bradshaw's help, P&M could expand their side-valve model range and transform P&M's image with sports and racing models with which to enter the increasingly important, and highly competitive Isle of Man Tourist Trophy. Bradshaw at first considered a conventional engine and frame design, but returned to the original P&M concept and developed an overhead valve version with an engine driven oil pump. This was claimed to be the first motorcycle engine with enclosed rocker gear and push rod tubes which allowed an oil mist supply to feed felt pads which lubricated the rocker spindles. The cylinder head also incorporated Bradshaw's patented spiral induction port (Pat No: 223324 laid 25th July 1923) to enhance atomisation and thereby improve combustion - in theory! The crankcase was a semi-dry design in which the eccentric rotary plunger pump delivered oil to the main bearings and also squirted oil at the big-end bearing; the piston and cylinder wall relied on splash lubrication. Surplus oil was scavenged by the flywheels and thrown over a high weir into a forward sump where it was cooled by the air flow before being recycled.

In his usual manner, general arrangement drawings were presented to P&M's draughtsman, Ben Hey (later Terry Smithems and Harry Asher) for preparation as production drawings.

Following its launch in October 1923, the new 84 x 90mm ohv 'Panther' became, if you'll forgive the pun, a roaring success. As more and more trials were being won on these wonderful new ohv models, it re-established P&M as the mount of choice for the sportsman and sidecar rider. For this ohv design, Bradshaw received a handsome royalty over five years, bringing him 'many thousands of pounds'.

The engine was further improved by P&M's new Development Engineer, Frank Leach. His first modification came in early 1924 by ridding it of the difficult to cast and machine, spiral induction port in favour of a conventional port which showed no loss in power. With these modified engines, P&M entered the TT for the first time in the summer of 1924, with Vivian Olsson and Tommy Allchin who, at the last moment, switched to a Douglas. It was not a successful event for P&M, but they returned in 1925 with Oliver Langton, 'Count' Ashby and an 18 year old P&M apprentice, Tommy Bullus, as reserve. Langton crashed during practice allowing Bullus to step in, coming 4th at 61.96mph - the race was won by H R Davies on his HRD at 66.13mph.

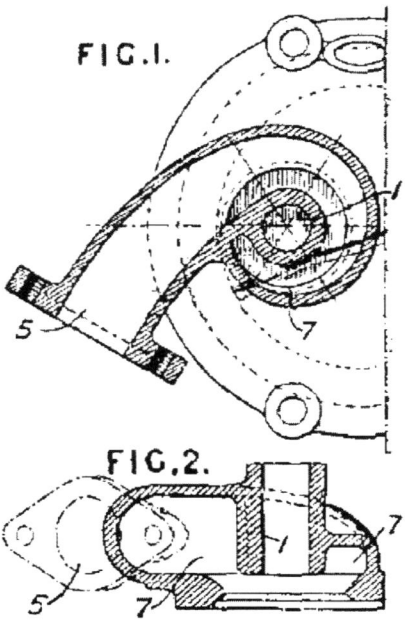

The patented swirl chamber

In 1926 Leach fitted a direct oil supply to the rockers but though P&M entered the 1926 TT, it was a disaster and they effectively withdrew from racing in 1927, though they continued to support several privately entered TT replicas in events at home and abroad. One engine was specially modified for Joe Sarkis of South Africa in 1927 using Bradshaw's laminated valve leaf spring arrangement in lieu of conventional coil springs. One spin-off from the single cylinder TT engines was their hall-mark twin exhaust pipes.

In the winter of 1934/35, a Panther 100 was be subjected to one of the most gruelling of endurance tests, the trans-Saharan crossing, en-route to Cape Town, by Theresa Wallach and Florence Blenkiron. Both ladies were Brooklands 100mph Gold Star award winners; Florence being the first lady rider to exceed 100mph. She was the secretary to the Technical Assistant to one of Hadfield Steel's directors. Theresa was a graduate engineer. Though Leach had designed a new, deeply finned cast alloy sump to improve oil cooling for the 1935 model range, this did not feature on the girls' 1934 Panther, christened 'The Venture'. The engine and motorcycle survived the journey despite Price's engine oil being reduced to the 'consistency of water'. On reaching Cape Town, Florence returned alone on a new 1935 Model 100, 'Venture II', shipped out to her by P&M - even though she was forbidden by the French authorities from crossing the Sahara alone. The full story of this incredible journey and the ruggedness of Bradshaw's engine and P&M's frame is told in *The Rugged Road* by Theresa Wallach, published posthumously.

Panthette

By 1925 Bradshaw had moved again to 6A Kensington Crescent and by 1927 to 5, Beauchamp Place, Brompton Road, London SW3 within walking distance of Kensington and the Royal Automobile Club. He clearly still had much influence with P&M for, despite Leach's several modifications to the ohv engine, in 1926 Bradshaw was called to P&M's works at Cleckheaton to discuss a new lightweight motorcycle. Tommy Bullus recalled picking up Bradshaw from Bradford railway station on one of his many visits and being told later that the new Panthette had been devised over that weekend.

The 60° transverse V-twin Panthette was an amalgamation of many of Bradshaw's ideas for a new, fault free, ABC and would provide P&M with a much needed model in the, then, under 200lbs taxation class. P&M had high hopes for it, as indeed did the eager motorcycle press when, following 2,000 miles of extensive testing by P&M's Joe Mortimer, the new Panthette was presented to the world

The Panthette unit construction V-twin (CMC)

The delightful 1927 Panthette V-twin, opposite. Below Panthette cylinder head showing leaf springs and valve gear arrangement

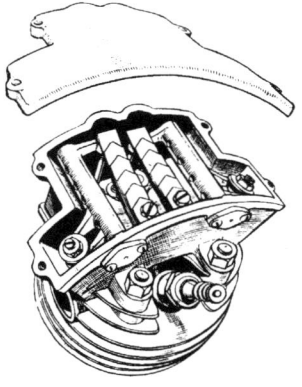

at the Olympia Show in October 1926. Like the earlier ABC, it was a show sensation with Bradshaw proclaiming, "It is the ABC of 1919 brought up to date, refined and improved with the benefit of the intervening years, to suit the market of 1927".

The 246cc engine was of a long stroke design (50mm x 65.5mm) with an enclosed valve gear using direct acting forked rocker arms to force open and close the valves by short leaf springs, rather than the conventional coil or hair spring. His decision was based on the leaf spring's supreme reliability and indifference to heat as well demonstrated in the Austo-Daimler in-line and Gnôme rotary aero-engines of the Great War.

The integral engine and 4-speed gearbox, with its enclosed clutch and lightweight flywheels, was split horizontally for ease of assembly. A car-type 'H-gate' gear-selector manipulated the gears while final drive was by chain from a bevel gear. The frame had a rigid rear with girder fork front suspension but only by lifting the tank was the frame's uniqueness revealed; it used a forged steel I-beam backbone, which incorporated the head stock, from which was suspended a flat steel strip engine cradle. BSA adopted a similar forged steel backbone and cradle frame in 1929.

Bradshaw's innovative design gave the Panthette a low centre of gravity and, with its narrow width, excellent handling and perfect cooling for the exposed cylinders - just as had his original ABC. Sales were optimistically projected at 3,000 a year and 'thousands' of forged steel backbone frames were ordered in anticipation. But the orders never came as the love affair waned soon after its launch, for come what may, P&M were unable to bring the overall weight down to meet the lower taxation class which lost them many potential customers. To make matters worse, the asking price, dictated by its advanced design, was too high for that class of motorcycle. Furthermore, the public were not enamoured with the angular fuel tank but, to their credit, P&M promptly redesigned it, transforming a frog into a very handsome prince. Nevertheless, the Panthette was a superb lightweight motorcycle and was much loved by their riders, factory staff and even the tall Joe Mortimer, who dwarfed the machine!

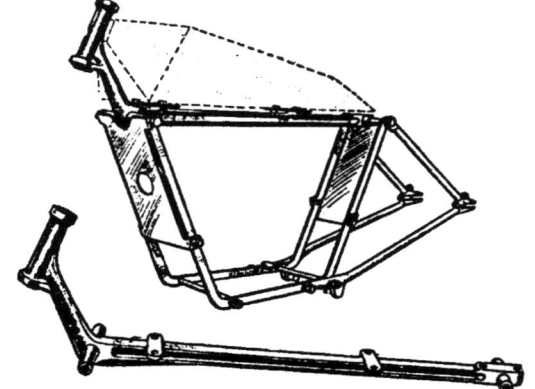

Cradle frame with forged steel backbone (CMC)

Panther-Villiers 250cc of 1929

However more serious problems were lurking under the surface. The oil supply proved inadequate, though riders overcame this by adding an extra pint to the sump allowing the big ends to dip in for extra splash lubrication. This of course caused adverse frothing and drag! The lightweight flywheels, combined with oil-immersed clutch plates, which dragged badly, caused the engine to stall far too easily, especially during gear changes. Being right-hand change with a right hand throttle, it was almost impossible to 'blip' the engine during changes to maintain flywheel momentum and this had the adverse effect of riders tending to set a high tick over and rev the engine too freely, leading to high fuel consumption. The engine bearings and bevel final drive also made the Panthette very noisy and, as it was later discovered, the kick start hypoid gear wore out at an incredible rate.

David Gow, a machine shop engineer at P&M between 1927-1932, was closely involved in improvements to the Panthette, which he too admired. He recalled the tremendous efforts made by shop floor engineers and the company to sort them out but there is strong evidence of how Frank Leach, who was far from enamoured with both Bradshaw and the Panthette, purposely helped bring about its early demise through badly engineered 'improvements' in order to bring forward a conventional 250cc motorcycle design of his own.

P&M were now in a very difficult position with low Panthette sales and these on-going development costs affecting their cash flow. By good fortune, Villiers were keen to promote their new two-stroke motorcycle engines and in May 1928 P&M gained valuable breathing space by replacing Bradshaw's V-twin with a Villiers two-stroke and conventional motorcycle gearbox. This rapid transition was only made possible by Bradshaw's forged steel backbone frame which readily allowed a simple engine/gearbox sub-frame to be fitted. And so it was that the new Panther-Villiers steadily consumed the huge stock of Panthette frames. More importantly, these new two-stroke lightweights fell well within the lower 200lbs taxation class. They sold remarkably well too, especially the sporty 250cc model and by 1929 production of the slow selling Panthette was brought to an end; only about 200 had been built and of these only a handful survive. By cruel irony, the 200lbs class was raised to 224lbs in 1930, but it came too late to save the Panthette. Once again, Bradshaw was ahead of the times!

By late 1929 the depression had begun to bite and P&M was soon perilously close to bankruptcy. Though blame for P&M's demise is always directed at Bradshaw's Panthette it has to be born mind that the depression also badly affected sales of P&M's profitable heavyweights whose production dropped to only 100 machines a month in 1930/31. Frank Leach now worked desperately on his new 250cc lightweight ohv model. Launched in late 1931 as the Panther Model 30, even these got off to a very slow start. By early 1932, heavyweight sales had plummeted to only 50 per month and as Britain's unemployment peaked at 2.8 million, P&M were forced to put staff on half-wages.

1932 saw P&M's trading profits cut by one half but by good fortune in the autumn of 1932, Pride & Clarke made P&M an offer too good to refuse. It saved P&M's bacon, but forced them out of the motorcycle manufacture's trade association - 'the Union'.

Pride & Clarke had upset the Union by undercutting their competitors and accordingly, union members were banned from supplying them. Those, like P&M, who ignored the ban were now faced with their own supply problems. For P&M, the ban was a very dangerous move indeed, but their situation was desperate and in early 1933 they delivered the first batch to Pride & Clarke of an exclusive, red liveried 250cc model, selling at the rock bottom price of £28/17/6d - it rose to £29/17/6d the next year and the 'Red Panther' became an instant success. As demand for P&C's Red Panther rocketed, P&M were forced to recruit local, redundant and unskilled, staff at minimal wages to assemble the motorcycles in a chain gang. The engines were still assembled by skilled fitters alongside P&M's own 250cc/350cc ohv model range. It was the cheapest, most complete 250cc motorcycle of the 1930s but although many thousands were sold, P&M made very little profit on these, but at least it kept them in business throughout the long recession.

Around 1940/1941, while living at Lovelace Lodge, Woodland Drive, East Horsley in West Sussex, Bradshaw returned to his Panthette concept. A drawing shows a 500cc V-twin (65mm x 75mm) with conventional overhead valve gear and a 4-speed gearbox with dry clutch, bolted to the rear of the crankcase; it appears to have shaft drive. Of greatest interest was the fact that the motorcycle's front down tube was bolted to the top of the crankcase between the cylinders; the saddle down tube was bolted to the rear of the gearbox making the engine an integral part of the frame. This was most probably designed for Phelon & Moore who, at the time, were considering various new ideas to compete with Edward Turner's Triumph twin. Was this 500cc V-twin intended to replace the Panthette, or to supplement the Model 90 500cc single of 1938, or was it an alternative to Bradshaw's in-line P&M twin of 1939? Whatever the intention, if one thinks of the Moto-Guzzi V-50 of the 1970s you have yet another Bradshaw design, years ahead of its time.

Bi-flex springs

Despite the Panthette's failings, Bradshaw's association with P&M remained strong and in 1938 he was instrumental in helping design a new range of conventionally framed, vertical single cylinder, sports tourers and an in-line twin Panther. For these he developed a 'Bi-flex' spring frame, first proposed in 1937 as described earlier which, in its prototype form, worked well and fully met P&M's criteria.

Drawing on his experience with the leaf sprung ABC, he produced a brilliantly simple, parallelogram, laminated leaf spring arrangement equally suited to front and rear suspension. Leaf springs offer a variable rate of suspension and, due to the leaves sliding against each other as the loading increases, have in-built damping, but in order to increase flexibility and reaction during cornering, Bradshaw drew on his earlier ABC patent (Pat No: 113345 of 1917) with 'waisted' leaves, separated by Ferodo friction material to dampen movement under load. All looked well and P&M had high hopes when Richard Moore instructed

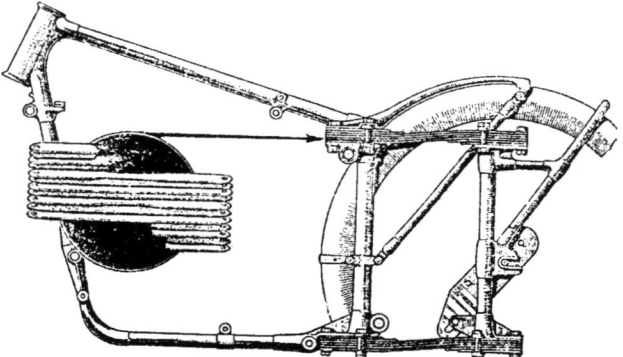

Bi-flex leaf spring rear suspension

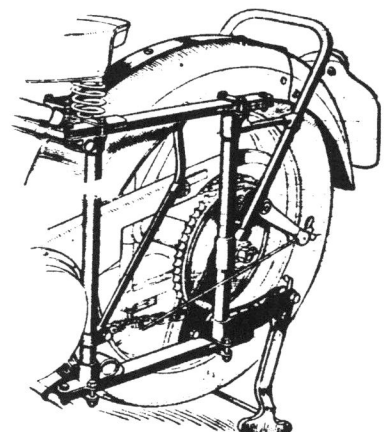

Bi-flex suspension on a Panther Model 90 (CMC)

Joe Mortimer to "try and break it" on a new 500cc Model 90. Unfortunately, Joe did just that, almost losing his life as he traversed a railway crossing 'at speed', snapping a lower leaf at the bracket, breaking the chain and bringing the machine to an abrupt halt. Mind you, Joe was quite used to such mishaps as on another occasion on a Model 100, the Enfield hub de-spoked during a road test whilst cornering - Enfield had neglected to tell P&M that they had changed the design.

Fortunately for Bradshaw the fault lay in the fabricated brackets and not in his springs; had he retained the proven ABC lower swing arm and upper leaf spring, all may have been well. Undeterred, P&M fitted the system experimentally to a 600cc Model 100 'Sloper', a lightweight model and the in-line twin, but with the onset of war, all further work was immediately halted.

In 1953, P&M developed a swinging-arm rear suspension of their own design, but so bad was the prototype that the usually compliant motorcycle press refused to publish details! Fortunately, (or unfortunately for its originator), P&M then pinched a rear suspension design off the former P&M machinist and engineer, David Gow, now living in Scotland, who had popped into the works one day on his modified Panther 100 when visiting his brother; he was mischievously whisked away on some false pretence by the P&M directors while the engineers took copious notes and measurements. P&M's 'new improved' swinging arm suspension proved a great success!

P&M in-line twin

In the summer of 1939, Bradshaw developed a 500cc vertical in-line twin, often referred to as the Panther 200, using the leaf-sprung Model 90 frame, with which P&M hoped to compete against Turner's Triumph transverse twin. Undoubtedly inspired by an idea Bradshaw had formulated in 1937 for a budget 300cc in-line engine, this new engine had two separate cast iron barrels with their conventional ohv gear in a shared cylinder head and rocker box; each cylinder's push rods were set either side of the engine; the spark plugs were at the front and back of their respective cylinders. It also adopted a common inlet manifold from the forward set Amal carburettor and a common cast iron, deeply finned exhaust manifold.

There was a separate Burman gearbox and chain final drive to the leaf sprung rear wheel. Everything appeared straight forward, but it was said that in engine testing, the inlet manifold appeared to upset the supply of the combustible mixture between the two cylinders, causing irregular running and, in turn, placed a major strain on the primary drive.

While there is no dispute over the two different methods employed in connecting the separate crankshafts, chain and gears, which came first is still uncertain, as the two people most closely involved give contradictory evidence. I shall start with the version, first given me by Joe Mortimer, which employed chain drive.

Bradshaw's stunning P&M in-line twin (CMC)

The crankcase was split vertically. Each crankshaft, of what was essentially a pair of 250cc engines, was fitted with a shrunk-on sprocket driving a single row chain over and under the crankshaft sprockets and idlers, producing a balanced contra-rotation with minimal vibration. This system worked well, but the irregular running attributed to the inlet manifold, caused a sprocket to slip with catastrophic results. The engine was duly rebuilt but now with

keyed-on sprockets. All went well until the erratic running at full throttle caused the chain to snap, whipping through the chain case, locking the rear wheel and causing the helpless Joe Mortimer to crash.

The engine sprockets were then replaced by a pair of 8" diameter keyed-on spur gears, again driving in contra-rotation. This proved so noisy, that they were replaced by helically cut gears which, though they reduced the noise, significantly increased drag, causing the engine to die above 1,500rpm. Special care had to be taken to minimise any backlash to prevent serious damage, but major problems were now encountered by the heat generated from the meshing gears within the crankcase, which thinned the engine oil so much so that it made it unsuitable for the high pressure exerted by the helical gears. Unfortunately, the available high pressure hypoid gear oil was unsuited to the crankshaft's plain bearings which required a high pressure oil supply.... the solution seemed to lie in chain drive.

The alternative version begins with spur gears, then helical gears and finally chain drive but in the absence of dated factory drawings, it remains uncertain as to which did come first. I suspect that Joe Mortimer's version is the more likely.

Slated in later years for using gear driven crankshafts for his in-line twin's 'failure', Bradshaw's design was in fact based on a well known and proven concept, first patented in 1895 by the great Frederick Lanchester! Engineers accept that the advantages of such geared-crank vertical twins far outweighed the problems of balance and whip in a conventional forged crankshaft, provided the major problem of gear backlash, friction and noise could be mastered. It is argued that had Bradshaw used a compression ignition engine (diesel) with full throttle air intake and metered fuel injection for a more consistently reliable and predictable 'bang', then his in-line twin may well have worked properly, as those involved on the project recognised that the problem lay almost entirely with the induction system and not the contra-rotating crankshafts. But unfortunately, the onset of war brought all motorcycle work to a halt and no further work was done on the in-line twin which was then relegated to the role of 'works hack' until the early 1950s when it was stored at Cleckheaton; and disappeared when the company closed.

It should also be borne in mind that a geared, double crankshaft, contra-rotating two-stroke twin cylinder concept, whose pistons rose and fell at the same time, was proposed in 1904 by Ralph Lucas of London, and redeveloped in 1909 with spur-gear drive, for the 'Valveless' motorcar produced for him until 1914, by David Brown Ltd of Huddersfield. A similar design was offered by a Mr Lamplough before the great war. The famous Trojan twin cylinder engine of 1912 however used a single crank with split connecting rod to the same effect, dispensing with the costly gears. All these engines were, however, twin piston 'single cylinder' rather than conventional, twin cylinder designs.

Furthermore, geared, double crankshaft opposed piston diesel engine designs were being developed in Britain by Doxford in 1912 as licensee's of Dr Hugo Junkers' opposed twin engines, pioneered by him in 1893. Intended for marine use, these massive diesel engines were not adopted by the Royal Navy but were successfully used in German U-boats in World War One. These were then developed by Junkers' famous aircraft company in 1925 into the Jumo 205, 6 cylinder, opposed piston, two-stroke diesel aero-engine which had two, six throw crankshafts coupled by idler wheels to a single output shaft. Granted, this used two pistons acting against each other to compress the charged gases and any irregularities in firing would have been smoothed out in each firing. The Jumo 205 developed 510hp at 2,100rpm and similar designs were developed in France and by our own Napier company, under licence, for the short lived Napier 'Culverine'. The Jumo 205 was used by Lufthansa in the Ju.86 airliner of the 1930s and in Blohm ünd Voss flying boats during the last war. The final development was the 207, but this ancient design was superseded during the war by the hugely successful, inverted V-12 Jumo 211.

P&M at war

Motorcycle exports to 'safe' countries continued for a short while but P&M failed to gain the much hoped for contract to supply the RAF, as in the first world war for the RFC. This was possibly because since the late 1930s all military motorcycle contracts were now handled by the army. Instead, P&M were contracted by the Government to produce components for bomb racks, bearing housings for Fairey Battle bombers, hinge pins, brackets and control columns for Fairey Swordfish torpedo bombers, as well as sub-contract work for Blackburn, Avro, Boulton & Paul, Dowty and Rolls Royce. They also made ships' stanchions which were almost certainly those designed by Bradshaw at Fairmile Marine, described later. Bradshaw also called upon P&M to consider the manufacture of an 'air-pump' which, apparently, was in prototype form at the Turner Manufacturing Company, Birmingham and for which he apparently had patent rights, though none have been found. It appears the 'air-pump' was developed around a Roots type blower. The story behind this and the toroidal rotary engine is told later.

Bond Scooter

During the later war years P&M drew up ambitious plans for a post-war motorcycle programme with the motorcycle press co-operating in teasing the public with subtle hints that the in-line twin, along with other exciting designs, would be in production once the war ended. But it was not to be, and after P&M's involvement with the toroidal engine and a brief involvement in the Bond scooter, Bradshaw's direct motorcycle work for P&M came to an end.

With P&M importing the French Terrot 'Scooterrot' from 1956, and Frank Leach developing a glass-fibre bodied 'Panther Princess' scooter, Granville Bradshaw introduced Phelon & Moore as technical advisers to Ewart Bradshaw's Sharp Commercials in the development of their glass-fibre bodied Bond P.1 scooter of 1958. The P.1 was designed under great secrecy during 1957 by, it is said, a close friend of Col. Reg Gray at their recently acquired India Mills in Preston. It was intended to be 'the best scooter ever made' and certainly featured many very innovative and advanced ideas.

The 148cc Villiers 31C powered Bond P.1, and the more powerful 197cc Villiers P.2, were launched in January 1958 and received rapturous applause at the October Motorcycle show. It is not known if Granville had an input in its initial design, but by April 1959, he had become engaged in redesigning the steering and frame geometry to improve its handling. Sadly, the Bond scooter proved a poor seller, mainly because they were considered too heavy and slow, much like other British designs compared to the sprightly Italian Vespa and Lambretta scooters.

Pleased with his one piece, moulded glass-fibre body shell for the Panther Princess, Frank Leach had overlooked one crucial point - how to extract it from the mould.... for there it remained, firmly trapped, much to the delight and amusement of P&M's staff, many of whom were not enamoured with him! Leach wisely dropped his design in favour of a simpler, steel panelled body-shell shared with the Dayton and Sun Cycle companies. Launched in November 1958, Princess production continued until October 1963. While production of the Bond scooters had ended in 1962 with some 1,680 of all types being produced, the Panther Princess fared even worse with only 229 examples.

The full story of the Princess is told in *The Story of Panther Motor Cycles* and the more comprehensive, *The Panther Story* while that of the Bond is told in *Lawrie Bond*.

A Bond P1 scooter being presented to the RAC/ACU scheme in 1958, with, on the left, the towering Col. Reg Gray looking on

13
ABC Cycle-car

Horse-power and taxation

The unit of 'horse-power' was established by James Watt in 1782. He observed a horse walking around a 24 foot circular track, operating a capstan pump at a working Cornish mine. Using a spring balance, he determined the horse was exerting a pull of 180lbs and completed $2^{1}/_{2}$ revolutions of the track per minute. From this, Watts calculated one 'Horse Power' was equal to 32,400 foot-pounds per minute; this was later standardised at 33,000ftlb/min. Both the French *Cheval-vapeur* and the German *Pferdestarke* were based on the power needed to raise 75 kilograms to a height of one metre, in one minute. Quite why, in their beloved metric system, they didn't chose 100Kg remains a mystery as does the way it was derived! Suffice to say one can visualise the 'power of one horse' irrespective of the science.

One HP = 1.014 CV/PS = 745.70 Watts. Simple really!

It is of course possible to calculate the theoretical power of a stationary engine by the indicated pressure on each piston in pounds per square inch (from the piston bore), and the productive force by its stroke; but this was wholly unrepresentative of the 'horse power' of an internal combustion engine given its wide speed range, inertia and thermodynamic characteristics. The truest representation of the actual, or effective, power an engine produces is at its output shaft where one can also measure the turning effort, or torque. To that end, engineers preferred to measure the engine against a braking force applied to the output shaft on a dynamometer. Hence it was quite common to see, say, a 3hp engine also described as 10bhp though the 'b' was often omitted.

Pioneering engines were thus described in perceived rather than actual horse-power, more in hope than truth, and as a general rule were categorised as follows:

HP	Equiv. cc
$1^{1}/_{2}$hp	250cc
$2^{3}/_{4}$hp	350cc
$3^{1}/_{2}$hp	500cc
5-6hp	750cc
7-9hp	1,000cc
12-20hp	2,500cc

Since 1st January 1904, taxation raised from motor vehicles had come from a fixed licence fee under the 1903 Motor Car Act, to which was added in 1909 a 3d per gallon levy on petrol, then selling for around 1/6d gallon. But from 1st January 1910, a new horse-power based road tax was introduced at £2/2/0d for up to 6 1/2hp; £3/3/0d for 6 $^{1}/_{2}$hp up to 12hp and in irregular intervals thereafter up to 60 hp. This was in addition to the petrol tax which rose to 6d/gallon in 1915. The new tax system gave impetus to designing lighter vehicles with less powerful engines and a new category of 'cycle-car' was created for those weighing under 7cwt (784lbs) with an engine capacity of no greater than 1,100cc.

A new category of motor tricycle (which included motorcycle and sidecar) was also introduced with the proviso that these were not fitted with a reverse gear. It is not surprising then that such legislation led to some truly incredible and rudimentary designs to avoid excessive taxes which, in turn, spurred the Chancellor of the Exchequer to introduce a simpler taxation system under the Finance Act of 1920, adopted on 1st January 1921, using an RAC scheme for motorcar taxation of £1 per 'RAC horsepower'; thankfully they also repealed petrol tax - Oh!, happy days! - only to be reintroduced in 1928 at 9d gallon at which level it remained until 1950.

The new RAC rating was based on the cylinder bore (bore squared x number of cylinders x 2.5); hence Bradshaw's 1,203cc air cooled engine, described below, yielded 10.3 RAC hp, but developed 35bhp. The RAC formula effectively imposed a penalty on engines more powerful than was absolutely necessary and, in turn, encouraged longer stroke, small bore engines which kept down tax revenues but saw ever more powerful, if slower revving and more sluggish engines. This rating system remained until the all encompassing Road Traffic Act of 1930 which introduced taxation classes by vehicle weight and category. Under the Finance Act of 1936, the Chancellor stealthily transformed the Road Fund Licence (intended only for road maintenance) into a Vehicle Excise Tax to feed his bulging coffers and be made available for squandering by any Government department. Clearly then, ever changing fiscal policies had a major influence on motor vehicle design.

Bradshaw - the motorist

ABC's success as aero-engine and motorcycle makers allowed Bradshaw the luxury of moving up from his lowly Triumph motorcycle to his first Rolls-Royce in 1913. This was a 1911 'London-Edinburgh' model Silver Ghost (he quotes chassis No. 1706) and was followed in August 1915 by one of the last Alpine Eagle models (chassis No. 36TB). In 1927 he took delivery of a new Hooper cabriolet bodied Rolls-Royce Phantom I (chassis 9LF) followed by a 1936 Model 20/25hp (chassis GBK80) with Park Ward saloon body. He also had an unidentified 1920s 40/50hp Rolls Royce chassis with which he experimented, but after his bankruptcy following the Hatry case, he was reduced to a Ford V-8 in the late 1930s and, post-war, a Ford Consul which he kept until the early 1960s.

He also enjoyed racing at Brooklands in his early days and circuited the course in a shortened Napier-Samson as well as Count Zbrowky's famous 1921 aero-engined 'Chitty'.

His first experiments for a peoples' car began around 1913 with a 4-wheel invalid carriage based on a motorcycle sidecar, powered by an accumulator battery which gave it up to 35 miles on a single charge. It is said one was produced with optional hand or foot control and that, at the end of the war, the battery was replaced by a 900cc version of his ABC horizontal twin motorcycle engine. Little is known of these experiments.

While Bradshaw had certainly planned a 5 cylinder 'Gadfly' radial aero-engine at the end of the Great War, he was also said to have worked on a design for a 5 cylinder radial engined motor car in the 1920s with a 4-speed gearbox and a novel 'hinged' rear axle supported by semi-eliptical leaf springs with half shafts from a central differential. It was apparently designed with a tubular steel chassis, but how close this exciting 100bhp sports car got to prototype stage is not known. He briefly proposed another 5 cylinder radial engined car in *The Autocar* of 1941 as described earlier.

ABC Cycle-car

His first major project after the Great War was for a cycle-car, announced in March 1919. The prototype was developed at Hersham. With ambitious plans for 3,000 cars per annum, sales would be handled by the 'New Service Organisation for Motorists' which was connected with Hubert 'Jack' Whitcombe's Motor Union Insurance Company.

Powered by the ABC twin-cylinder, air-cooled engine - possibly a 900cc but also described as a 1,198cc (91.1 x 91.5mm) 35bhp engine (rated at 10.4 RAC hp) - it had the usual machined steel cylinders, overhead valves and hemispherical head design. It was fed by a pair of ABC carburettors. Transmission was through a 3-speed unit construction gearbox to a bevel final drive.

The cycle-car was not at all refined and, given its war-time ancestry, showed signs of having been designed for rapid assembly at least possible cost, rather than for longevity. Free floating gudgeon pins for the lightweight cast alloy pistons and lightweight con-rods were retained by press-fitted domed copper buttons, but for strength, the big end was in one piece and was fitted by sliding along the crankshaft, over the webs - 14 bearing rollers were then inserted into the big end bearing shell through a small slot machined in the web; once installed they were retained by split rings. The many assembly bolts in the crankcase had fluted, milled heads which allowed them to be pressed home into the soft alloy casting, preventing them from turning, in keeping with an earlier patent (Pat No. 133093).

The transverse twin's barrels and crankcase were cooled by air ducted from a dummy radiator, drawn through by a large fan. Apertures in the bonnet's side panels allowed cooling of the cylinder heads. To the unwary, the filler cap on the dummy 'radiator' could all too easily be removed and topped up with water - it was in fact the petrol filler cap to a tank contained within! Oil for the separate engine oil reservoir was accessed from under the bonnet, being pumped direct to the bearings and by jets to the cylinder walls. The timing gear train and tappets were lubricated by oil mist.

Uniquely, the engine, cone clutch (later single dry plate clutch) and mid-mounted gearbox, were fitted to a large steel platform sub-frame which was then fitted into the open ladder framed steel chassis which was braced by tubular steel cross members. Sub-frames are standard practice in modern motor cars; a major bonus is that the engine and transmission could be bench tested prior to fitting. An interesting clutch withdrawal mechanism was used in which a spring counterbalanced the pedal, but this required a long travel before engaging the Skefco thrust bearing and clutch withdrawal collar. Gear selection was by a straight through gate. The drive shaft was contained within a torque tube, supported by bearings, driving the axle's fabricated bevel gear crown wheel and differential unit.

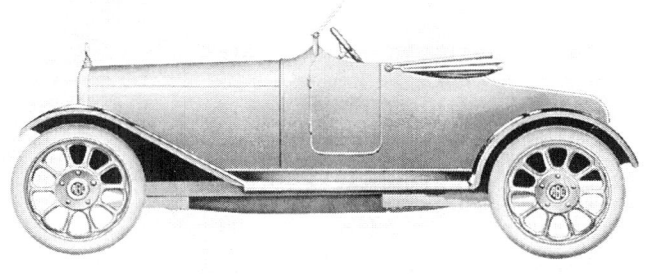

Top, 1920 ABC 12hp Standard 4-seater. Bottom, 1920 ABC 2-seater Light car (GB)

Left, the dummy radiator for the air-cooled 1198cc (the Autocar), and, right, the 2-cylinder engine. (GB)

The final drive assembly was held in a cast alloy housing, connected to the wheel hubs by 14G rolled steel cones, flanged at both ends, by which it was riveted to the hub and differential housing. The hubs had brake drums actuated by a compensated link to the Ferodo lined internal expanding hand-brake shoes and the external contracting foot-brake band. There were no front brakes.

The I-section drop forged steel front axle and the tubular rear axle were suspended on the tips of, undamped, quarter-elliptic cantilevered leaf springs. Wheels were of the 5" Sankey artillery pattern fitted with 700mm x 85mm 'voiturette' tyres. Steering was by simple worm and nut.

Both the first and the longer chassis second prototypes were exhibited at the spring 1919 Motor Show, priced at £195 each and with a launch promised for June 1919. But June came and went, as did November when the price rose to £295. Meanwhile the prototypes were being tested in trials by S C H 'Sammy' Davies and Jack Falahee; both did well, often pitched against cars of much higher power in the under 1,500cc class. Sammy Davies later won a Gold Medal in the 1920 London to Land's End Trial in his alloy bonneted ABC cycle car, registration number PB 5551.

ABC Motors (1920) Ltd

ABC's assets included the $1^3/_4$ acre Hersham works, a 20 acre development site opposite (which included Hersham Lodge into which they had moved their offices) together with 50 Queens Road, Hersham (most probably Faulkner's original forge of the 1790s). Their designs and patents (except the aero-engines which were passed to Walton Motors Ltd, a subsidiary of ABC) received royalties of £3/10/0d from each Sopwith-ABC motorcycle built in the UK and from the Société Francais de Moteurs ABC in France, plus 15/0d per Skootamota built by Gilbert Campling. In late 1919, the rapidly expanding and over ambitious Harper-Bean Ltd car company, co-founded in November 1919 by Hubert Whitcombe, took a keen interest in the ABC cycle-car and soon offered the company a substantial investment programme, promising inexpensive supplies of castings and forgings from the various Harper-Bean companies, as well as a new factory in which to build the ABC cycle-cars at a rate of 5,000 per annum and at a projected profit of £20 per car! After due consideration of this exceptionally attractive offer, ABC's directors agreed at a meeting in March 1920 to sell for £200,000. ABC Motors Ltd was duly put into voluntary liquidation and reformed as ABC Motors (1920) Ltd.; they also gained new telephone numbers: Esher 306/307.

Harper-Bean's chairman, Hubert 'Jack' Whitcombe and Managing Director, John Harper-Bean, replaced Ernest Noel as ABC's Chairman. (Ernest remained Chairman of Mercantile Investment &

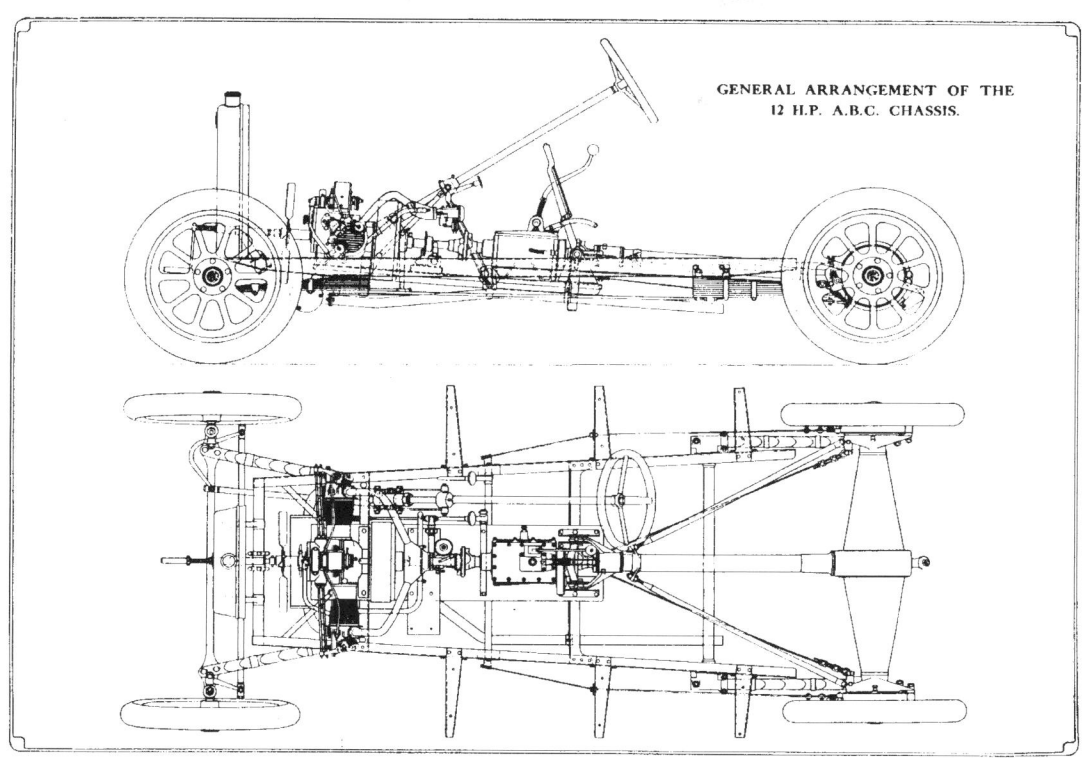

General Trust Ltd and a director of ABC). Ronald Charteris became Managing Director (remaining MD of Walton Motors Ltd and a Director of Société Francais de Moteurs ABC). Maurice Yorke (former Director of ABC Motors Ltd) and T A Dennis (who had been company secretary since 1915), also remained. It appears that John Harper-Bean intended to use the ABC as a means of using up major transmission components ordered, in excess, for his poorly selling Bean cars to which Granville took grave exception and was duly given notice of termination of contract (most probably with mutual agreement). Thereafter, Granville was retained as a design and consulting engineer to ABC, working from his home at Darby House, Sunbury on Thames.

Harper-Bean Ltd

The original business was formed by Absolom Harper in Dudley at the turn of the 19th Century as iron founders. Absolom had two sons, Edward and John. John Harper's only daughter, Mary, married George Bean in 1879. George later joined the family firm as principle shareholder and in 1907 the firm became 'A Harper & Sons, Bean'. As an important manufacturing base for government war-time contacts, they became a limited company during the Great War and, like Sopwiths, sought to expand into new areas to secure their future in the post-war years.

In late 1918, George and his son, John Harper-Bean (also known as 'Jack') bought the assets of the ailing Perry Motor Co. of Tyseley who, like Star, was a former cycle component maker and had produced, between 1915-1916 a very sporty 12hp car. Under John Harper-Bean, this was briefly relaunched in 1919 as the 'Perry-Bean' and then as the 'Bean' at a new Tipton factory, using American production line methods which had allowed Henry Ford to become the dominant force in motorcar manufacture.

Against his father's better judgment, John teamed up with Hubert 'Jack' Whitcombe of the Motor Union Insurance Company who had very recently formed the British Motor Trading Corporation. Whitcombe duly co-funded Harper-Bean Ltd in November 1919 with an outlandish capital of £6 million. They immediately went on a rash buying spree acquiring major interests in Swift Cars (a 51% share; these were built by the former Coventry Sewing Machine Co, Britain's oldest cycle maker), ABC Motors, Galley Radiators Ltd and Hadfield Steel of Sheffield. They also gained share holdings in the Regent Carriage Company of Long Acre, London, Aeromotor Components Ltd, Coopers Mechanical Joints Ltd (gasket makers), Rushmores Ltd and Jigs Ltd. They also briefly held a 75% share in the Vulcan Motor Manufacturing & Engineering Company who had been contracted to build ABC Dragonfly engines, but Vulcan quickly pulled out of the consortium as the depression set in.

In theory, Harper-Beans' unrealistic assessment of 100,000 cars per annum from their several companies, all marketed through their 50% holding in the British Motor Trading Corporation would, through economies of scale from their own engineering and components companies, ensure low unit component costs allowing Harper-Bean to take on the likes of Henry Ford. However, unable to meet demand for bodies from Regent Carriage, they now contracted the Grahame-White aircraft company at Hendon as their body builders. Grahame-White were already building luxury bodies on chassis such as the Rolls-Royce, but when Claude Grahame-White asked for part-payment, John Harper-Bean immediately cancelled the contract and awarded it to the Handley-Page aircraft company! This kept Handley-Page's Somerton Road, Cricklewood, works alive in the post-war years and was ultimately worth £750,000. Handley-Page's profits on war-time aircraft work was £51,000, of which £30,000 had to be set aside for the Government's Excess Profits Tax. H-P later took on the body for the short lived Eric Campbell sports saloon.

Claude Grahame-White, was a pioneering aviator who first flew in 1910 and then promptly formed a flying school at Hendon in 1911. During the war, Hendon airfield was requisitioned by the Royal Naval Air Service and much use was made of his flying school, while his workshops turned to contract building aeroplanes for AVRO, Handley-Page and others. At war's end he offered joy-riding flights in DH9s, but despite briefly opening Hendon as a civilian airport, he too was forced to rely on high class furniture and car bodies, yet remarkably, Grahame-White also designed a very cheap and cheerful cycle-car to employ his workers; the first used a 348cc single cylinder Wall 'Auto-wheel' in their 4hp 'Buckboard' of 1919 - he claimed a production rate of 100 cars a week! This was followed by a twin cylinder 7hp friction drive 'Wondercar' and for 1924, a prototype 4-cylinder 1,094cc Dorman engined car. After this, Grahame-White moved to a new factory as furniture makers and sold his Hendon works to the Royal Air Force in late 1924; these were temporarily used for production of the Angus-Sanderson car. Parts of the old Hendon works survive at the Royal Air Force Museum.

ABC Sporting Light Car

Harper-Bean's immediate objective was to maintain cash-flow through continued production of spare parts for the ABC motorcycle and the ABC Firefly auxiliary engines. Meanwhile with funding from the British Motor Trading Corporation, they built a large new factory, with North-lights, behind the Lodge with a separate long, narrow, single storey works, believed to be their paint-shop. Production finally got under way at around two cars a week in late autumn 1920. New models were duly shown at the Motor Show: the ABC 'Regent' with 4-seat bodywork by the Regent Carriage Company and the 2-seat ABC 'Sports' at 370 gns with a factory built open sports bodywork, *sans* doors. They also offered a standard chassis for coach builders, such as Compton & Herman of Brown's Yard, Hersham who produced an attractive aluminium sports body with a rear dickey seat and a folding fabric hood.

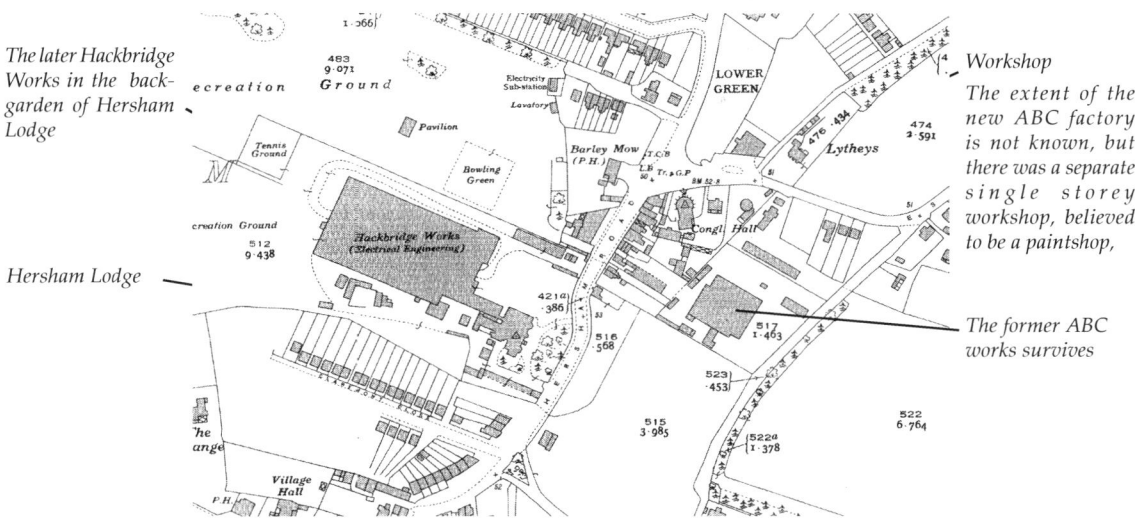

The Hersham site in 1933

Its closest competitor was the Jack Sangster designed Rover 8hp, air cooled horizontally opposed 998cc (85mm x 88mm) transverse twin of 1919 which used a worm-gear axle borrowed from its 12hp sister. Both the ABC and Rover were regarded as transition models between the out and out lightweight cycle-car, such as the 'Buckboard' and a proper motor-car. The Rover was built exclusively at a recently acquired former munitions plant at Tyseley. A sedate runabout, it initially sold for £300 but dropped to £220 in 1922 and down to £145 in 1924. Its engine was later enlarged to 1,130cc.

One early ABC prototype, which had already covered 9,300 miles mainly in the hands of *The Autocar* staff in late 1920, was given a favourable report in April 1921. They praised the engine's tractability and performance: 'there are few small cars which are more pleasant to drive, and the little machine has a feeling of sheer power which is very delightful'. But while the engine was undoubtedly noisy, it became less wearisome to the driver and passengers 'at high speeds'. Most of these unpleasant noises were due to the sounding board effect of the engine's platform sub-frame, bonnet and, particularly, the large bulkhead board - but these were systematically modified by ABC to reduce their impact.

The handling was described by *The Autocar* as the 'first small car in the driver's experience which is not tiring, for not only is the steering perfect but the gear change is exceptionally easy'. Maintenance and decarbonising was also easy as a single BSA 4-way spanner was adequate for most tasks. Concerns over air-cooling were dispelled when vigorously tested up the steep Kirkstone Pass, where it showed no ill effect 'while its water cooled rivals have at times been passed in sore distress, emitting clouds of steam'. Oil consumption was also low but there was a tendency for the crankcase to become pressurised causing oil loses; this was cured by a larger breather.

After *The Autocar* test, the car was entered in the London to Lands End trial but suffered a seized piston ascending Porlock Hill - given its high mileage none were worried by this minor mishap. Overall, *The Autocar* was very impressed, but sales proved disappointing as the plunging economy and industrial strikes took their toll. Already by mid 1920, the Harper-Bean consortium was facing a deep financial crisis resulting in both Swift and Vulcan withdrawing from the group. The Bean car company went into receivership in October 1920 due to unpaid bills, causing John Harper-Bean to resign in November whereupon Bean's Tipton works were permitted to reopen.

Granville Bradshaw in 1954 with an ABC Super Sports

John Harper-Bean and H J Whitcombe both resigned from ABC in April 1921, followed by Ernest Noel in December. Faced with this change in circumstances, the decision had been taken during 1921 to build bodies in-house and with cheaper steel disc wheels. ABC now offered a short lived deluxe 'Surbiton' 4-seater to replace the Regent, a Standard 4 seater at 365 gns, and a Standard 2-seater which was essentially a Sports with doors and dickey seat. There was still of course the ever popular Sports at 330 gns.

For 1922, from chassis number 203, ABC adopted a new one piece spiral bevel rear axle crown wheel within the differential, making for a quieter and smoother ride; it also had a modified gear box. The model range was also reduced to the Standard 4-seater and Sports both with aluminium bonnet and French-grey coloured, fabric covered body. Depending on needs, there was a CAV 6v or a Smiths 12v dynamo system and from chassis No.400, an optional CAV SD 6v starter motor. From chassis number 650, the exhaust was moved from behind the engine to under the seats. De-luxe features included a honeycomb grille and bonnet apertures while an external battery box was fitted to the Standard model. The crucial chassis dimension were 8'6" wheelbase, 47" track, $8\frac{1}{2}$" ground clearance, 12'0" overall length and 5'0" width. The rolling chassis weighed $8\frac{1}{2}$ cwt.

Automobile Engineer magazine of April 1922 was intrigued by the unorthodox design but, equally, praised its logic. They summed up with the observation that: 'The whole design shows evidence of careful thought, tending, perhaps to a departure from simplicity. In practice the car is not very quiet, but is remarkably fast and is built to a better standard of design and workmanship than some of its class'. But all that criticism was forgotten as the lightweight car offered a handsome performance with a top speed in excess of 60mph and indeed, with several being exported to India, they did well in the notorious 120mile Bombay to Poona race returning an average 40mpg.

Production was now running at around 200 cars a year, still well short of their ambitious plans. Despite the minor improvements, it remained a noisy car mainly caused by the metallic ring from the machined steel barrels; it was still difficult to start and poorly lubricated despite its large capacity worm gear driven, eccentric plunger pump. But worse, like his earlier ABC motorcycle, Bradshaw's freely revving engine developed a habit of shedding its push rods and tappets. Though this was not a problem unique to the Bradshaw engine, unlike the motorcycle with its wide under tray, these usually disappeared from the car into a dense roadside hedge. However of greater annoyance was the hand cranked starting which required the cautious driver to 'bounce' the engine into life against the compression of the opposing cylinder. Backfires were, painfully, common-place and many a wise owner parked facing downhill or waited until a crowd was to hand to offer an embarrassing, push-start!

Brooklands

Another pioneering aviator, turned racer, was Eric Gordon England whose father, George, was an ABC dealer. Born in 1887, Eric had built a Weiss glider at the age of 19 and launched himself off Amberley Mount in Sussex - and survived. However, he had mixed successes at Brooklands crashing his Hanriot monoplane into the pond behind The Blue Bird Cafe in 1910 followed the next week by his, now ENV powered Weiss glider, dropping into the famous sewage lagoon.

He had better success racing ABC cars and his graceful, Compton bodied, bullet shaped aluminium sports model, with its detachable wings and Rudge-Whitworth wire wheels, was entered into the September 1921 Nailsworth Ladder trials. However, its reduced ground clearance from an extra protective under tray, caused the ABC to ground. In November 1921, the Junior Car Club held its first post-war 200mile race at Brooklands with Eric Gordon England competing in a rebodied 1,198cc ABC flat twin racer. The original Compton body had been replaced over a weekend by a lighter, squarer aspect body and to lighten it further, the cast iron pistons were replaced by aluminium ones, the cooling fan removed and an oil mist supply rigged up to prevent the exhaust valve overheating; even the foot-brake connecting rods were removed! The rear axle ratio was also changed to 3.25:1 and, now breathing through a pair of massive Solex carburettors, reached 80mph in tests and competing against cars of much larger capacity, completed the race in 2hrs 51min 5sec; his 'smoky' ABC came 14th overall.

1922 saw three more ABCs entered at Brooklands by W L Kenning, Keith Mason and a Lt. Grey who entered a standard 2-seater with dickey seat in the August meeting. Gordon England entered in the Lightening short scratch meeting followed by the Whitsun race in which he was forced to retire. He then entered the August JCC 200mile race rebodied with the streamlined Compton staggered 2-seat body, now resplendent in Saxe blue with red stripes. The engine now featured hollow push rods, Philbrin battery and coil ignition, and Higgs 'Specialoid' alloy pistons. It lapped beautifully at over 80mph, requiring only new spark plugs in a pit stop, coming 6th in class.

In the Easter of 1923, the car was entered by J Campion in what was to be its last race with an ABC engine for, during the summer, she was re-engined with a Bristol Cherub aero-engine and entered by Gordon-England in the 200 mile race that autumn. However in a practice run, the Cherub's main roller bearing showed signs of failure; promptly repaired, all then went well and he came 4th in class at 69.75mph. The race was won by Kenelm Lee Guiness (of KLG Spark Plug fame) in a Talbot-Darracq. This was to be Eric's last race in the ABC as he swapped allegiance to the Austin 7 while his ABC passed to R Malcolm to be re-engined with a 1,500cc ABC Gnat aero-engine and entered into the Whitsun 1925 event, now repainted black with white stripes, but it was an uneventful entry for him as were two further races. At the end of the 1925 season, the faithful ABC was retired.

In spring 1985, Geoffrey Bradshaw presented Doris, Eric's 91 year old widow, with a long lost *Autocar* photo of Eric in the ABC at Brooklands. Writing to thank Geoffrey for the kind gesture, she remarked that for many years she had presented the Gordon England Cup at Beaulieu, but at the annual Brooklands reunion, ever fewer people knew her, yet when she was presented with the photograph, everyone flocked to meet her. For her 91st birthday, she went up in a glider to celebrate Eric's maiden flight of 1906. Immediately afterwards, she eagerly booked up for a flight on her 92nd!

The end of Harper-Bean

Neither Harper-Bean's hopes of benefiting financially from ABC's patent royalties, nor their projected annual sales of 5,000 ABC cycle-cars came anywhere near expectation for, almost immediately, ABC returned an annual loss of £100,000. During 1921, most of the Harper-Bean board members resigned, leaving only Ronald Charteris, Maurice Yorke, T A Dennis (with technical assistance from Bradshaw) to gallantly fight a losing battle for survival. In a desperate bid to cut overheads, wages were slashed and many staff were laid off but by spring 1922, the list of creditors was mounting rapidly. Maurice Yorke resigned in April 1922, followed by Ronald Charteris in July.

In September 1922, the Regent Carriage Company of Long Acre, London petitioned to have ABC Motors (1920) wound up and a receiver was duly appointed that November. Now severing all ties with ABC, Granville Bradshaw concentrated on his engineering consultancy work for his Bradshaw oil-cooled engine and the Belsize-Bradshaw car.

Meanwhile, frantic efforts were being made in 1922 by A Harper, Sons & Bean, George Bean, their bankers and Hadfield Steel to restructure Harper-Bean Ltd. The newly built ABC factory and Hersham Lodge offices were sold to the Hackbridge Electrical Construction Company for a new factory for the manufacture of transformers while surviving operations at ABC retracted to their old works, opposite. (In 1967 Hackbridge became part of English Electric and on their merger in 1968 with GEC, production was gradually moved to Stafford causing the 'new' Hersham works to close in 1972. They were demolished in 1978 for a new Air Products Ltd works and offices).

Bean Industries

Despite their efforts, these were far from happy times for Harper Bean and in March 1925 Harper Bean Ltd went into voluntary liquidation but was soon bought in its entirety from the bank by Hadfield Steel and in June 1927 they dissolved Harper-Bean Ltd and reformed it as Bean Cars Ltd at which point John Harper-Bean left to join Guy Motors. Bean motorcar and commercial vehicle advertisements later proudly boasted the fact they were 'built from Hadfield's steel', but unable to compete, vehicle production ceased in 1931 in order that Hadfields could reinforce Bean's position as component makers and drop forgers; but in January 1937, Hadfields sold their interest in what was by now Bean Industries Ltd, and John Harper-Bean returned as Managing Director. After an abortive attempt in 1940 by GKN to gain control of Bean's drop forge business, John Harper-Bean left to join Guest Keen Nettlefolds as Manager of their Garrington drop forge business; he later became a director of GKN. Meanwhile, Bean Industries had passed to the Thomas Ward company, who in 1939 gained control of the Triumph Motor Company which had separated from the motorcycle business in 1936. Triumph later passed to Sir John Black who, being born and raised in Kingston, had been given his first job by Granville Bradshaw at ABC Motors. What a small world!

After service in the Great War, the then Captain J P Black, trained as a lawyer and joined Spencer Wilks in the 1920s to run the Hillman car company; both he and Spencer were William Hillman's sons-in-law. In 1929 Black joined the Standard car company, becoming joint Managing Director in 1933, and in 1940 was appointed Chairman of the Joint Aero Engine Committee in charge of shadow factories, for which he was later knighted. In 1944 he bought the Triumph car company from the Thomas Ward Group followed by Beans in 1956 which then became Standard-Triumph's component manufacturing subsidiary. When Standard-Triumph was taken over by Leyland Motors, Bean Industries passed into the Rover Group, but regaining their independence in 1988 as Bean Engineering, they bought the Reliant Motor Company from the receivers in 1991 and thereafter built the famous Reliant 850cc all-alloy engine. Reliant was after all one of their longest standing customers; they also owned Bond cars and since the early 1960s, they had used many Standard-Triumph suspension parts from the '8', Herald and TR series for their 4-wheel Rebel and Scimitar models. By 1994 Bean and Reliant Motors were once again in deep financial trouble and the group finally went into receivership in November 1994.

ABC Super Sports

With retrenchment into their old ABC works at Hersham, cycle-car production trickled on with short-time working using all available parts while the company concentrated on improving on what they had. In gallant support, sympathetic engineering businesses began to produce kits for ABC owners to overcome inherent flaws in the big end bearing assemblies with their split retaining plates, as well as improve lubrication to the exposed rockers. Among them was GT (Geoffrey Taylor) of Kingston, Jarvis Ltd, R S Inglis Ltd and BEW (Brookland Engineering Works) of Cobham, many of whom also built kits for the Sopwith built ABC motorcycle.

In July 1923, T A Dennis negotiated the acquisition of ABC's assets from Harper-Bean resulting in a new company being registered in December 1923 - ABC Motors Ltd. In January 1925 the former joint ABC-Bradshaw owned Walton Motors Ltd engineering research company was dissolved and in October 1925 the final remnants of ABC Motors (1920) Ltd were wound up. (The later Hersham & Walton Motors Ltd is totally unconnected).

By late 1924, improvements to the engine and its lubrication had transformed the engine's refinement and durability resulting in a 1,326cc (96mm x 91.5mm) cast iron engine with minor valve gear improvements and a larger crankshaft having double roller big ends. Breathing through a pair of Zenith carburettors, it now produced 40bhp (12 RAC hp rating) and was fitted to their new ABC 'Super Sports'. The original 1,200cc 'Sports' model was finally abandoned though the smaller engine remained an option and even the double roller bearing crankshaft was offered for retrospective fitting.

The Super Sports was a particularly handsome two seater Compton bodied sports car, boasting steel disc wheels and an aluminium bonnet. It had a brisk performance through ABC's separate 4-speed gearbox fitted with its unusual vertical gate-change; first was down and forward, second full back, third into neutral then up and forward into top gear. Suspension was still by the undamped quarter-elliptic springs but with the narrow, high pressure beaded-

Top, Compton bodied, artillery wheeled Super Sports. Bottom a 1925 Super Sports advertisement

edge motorcycle tyres of the day, and with only rear wheel braking, driving on wet roads was quite hazardous!

Offered at £275, the first Super Sports built was much modified for a South African enthusiast, but he died before its completion so the car was registered PD 4837 and used personally by T A Dennis for several years. Sadly, ABC's profits did not allow them the luxury of exhibiting at the Olympia Motor Show and only a further 41 Super Sports were built to order over the next two or three years during which time it is believed they even experimented with a water cooled Anzani engine and an ABC Scorpion Mk.II aero-engine in a 'Super Sports' chassis. But the ABC, and indeed other light cars such as the Rover 8, had had their day and it is reckoned only around 1,500 ABCs of all types were built at Hersham up to 1927 as the company once again turned its attention to aero-engines.

End of the cycle-car

Often regarded as the Edwardian equivalent of the Citroen 2CV, over 17,000 Rover 8s were produced before production ceased in 1925 as it struggled to sell against the outstandingly successful water cooled, 747cc 7hp cast-iron side-valve 'Austin 7'. Launched in 1922, the 'baby' Austin almost single handedly killed off the cycle-car industry. Over 300,000 Austin 7s were eventually produced until 1938. Their 7hp engine was adopted in 1937 by Tom Williams for his Reliant van, replacing the JAP V-twin. A contemporary of Bradshaw, Williams had designed Triumph's first chain-driven motorcycle before moving to Dunelt and then to Raleigh, where he designed their famous Safety-Seven three-wheeler of 1933 but with the decision to axe it, Williams bought the rights and set himself up as the Reliant Engineering Company in 1935 and when Austin ended production of the 7hp engine in late 1938, Williams bought the rights to the engine. The first Reliant engine burst into life in 1939. Twenty years later, Reliant launched their own, indestructible, all-aluminium 600cc ohv version of the Austin 7 unit which, like Bradshaw's ABC engines, proved to be highly versatile, powering everything from motor-cars to motorcycles, and portable fire-pumps to outboard motors!

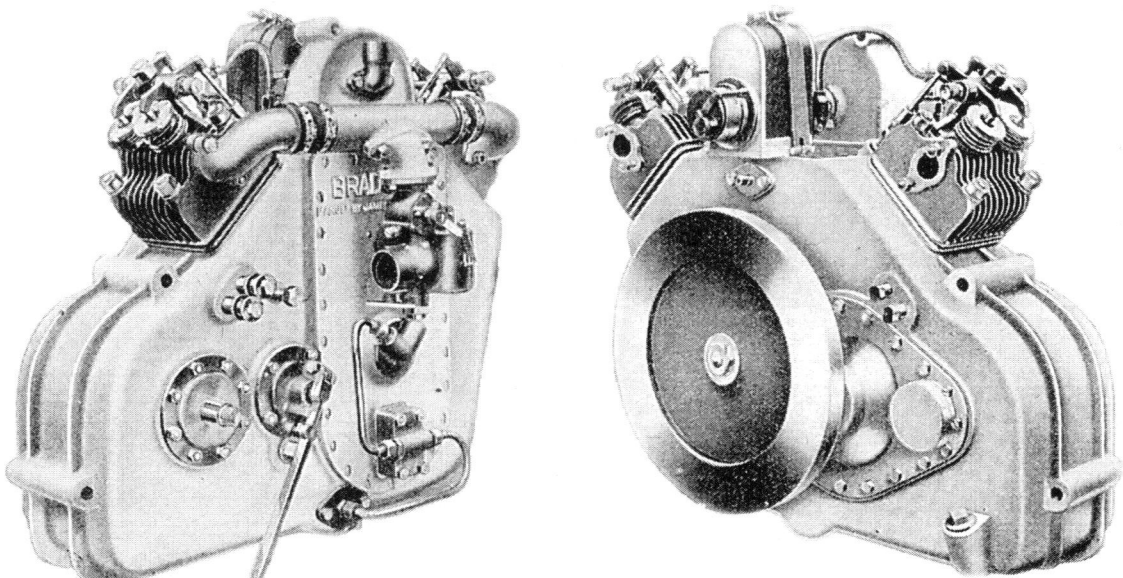

Bradshaw's 1921 1094cc oil cooled ohv V-twin

14

Belsize-Bradshaw

The Marshall bicycle company was based in Clayton, Manchester and built their first motorcar in 1896 based on a French 'Hurtu' *vis-à-vis* design; Hurtu were also bicycle makers. Initially marketed as the 'Marshall', in 1901 they adopted the 'Belsize' name and formed the Belsize Motor & Engineering Company later that year to build a wholly British car, their first being a twin cylinder 1,728cc water cooled Model 12 of 1901, which soon gained them a reputation as makers of solid, well built and reliable models. They offered 28 models between 1901 and 1915, covering the entire range from runabouts to landaulettes, taxis to light commercials, all using multi-cylinder variations of the same basic engine to reduce manufacturing and maintenance costs. By 1914, the company had grown to some 1,200 employees producing 60-70 cars a week, requiring expansion into several workshops between Cycle Street and Wilson Street, off Clayton Lane, Manchester. Parts of these buildings survive despite extensive clearance by the Clayton Aniline Co., whose dye and tannery business dominates the area.

During the First World War, Belsize were contracted to produce munitions, light trucks and an order for 1,000 ABC Dragonfly radial aero-engines; the order was cancelled in late 1918 and only 48 had been completed with some 300 or so still in progress. The post-war years saw consolidation of their model range as they concentrated on the medium car market with 4 or 6 cylinder ohv models, competing with companies such as Star and Rover, but sales of their light lorries and particularly their once popular taxis, plummeted and the company found itself in a perilous financial state.

Belsize-Bradshaw

Bradshaw's air-cooled horizontal twin engine and his proposed oil-cooled engines, together with his work on the ABC cycle-car, appealed to Belsize, so they contracted him in 1920 to design in its entirety a new 9hp motorcar, using a new 8'0" wheelbase chassis and a two or four-seat body. With the bonus of a £2/10/0d royalty on each car, Granville spent much time promoting his new design among motor dealers to get the project underway with firm orders.

By June 1921, Bradshaw had developed a new 1,094cc (90mm x 86mm) 90° oil cooled ohv V-twin engine (see left, opposite) as well as one with integral multi-plate clutch, clutch-stop and 3-speed gearbox complete with kick-start. There was also a planned short stroke (90mm x 77.5mm), 1,000cc V-twin motorcycle engine. These engines were built by James Walmsley of Preston alongside the oil cooled motorcycles engines described earlier. They were to be marketed by Gilbert Campling from his and Bradshaw's offices in Jermyn Street, but they found little success other than with the Belsize-Bradshaw

Bradshaw's Belsize-Bradshaw coupe

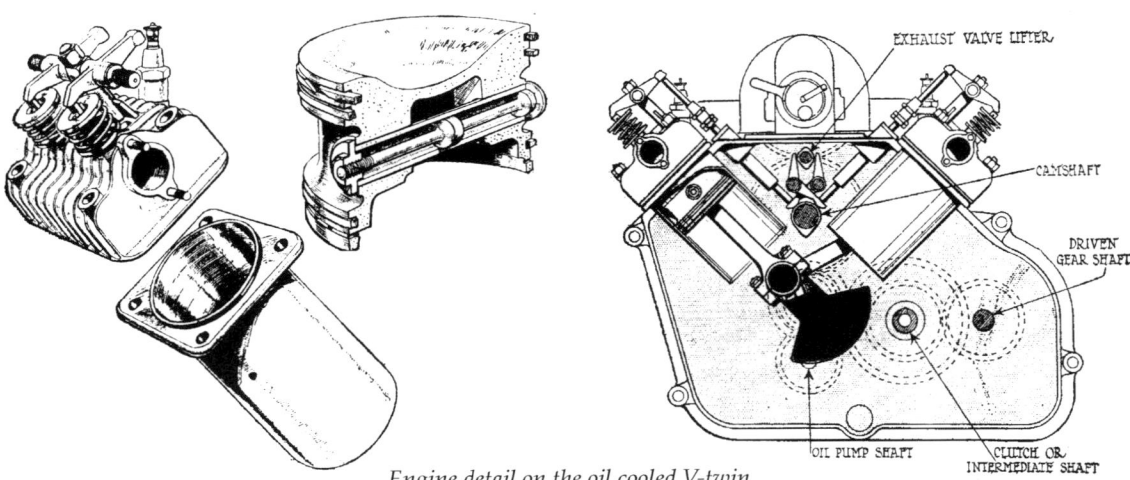
Engine detail on the oil cooled V-twin

and a prototype Lea Francis. For those living in London, eager for the new model, Belsize later opened an impressive showroom at 2-3 Duke Street, St James's SW.1 and a repair shop at 33 Cumberland Market, London, NW.1.

Fitted with 3-ring, waisted aluminium pistons, using a unique bolt type gudgeon pin, the cylinder barrels, camshaft and tappets were enclosed within the crankcase, cooled by oil spray pumped from the $1\frac{1}{2}$ gallon (later 2 gallon) sump, circulating at half a gallon per minute. The smoothly finished unit construction crankcase sump also supplied the worm and nut steering box, bolted to the crankcase. The integral multi-plate clutch and 3-speed gearbox, was some 30 years ahead of Alex Issigonis' 'revolutionary' unit construction Mini engine/transmission. The exposed ohv gear used duplex coil springs for safety. A single Claudel Hobson (later Zenith) carburettor and Fellows magneto (later coil) ignition was fitted.

In August 1921, Bradshaw lodged a patent for an improved gear change mechanism (Pat No. 188087), fitted between the engine's heavy external flywheel and conventional clutch and gearbox. This was in effect a 'clutch stop' which comprised a friction disc (item No. 327 opposite, top) on the end of the drive shaft (308) which engaged a cone on the clutch housing (304), matching the main shaft and layshaft speeds and, thereby, never allowed the clutch to spin slower than the propshaft, facilitating a smooth, crash free gear change without the need to double de-clutch through any of the gears, 'which makes it practically impossible for a bad gear change to be made, even by a novice'. The design also allowed for the further movement of the main shaft (308) to engage a transmission brake (321). Having sold the design to Belsize, Bradshaw allowed the patent to falter through non-payment of the sealing fee - a practice which he followed more often in later years.

Belsize-Bradshaw ohv V-twin

With this clever clutch design, a heavy flywheel and a responsive engine from a slow and reliable tick over, it allowed for a smooth take off and superb pulling power from very low speeds right through to 60mph. A central bearing supported the leather disc universal joint and the long torque tube drive shaft to the bevel final drive.

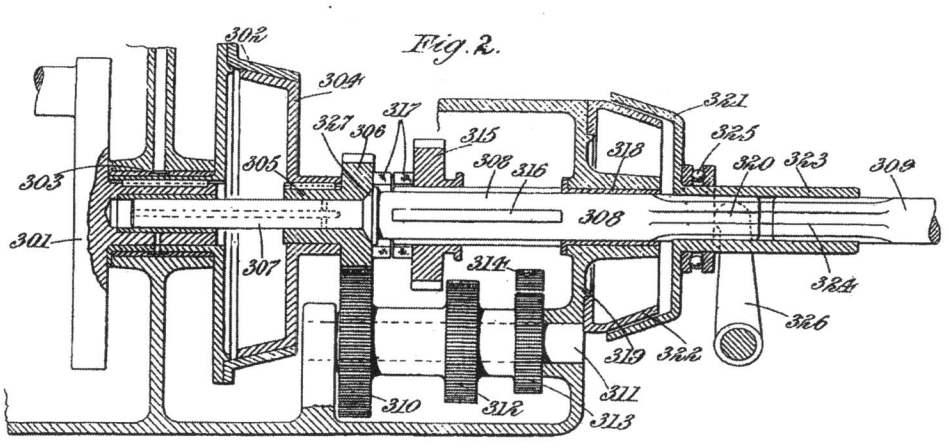

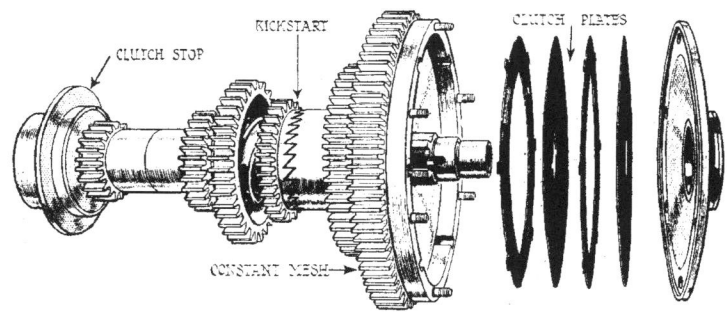

Top, Bradshaw's patented clutch-stop (Pat No 188087)
Opposite, Integral oil-cooled gear, clutch and clutch-stop
Bottom, cross section of later Belsize-Bradshaw 1370cc side-valve V-twin

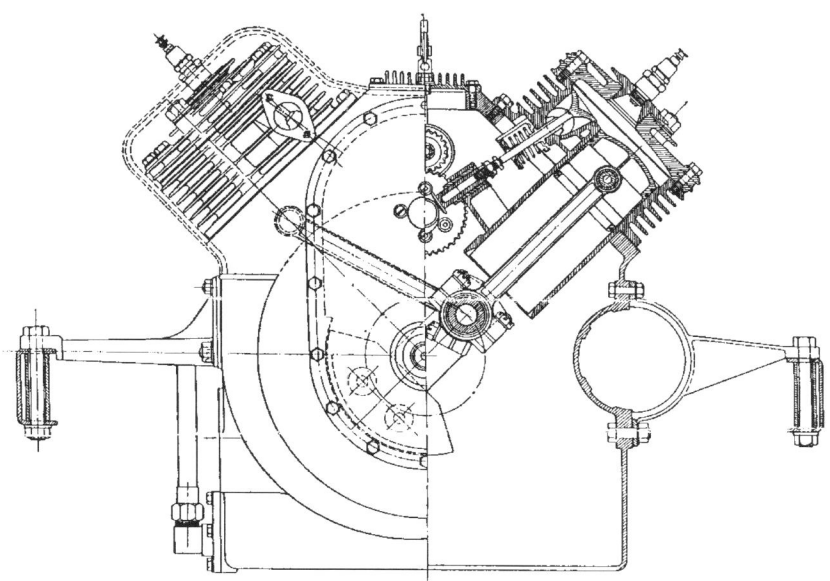

The rear axle design was similar to the ABC cycle-car with conical steel half-shaft housings. Rear brakes also followed the ABC design with external contracting hand-brake and internal expanding foot brake.

With the compact unit construction engine/gearbox set well to the front, all routine maintenance was possible through a large bonnet. Furthermore, as the engine had forward facing exhausts, it helped keep heat away from the passenger compartment.

The car was built in its entirety in the main assembly hall at the Clayton works, hidden behind security screens. Bradshaw regularly called in to check on progress.

The Autocar were very impressed by this quiet and smoothly running 1,094cc engine, whose unit construction 3-speed gearbox, bevel final drive and oil cooling made the Belsize-Bradshaw a much quieter car than many of its competitors. However, the 1,094cc ohv engine was replaced for the Olympia Show of October 1921 by a new 1,289cc (85mm x 114mm) V-twin side valve unit which attracted a £9 annual road tax.

Motor magazine heralded the Belsize-Bradshaw as 'a show sensation'. Road test reports in *The Autocar* of the 1,289cc model gave praise to the smooth and relatively silent engines as well as the excellent suspension from the four quarter elliptic, cantilevered leaf springs, ABC fashion, from the wide, C-section rolled steel open framed chassis. The track of the front wheels was 8" wider than the rear which, together with self-centering castor action on the front hubs, helped control both directional stability and roll.

By 1922 the side valve engine had been enlarged to 1,370cc (85mm x 121mm) and was fitted with vanes cast into the flywheel which fanned cooling air, via ducts, to the stubby cylinder heads. This new, stumpy engine was of a unique shape, well hidden from view under a normal bonnet and a larger curved brass dummy radiator.

Initially offered as a 2-seat coupe at £280, the simple open fabric covered body, finished in blue, had a wide foot-well for a three-abreast, blue leather seat. There was also a slightly narrower, door less, 2-seater Sporting model at £235, with access to an enclosed boot by a hinged seat back - for an extra £5 it could be 'tuned' using lighter aluminium pistons.

'Oil boiler'

Like most engines of its day, Bradshaw's oil-cooled engine tended to leak oil (which was eventually cured). It also gained the unfairly mischievous nick-name of 'oil boiler' yet, like his oil-cooled motorcycle engines, they never did - despite road tests up the nearby Snake Pass in the Derbyshire Peaks. It was however a devil of job to start by hand and those many who bought the car 'on the cheap', avoiding the extra cost of the optional electric start (£15), soon regretted their decision and learnt to park on a downward slope to avoid any embarrassment from having to swing the engine!

The industrial archeologist and chronicler, L T C Rolt, remembers well his father's Belsize-Bradshaw, describing it as "an opportunist and unsuccessful bid by the Belsize Company to conquer the British light car market with what was really a refined cycle-car... the oil cooled engine was a beautiful piece of engineering and when running was as smooth and silent as any four cylinder engine of the day." E S Chapman, an apprentice at Belsize between 1919-1924 (later Works Manager at Dagenite batteries), recalled how as apprentices they often looked out for cars broken down at the roadside, offering a speedy repair. Owners were often left bemused on seeing the engine being quickly lifted out of the chassis for better access.

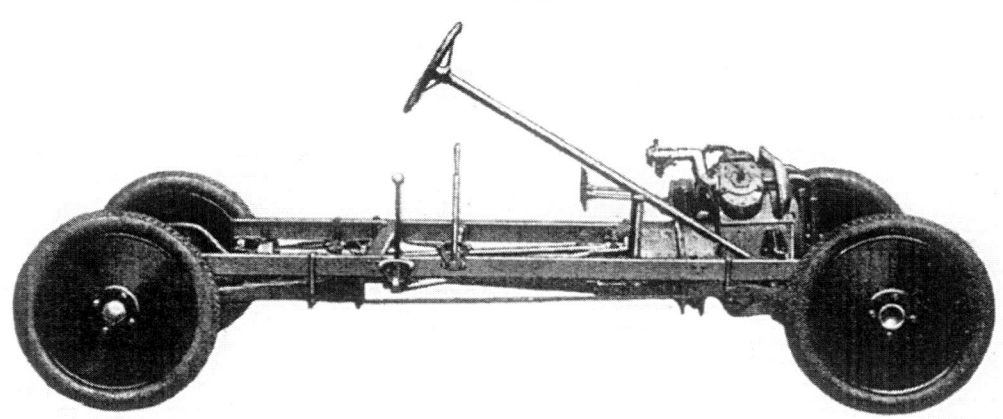
The Belsize-Bradshaw chassis showing the compact side-valve engine neatly stowed behind the front axle

Although Bradshaw had optimistically projected production of 50 cars a week, tragically, just as work was getting under way, industrial strife arose following the devastating three month national strike in late 1919 by foundry moulders, severely affecting the supply of castings to the entire motor, motorcycle, engineering and electrical industries and led to a series of almost suicidal national labour disputes and random transport strikes from 1920-22. This caused the Federation of Motor Manufacturers to order a series of wide spread lock-outs of their engineering and motor-company members, the longest of which was 11 weeks, which had a devastating effect across the whole British engineering industry.

Sales were also affected by the very unfair perception of unreliability. As Belsize were eager to recover from the losses caused by the recent industrial strife and worsening economic climate, they were forced to drop prices further. By October 1922 the 2-seat tourer was offered at £235 and then, in 1923, at £210; the enclosed coupe with dickey seat also dropped to £210; the four seater saloon cost £260. But with no signs of recovery, Belsize was forced to consider replacing the oil-cooled engine with a conventional 4-cylinder, 12hp motor, but by now the company was facing insolvency following a petition in April 1923 by one of their creditors, S R O Ball Bearing Co. Ltd., to wind them up.

It is evident that Belsize's finances had been a shambles for quite some time and drastic steps were taken in the months after the petition to restructure the company's finances and share capital (in which Bradshaw had invested £500 in £1 shares). As a result, their capital was reduced from £300,000 to £102,000, but this was insufficient to save them and a receiver was appointed in 1924 at which point the oil-cooled Belsize-Bradshaw was immediately dropped. Within a year, production of all their other models ceased; the company was declared insolvent and put into liquidation in April 1925.

Serial numbers are not certain indicators of production levels, but chassis numbers suggest in the region of 2,000 cars may have been built. Thankfully for these Belsize-Bradshaw owners, a supply of spares was secured by the Belsize-Bradshaw Car Club, hastily established in 1923. Several engines were also apparently sold for industrial use, but little is known of their application.

There is remarkably little on the Belsize-Bradshaw in Granville's archives and one wonders how much he was enamoured with the project. Neither is there any reference to a claim that he designed an ambitious 2-ltr, six-cylinder oil-cooled engine for Belsize, nor that he may have been involved in the development of Bert Houlding's Moveo Car & Engineering Company's six-cylinder, luxury Grand Tourismo coupé chassis.

Lea-Francis

It appears the oil-cooled engine featured in only one other car, a 1922, 7hp Lea-Francis cycle-car prototype of which only one was made. The company already had a somewhat chequered history since 1904 as makers of bicycles, and a light car which R H Lea (who had earlier worked with George Singer) almost immediately granted under licence to Singer. From 1911, Lea-Francis made motor-cycles and from 1920, motorcars..

Their renewed attempt at motorcars yielded very few sales but in alliance with Vulcan Motor & Engineering and the British Motor Trading Company they began work on a range of cycle-cars and light cars for 1922 which included an 8.9hp 1,074cc sv Coventry-Simplex powered model (rapidly replaced by 10hp ohv Meadows which remained in production until 1925) and a cycle-car with a 7hp, 690cc (76mm x 76mm) Bradshaw oil cooled horizontal twin. Shown at the November 1922 Olympia show, it featured a 4-speed direct coupled gearbox, quarter-elliptic springs and a sporty 2-seater body. Offered at £190, it failed to sell and it appears that, like Alan Lea's first cycle-car of 1919 with a Coventry-Victor twin-cylinder air-cooled engine, the Bradshaw powered prototype was despatched to Vulcans and never heard of again. The 690cc engine was almost certainly developed from that used in the Zenith motorcycle, but no other reference to it has been found. They have undergone five major revivals, not all successful, but they survive today.

Air, oil or water cooling?

Bradshaw's interest in oil cooling, however, persisted and in late 1941, *The Autocar* published several letters from correspondents on the idea of using oil in a car's radiator rather than water. Bradshaw was himself by now toying with the idea following an experiment in which he drained the water from both his Rolls Royce 40-50 and Ford V-8; he discovered the engines ran quieter without a water coolant and reasoned that the denser glycol, used as an anti-freeze additive for high altitude fighter aircraft, would transmit less noise than water in the cooling system. He duly experimented with four new glycol and oil formulae in the cooling system but all to no avail.

Of interest, in April 1946, *The Autocar* carried the story of large oil-station pumping engines in the Arabian desert using oil rather than water as a coolant. They also reported a new engine designed by a French engineer, M. Viloet, using light oil as a coolant, circulated on the thermo-cycle principle via large finned alloy radiators, but little else was heard of the engine.

15
Inventor Extraordinaire

After the many successes at ABC, Bradshaw broadened his horizons in the late 1920s and 1930s, producing some of his most interesting, if weirdest, ideas. Many of these came from opportunist money making schemes which brought him great wealth, followed by bankruptcy.

Granville certainly enjoyed his millionaire's lifestyle; his Rolls Royce Phantom, the magnificent Lowfield Park manor house, a town house and well manned offices in Central London from where he conducted his business affairs. His decaying, hand written notes suggest a playboy's life at The Cosmo Club in London, of Paris and Dusseldorf, of financial deals going wrong and of 'the need of a good solicitor who won't let one down'. His incomplete memoirs even recall Percy, the giant pig, of which no further mention is made! He was a keen follower of cricket, football, tennis, snooker, motor and motorcycle racing but, although he regularly did the football pools and laid bets on the horses, he rarely had a significant win. There is a strong suggestion he had invested in some racing stables.

His versatile and flexible mind extended to writing two novels, one of which was *The mystery of the old Manor clock*. He also prepared a book to assist other patentees as well writing his autobiography which he had almost completed by June 1960, but despite help in February 1959 from a literary agent, Commander Wilfred Granville and submissions to Temple Press (who published *The Autocar*, *MotorCycling* and *Flight*), he remained unpublished.

Company promoter

The exact chronology of events in these inter-war years is unclear and what little is known is found mainly in letters to his solicitor in 1960 when, at the age of 74, it is clear he was in ill health and his memory was fading, causing him to often exaggerate his successes somewhat by 'gilding the lily'.

On being ousted from ABC in 1920, it appears that Bradshaw was offered a design post with Armstrong-Siddeley – whether this was for their radial engine is not known - but he clearly rejected it. Bradshaw worked from his Darby House home in Sunbury-on-Thames as a consultant engineer, primarily to the ABC and Belsize motor car companies but by 1922, he had established a London office at 90 Jermyn Street, off Piccadilly, as Granville Bradshaw Ltd (automobile engineer) from where he worked closely with Gilbert Campling. After this, Bradshaw became more of a 'Company promoter', making his money by selling his ideas, often only on a provisional patent, or forming companies to exploit their potential. Many patent applications in these post-war years simply faded away due to the increasing time it was taking the Patent Office to fully assess and grant patent rights, by which time technology, or his interest, had moved on.

His earliest such patent was for a smoker's ash tray (Pat No: 167923 laid on 18th June 1920). This simple device comprised deep compartments in various decorative patterns so that when cigars or cigarettes were placed on it, they would safely burn away allowing the ash to drop into a tray, hidden from view.

The next (Pat No: 171810 laid on 6th September 1920) was for a 'wardrobe', jointly patented with a Robert Rood. It was ludicrously simple! The 'wardrobe' comprised a pair of brackets and a rack on which were hooks and sliding hangers, hidden by a curtain. The patent allowed for this to be made 'portable' by having it hang off a door or picture rail. Today, such simple ideas would not warrant nor likely be granted a patent, though many such 'innovations' are now regularly offered as new and exciting ideas in magazine catalogues which appeal to a gullible public.

Advertising machine

One device however had a serious commercial purpose. Pat No. 260641, laid on 6th May 1925, was for a combined games machine and advertising device. On the premise that one remembers an advertisement or product by association with an event, he designed a table top 'skittle alley' in which projectiles were aimed at targets which, on being hit, revealed an advertising slogan mounted on a rotating drum. Striking other targets revealed the next advertisement and so on. By incorporating a counter to record the number of successful strikes, prizes associated with each advertisement could be released at predetermined intervals, much to the delight of the player. It is not known for whom this was developed, but the British Automatic Company (who had patented several coin-feed machines), appears a likely candidate.

'Little Stockbroker'

One of his more interesting ideas was developed while at Jermyn Street and came about through a 'ten bob bet' in 1926 from an unnamed social acquaintance at the RAC Club. It appears this may have been a Mr White, an engineer who ran amusement arcades across Britain and who imported 'fruit machines' from America, where gambling was frowned upon. At the time, this gentleman was embroiled in a lawsuit and, having been fined a considerable sum, intended challenging the case in the High Court and, if necessary, the House of Lords. The best legal brains were of no use to him for he needed someone to overcome the technicalities in the wording of the law and having heard of Bradshaw's reputation as a genius with a uniquely fertile mind, considered him to be his likely saviour.

Bradshaw was presented with a tricky problem relating to non-random 'fruit machines'. These could be legally operated with a guaranteed payout at certain times, but which required the 'skill, observation and judgment of the player', pitted against others, as if in a card game, yet which would generate an income to the amusement park. English Law prohibited gambling in public including racing bets - though of course everyone did it - so it was nationalised as The Tote! Any device which operated or stopped in a purely random fashion was considered illegal, for the game had to involve 'skill and judgment' and be based on calculable probabilities. How then to make them legal?

Ever keen for a challenge, Bradshaw studied the law and solved the problem by gearing the mechanism so that it paid out in a set pattern but retained a small percentage for the promoter. The mathematical permutations required were complex and, with a slide rule always to hand, ratios of gearing were worked out at great speed. The gear ratio of each machine was different, making it impossible for a player to know a machine's pattern before it paid out requiring the punter to observe and then play the machine at just the right time to secure a win; of course others were also keen to play at just the right time too and deprive the hopeful punter of his winnings.

An application was laid on 24th January 1927 and granted as Pat No: 288744 on 19th April 1928. Dubbed the 'Little Stockbroker', the 'game' relied on a predetermined gearing of several discs in repeated sequence of cycles and was modelled on a popular Stock Exchange card-game. The player would know that a 'dividend' was due, but would not know at what precise point in the machine's cycle, nor its value; if he played too soon after a previous winning player, when the 'market' was rising, he would get only a small dividend insufficient to recover his input, but if he left it too late, he might loose to another player.

By now living at 5, Beauchamp Place, Brompton Road, London SW3, he patented a variation to the above (Pat No: 306858 laid 1st November 1927) which issued irregular sums of money at irregular intervals, but which ensured the machine would not exceed its float or reserve of monies. Once again, observation by a keen player could determine when a payout was likely.

His interest in gaming machines resurfaced at the end of the war when he set about a new design and took an initial order for 50 to an engineering company, but he then suddenly realised halfway through its manufacture, that a war-time Government ban of late 1939 prohibiting the manufacture of any form of amusement machine, was still in force as part of the war time rationing of steel. The project was promptly cancelled for fear of a conviction and imprisonment. By the time the law was repealed, he had moved onto new projects.

The abondoned post-war gambling machine under construction (GB)

Amusement machines

The 'Little Stockbroker' prompted Bradshaw, Herbert Taylor White and Ben Dawson to form a public limited company in 1928, Coin Operated Machines Ltd, based at Riverplate House, Finsbury Circus, EC.2, to exploit their ideas and Bradshaw's patents, particularly his "semi-automatic photographic studio", described later. A major investor was Stephen Walter (of *The Daily Telegraph* family); they all made a small fortune.

He developed more amusement machines starting with a target game (Pat No. 317136 laid 15th May 1928) in which a projectile is spun in a curved path at a target, which would change at a predetermined sequence in degrees of difficulty or value; those that succeeded in being struck would generate a prize of variable value. In November 1928, Coin Operated Machines Ltd was absorbed into the Associated Automatic Machine Corporation, based at Trafalgar House, 11-13, Charing Cross Road, SW.1, from where Bradshaw devised a pinball machine (Pat No. 325353 laid 14th January 1929) requiring two coins be inserted. Depending on how successful the player was in pocketing a ball along a heart shaped, two track course, the coins were either returned singly, in full or not at all.

Pat No: 328749 (laid on 23rd March 1929) was for a coin operated 'Noughts and Crosses' game featuring a matrix of illuminated windows each with two bulbs indicating a '0' or 'X' as each player took their turn; solenoids prevented the opposite player from selecting the same window.

Incredibly one idea for a death defying scenic railway was patented in Britain (Pat No. 329789 laid on 22nd April 1929), France, Germany and America! This featured a gravity railway using wheeled sleds which descended a spiral course around a mock mountain. The wheels appeared to run along a railway track, but hidden within the mountain was an inner track along which the extended axle, carrying the sled, actually ran. At various places along the outer, visible track, sections of rail were missing as if removed by landslide. As the sled hurtled down the track, it appeared to jump the gap! Whether this was intended as a child's toy or a life-size death defying helta-skelta at a fair ground, is not known.

From Charing Cross, Bradshaw moved to 11D, Canton House, Regent Street where he devised a shove ha'penny board employing the theoretical principles of a hovercraft, Pat No: 345960 (laid 27th November 1929). It had a polished cast iron deck marked by transverse score lines. A cast iron puck had a shallow scalloped underside. As it was shoved from the start line, it passed over a slight lip allowing the puck to trap a cushion of air under it and thus glide, fiction free, along the polished surface until the air had been expelled - which it promptly was!

His final patent for an amusement machine, (Pat No: 380896 laid 29th July 1931) came while living at 59, Park Street, Camden Town. This was for a bowling game in which the ball was rolled at a target in order to knock it down. However, the target bobbed up and down on a sprung column operated by a rotating cam and only at certain points in its movement, could the target be successfully hit and knocked down.

His extensive wealth from these amusement machines accumulated rapidly and allowed him to form the Eastwood Automatic Company, the General Amusement Company and a further subsidiary, Auto Rides Ltd which secured the UK rights to the American dodgem cars. Originating in America in the early 1920s, these 'dodgems' were contemporary motor car designs but of a 'midget' (reduced) scale. Seating one or two people, and powered by a small petrol engine or batteries, they were surrounded by a steel car-bumper much like today's dodgems, to be driven around circuits of various shapes; these soon became immensely popular at sea-side resorts and fairgrounds. The famous Rytecraft single seat car of the 1930s was originally built as a 'midget motor'.

Formed in March 1929, Auto Rides Ltd acquired several small businesses involved in public amusements, one in particular was Laurence Delaney's Midget Motor Ride Ltd which held entertainment rights at the Eastern Parade Bandstand, Southend-on-Sea. (Delaney had earlier secured a patent (Pat No. 259868 of 1926) for a novelty bowling game machine). Bradshaw claims in his letters that he supplied the dodgem cars to Billy Butlin's holiday camps; while this may be so, it was more likely to have been through Delaney's Midget Motor Ride Ltd, for this is at variance with Butlin's account. It appears that Billy Butlin first saw them in Toronto where he was then living. Although he was British and had been born in South Africa, he served in the Canadian army and moved to Britain in 1921 where he spent the next five years in travelling fairs before setting up his own permanent site in Skegness in 1927 where, 'after many months of negotiations with the manufacturers', he installed the first dodgem cars in 1928, followed by his other sites. His famous holiday camp opened in October 1935. Auto Rides' directors were Granville Bradshaw, Laurence Thomas Delaney, John Raymond Gray and William Ouchterlony Hunter. In his capacity as Auto Rides and the General Amusement Company, Granville then bought the Teignmouth Pier for £9,000 and, after improving its profitability, sold it two years later for £25,000. Midget Motor Ride Ltd had been dissolved by 1935.

Heart monitor

The excitement generated by these amusement machines may have brought about his patented heart monitoring device, (Pat No: 322503 laid 9th July 1928). In observing how a spring balance, or scale, fluctuated with the slightest vibration of a body, he devised a very sensitive set of scales on which a patient stood and to which was added a finely balanced spring loaded bar which fluctuated in sympathy with any movement of the body. By pressing one end against the patient's beating heart, the bar magnified the minute pulses by exciting a liquid filled U-tube manometer which, by simple ratio, indicated the relative blood pressure. By assessing from the patient's weight, heart beat and relative pressure, the doctor (or gambler!) could determine the patient's health. Though it would be unlikely to work in practice, its merits were sufficient to persuade the famous W S Avery spring balance company to buy the patent, but no more was heard of it. Perhaps it was a move by Avery to thwart their competitors?

Photography

It is ironic that Bradshaw's wide repertoire of inventions opened him to such ridicule while other pioneer engineers, such as John Alfred Prestwich and Prof. Low, had also demonstrated as wide a variety without ridicule. Prestwich, for example, trained as an electrical engineer at S Z de Ferranti and also worked with his father at W H Prestwich Ltd making and patenting photographic devices, projectors and tripods long before his famous JAP engines; one of these included a joint patent for a motion picture projector with William Friese-Green, the British pioneer of colour cine-photography.

Automatic coin operated photo-booths which took and developed self portraits, date back to the 1880s in America but they were far from reliable and consequently required an attendant to be on hand, but in 1925 Anatol Josepho, a young Siberian photographer, who emigrated to America in 1912, devised an automatic photo-booth which took and developed 8 frames in as many minutes. Rasing $11,000 from American sponsors for its development, his 'Photomaton' gained for him overnight success when in 1927 a consortium of businessmen paid $1 million for the patent rights and formed Photomaton Inc. The UK rights were obtained in 1928 by Clarence Hatry who established the Photomaton Parent Corporation Ltd at 14, Regent Street, London and installed coin operated photo-booths on many railway station platforms. Hatry had a passion for things mechanical and the Photomaton business was certainly his favourite. In early 1929 Ben Dawson co-patented with Photomaton a prism system to produce a reversed optical image from the lens onto the sensitised photographic reversal process paper used in these automatic negative-free photo-booths.

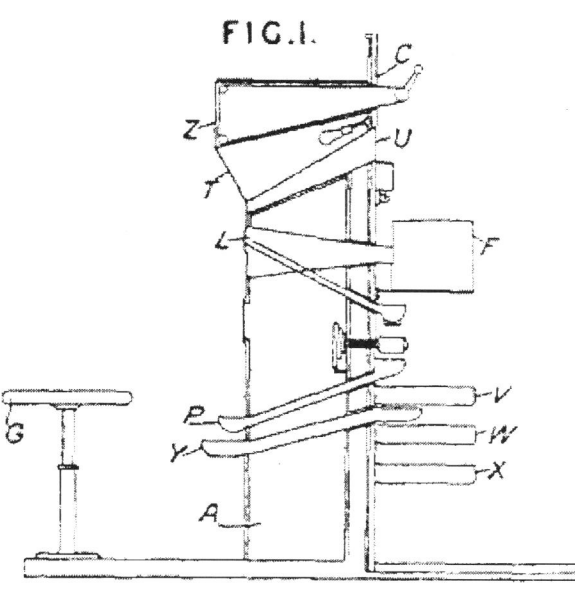

Bradshaw's photobooth! The photographer worked under cover on the right.
Z - instructions; T - photolight; U - photographer's viewer; L - coin shoot; P - coin return; F - camera; V,W,X - processing trays; Y - photo delivery shoot

Now, have you even wondered why photo-booths are so large, or wondered at the miracle of a 'freshly brewed' cup of tea from a vending machine?! Well, one of Granville's most amusing inventions had a very serious side to it and it appears may have been connected with Hatry's interests. Many people found, just as today, that having their photograph taken was a most embarrassing moment, yet Bradshaw's solution was so, embarrassingly, simple!

Pat No: 296511 was laid on 11th May 1927 and refers to a coin operated photo-booth, inclusive of a revolving, height adjustable, seat, just as one would find today in a railway foyer. The client entered, adjusted the seat, checked themselves in the mirror, inserted a coin and followed the instructions displayed, mechanically, as the process evolved.

Illuminated by lamps, a series of photographs were then taken, immediately processed and then presented to the client through a slot moments later. The really clever bit, to which the patent relates, is the use of a real live photographer hidden behind the facade who takes the money, rejects dud coins, operates the instruction flags and finally takes and processes the photographs! The patent even covers the accompaniment of authentic mechanical noises to add to the illusion.

Bankruptcy

Bradshaw claimed he was grossing £250,000 per annum from his company promotions and employed a staff of 20-30 people in the late 1920s/early 1930s, but all was soon to go badly wrong for him.

Having recovered from the General Strike of May 1926, Britain's economy teetered slowly back onto its feet, only to be knocked back down into a long depression following the first Wall Street crash of 1927 and its knock on effect from the second major crash, which badly affected the London Stock Exchange in 1929.

These events were coincidental to Bradshaw's huge personal financial loss through the Hatry fraud case. Bradshaw's penny slot machine patents and interests had attracted the attention of Clarence Hatry, the City's 'God of finance', who controlled a large holding group which included, among others, the Photomaton Parent Corporation.

From humble beginnings, Hatry had made his fortune buying and selling small companies, 'improving their position' then selling them on for a quick profit. With his accumulated wealth, he expanded into insurance and business banking and became ever more ambitious in his plans. In 1920, he persuaded the many small scale Scottish jute farmers to merge and, thereby secure and strengthen their industry. He applied the same skills to other industries which lead to the successful creation of Allied Iron-founders Ltd in 1928 - today a very famous business. His secret lay in offering to buy all the shares in one business, to be paid for in instalments; these shares, though not yet paid for, were then used as collateral to buy the shares of the other businesses and so on. By working quickly and skillfully, he was able to sell the newly merged companies as a single concern at a sensible profit long before final payments on the outstanding shares were due. His skill earned him the reputation as the 'Golden Boy' of the City of London and investors and bankers alike thrived off his investments. But he also made many enemies who later sought their gleeful revenge. Central to Hatry's operations were an Italian financier, John Gialdini, and the Corporation & General Securities Ltd (with an unwitting Lt Col Sir W E G Archibald Weigall as a director) set up in 1926 to carry out Hatry's corporation loans and share dealings. As part of Hatry's plans to amalgamate other businesses he formed the Associated Automatic Machine Corporation on 1st November 1928, at Trafalgar House, to acquire Bradshaw's Coin Operated Machine Company and Walter Dailey Morgan's Visible Writing Machine Company. Associated Automatic's other directors included Herbert White, Ben Dawson, Gialdini and Weigall.

Hatry proposed that Bradshaw purchase a rival company, Sir George Touche's British Automatics Ltd, which had secured many valuable railway foyer sites, and merge it with their Associated Automatic Machine Corporation. This Bradshaw obligingly did on 18th January 1929, using his own money and, naively it now seems, handed over all title to this and his other companies to Hatry to form the new company in exchange for shares valued at £1 million; the assets included £500,000 million in Gilt Edge securities held in Hatry's bank. Hatry himself took 300,000 Associated ordinary shares in exchange for preference shares, for financing the deal. However, as Bradshaw could not trade Associated's shares on the stock exchange for six months, they were duly deposited at his Piccadilly bank.

Meanwhile, Hatry was using these, and other shares, as collateral in a major venture to merge several small heavy iron industries, in a move similar to his successful earlier formation of Allied

Iron-founders Ltd. In the spring of 1929, he duly bought, in cash, the entire shares in the United Steel Companies Ltd, a consortium which was already in financial problems. Then, at a critical stage and in urgent need to quickly raise £8 million, the first effects of the Wall Street crash took their toll on investor confidence in both Britain and America, as many major British companies were at that time American owned. Following a general election in May 1929, and the installation of Ramsey McDonald's new Labour government, investor confidence in Britain collapsed and, with it, support from Hatry's favoured backers. In desperation, he called upon the Bank of England for a bridging loan, but in this now very uncertain economic climate, they became suspicious of Hatry's dealings and refused.

With the deadline to pay for the shares looming and still lacking funds, Hatry and his fellow founding directors John Gialdini, Edmund Daniels and the unwitting Company Secretary to Corporation & General Securities, John Dixon opted for a foolhardy ruse by printing and issuing more temporary stock certificates in his businesses than existed and, by carefully moving simultaneously several bank accounts and begging new investors not to register these temporary stock certificates until the transactions were complete, 'to avoid double stamp duty', he and his partners would in theory be able to raise the funds and, with a quick sale of United Steel, use the profit to repay the payments against the original temporary stock certificates.

Unfortunately, it did not go to plan as one major investor got cold feet and registered his non-existent stock certificates. On October 21st, 1929 traders at the Stock Exchange immediately suspended share dealings in all of Hatry's businesses, leading to a major crash as shareholders discovered their investments were worthless; the Hatry group collapsed with debts of £14m, the highest then recorded. This led to wide scale bankruptcies, made worse by the Bank of England having to raise interest rates which, it was said, helped trigger events leading to the Wall Street crash of late 1929.

At the time, Bradshaw naively blamed the Wall Street crash for Hatry's failure, but then reality dawned that he had fallen victim to a major fraud. Hatry, ever the gentleman that he was, immediately owned up to his utter stupidity, but rather than being correctly tried under civil law for fraud, the cloud of vengeful anger which now existed, called for him to be tried under the far graver criminal offence of forgery, for which he was duly sent down in 1931 for 14 years hard labour. Daniels received seven years, Dixon five and after much political pressure the chief fraudster, Gialdini, was prosecuted under Italian law but on a lesser charge of 'falsity of a private document', sentenced to 26 months and was out after twelve! Hatry was now being blamed for everything under the sun and it was only years later that his gross crime came to be seen more as a stupid gamble which didn't pay off.

The knock on effect was a major loss of income for Bradshaw from shares and royalties from such businesses as The British Photomaton Trading Company, which marketed the photo booths, and which went into voluntary liquidation in April 1932. With confidence gone, many smaller businesses associated with Hatry also collapsed. Many others entered into voluntary liquidation to salvage what they could.

Granville Bradshaw lost a 'small fortune', later estimated by Counsel at £160,000, from which he never fully recovered. In a desperate bid to stay solvent, his son Peter was withdrawn from preparatory school and his hopes of entry to Harrow School dashed. Gradually the furniture at his 16th century Lowfield Park manor house was auctioned off and by 1933, Granville was forced to sell Lowfield to pay his creditors. In the meantime, he had secured a concession for end-of-pier amusement machines and having Peter at home, put him to good use servicing them, often late into the night. But, even selling Lowfield was insufficient to solve his dilemma and with liquid assets of only 19/6d against debts of £19,353, the former millionaire was finally declared bankrupt on 20th July 1936 by the Official Receiver who even, audaciously, tried to take his wristwatch! At the time he was still renting 16, Sloane Avenue, Kensington, and relying heavily on the generosity of his brother, Ewart, to tied him over. He was released from bankruptcy two years later, on 23rd March 1938.

War against crime

The only war Granville fought was against the taxman and thieves! There is a subtle difference. One day in the 1930s, when living at Sunbury Court, following a chat with a railway official at East Croydon station, he pondered the fate of the missing light fittings and screws in his railway carriage and came up with the idea of a thief-proof wood screw. He promptly made a working example before forgetting all about it until 1937, when living at 12, Wigmore Street, London W.1.

He now proposed many variations of a one-way, cam-action or serrated tooth cone clutch, self-tapping drive insert, by which to drive home a wood or machine-screw, but whose design also made it impossible to remove without a special device. Some 13 patent applications were submitted to British, German and Australian patent authorities between September 1937 and March 1938.

The first (Pat No. 501548, laid 3rd September 1937) was for a countersunk screw in which the head had a softened, shallow depression. To drive the screw into the wood, a screwdriver with a self-tapping quick-threaded spigot end was inserted into the depression and screwed home; by reversing the screwdriver, it withdrew the self-tapping spigot leaving the woodscrew in situ with no means of conventional extraction. This was followed by a design (Pat No: 502348 laid 13th September 1937), for an Archimedes spiral screwdriver, of the Stanley 'Yankee' pattern, using special spigots for the above driver inserts. By incorporating a sprung ratchet which allowed for pump action, up to 15 screws a minute could be inserted, but if the handle were suddenly pulled back, it locked the ratchet, allowing withdrawal of the insert in the usual anti clockwise manner. The same patent also included a left-hand threaded extractor insert which cut into the tapped centre during 'normal', anti clockwise, unscrewing action, enabling the screw to be removed.

From a manufacturer's viewpoint, two major advantages of these screws were that by making the screw and screwdriver as one, a far greater positive driving force was possible eliminating the need

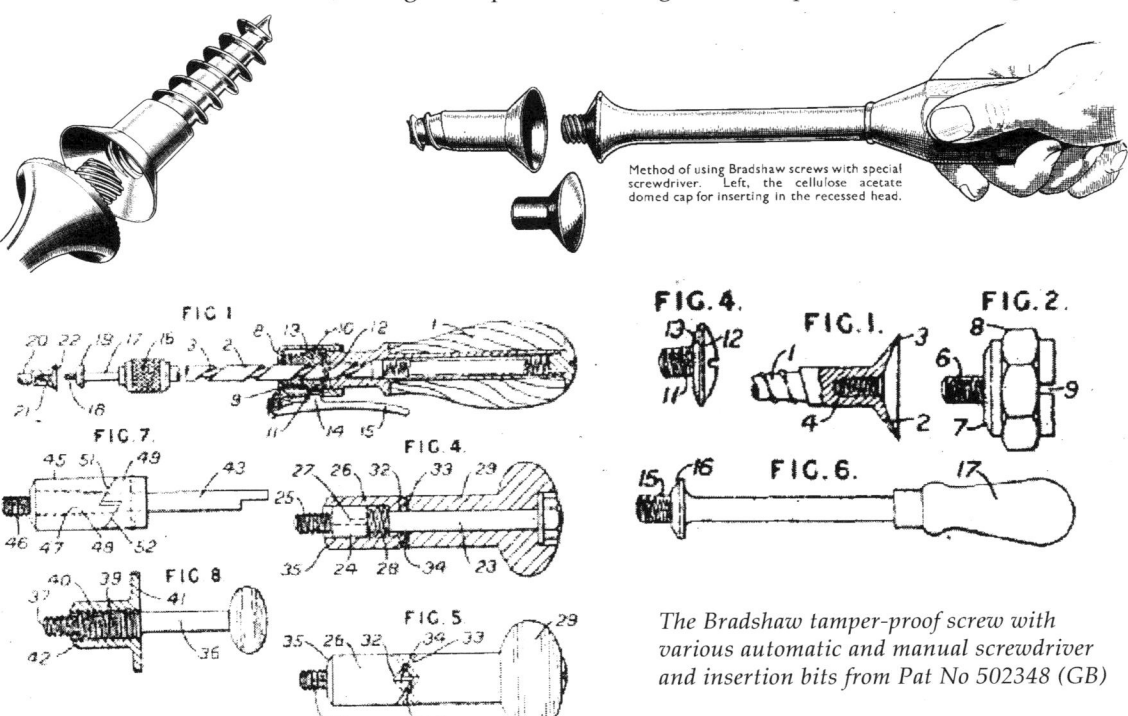

Method of using Bradshaw screws with special screwdriver. Left, the cellulose acetate domed cap for inserting in the recessed head.

The Bradshaw tamper-proof screw with various automatic and manual screwdriver and insertion bits from Pat No 502348 (GB)

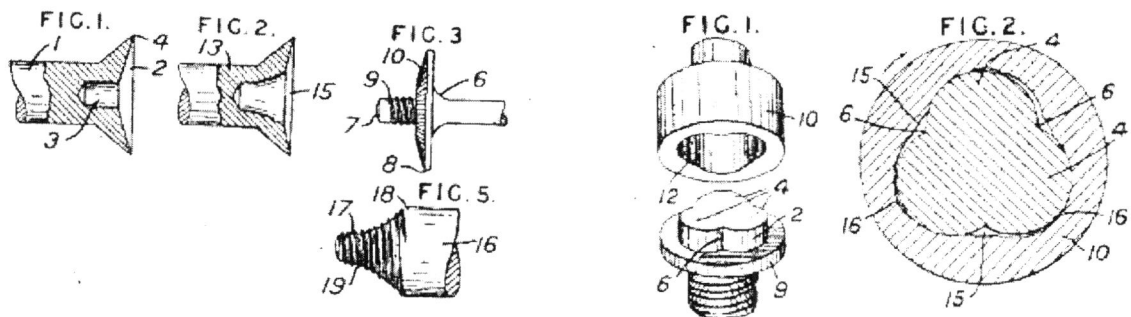

Variations of clutch drive and self-cutting drive inserts (Pat No 511033) and lobe headed screw (GB)

to bore a pilot hole, and secondly, there was absolutely no risk of the driver slipping and damaging the surrounding material.

With these patent applications lodged, and although still bankrupt, he announced in *Autocar* of September 1937 that a range of screws would be made available from the Tamper-proof Screw Company of 82, High Street, Camden NW.1. Confident with the potential for his screws, he formed a new company in December 1937, to manage the patents, The Bradshaw Patent Screw Company, based at his new home at Hanover House, 14, Hanover Square, London W.1.

A further variation was described in Pat No: 511033, laid in 7th December 1937, for a wood screw with a drilled centre into which the special screwdriver insertion bit cut its own thread and self released on withdrawal. His final design was for a lobed, raised head machine-screw, requiring a matching socket head driver; the rounded lobes prevented normal spanners or wrenches from gaining purchase. Pat No: 513686 was laid in 17th March 1938 and he secured patents in France, Belgium and America where it gained much publicity in the *Scientific American* journal of September 1938.

The question of fitting a suitable cap to prevent vandals inserting a withdrawal tool was easily resolved by the newly discovered cellulose-acetate plastic, derived from a by product of wood-based artificial silk (Rayon). This new 'wonder material' was electrically non-conductive, highly resilient and above all, a remarkably strong mouldable and machinable engineering plastic. It is still much used today for shatterproof screwdriver handles. Caps moulded from this material could easily be driven home into the screw's tapped head. The acetate could also be colour matched to surrounding surfaces and productivity could be greatly improved by simply rubbing it down for painting without the need to counter bore, stop and flatten, as was required with brass or steel screws. Furthermore, by using a cap as a visible decorative feature, there was no need for expensive chromium plated screws whose manufacturing process and barrel-polishing, both reduced the thread's tenacity and frictional grip in timber. Such grip was an essential feature of unplated screws which prevented them working loose though vibration, or through the expansion of the wood by moisture.

Both The Bradshaw Patent Screw Company and the patent rights were bought by the leading British screw manufacturer, Guest Keen Nettlefolds Ltd, for £12,000 and were marketed as 'The Bradshaw tamper-proof' self-attaching screws. However, it appears that fearful of erosion of their lucrative market, GKN promptly shelved the idea to which Granville later commented that while vandalism went unchecked, sales of replacement slot headed screws rose - along with GKN's profits - and also kept the Police in business for, as a senior police officer told him: "theft was a job for the Police since they were employed for this class of work". How times have changed!

Today of course, vandal proof-screws are common place and many 'modern' designs are evident in Bradshaw's original patents. It should be borne in mind that the original cross-head (Philips) design, developed in America by the Fitzpatrick Corporation in 1933, would only withstand casual tampering.

However the nature of the Philips design does not allow safe high torque application as the convex internal shoulders allows the screwdriver to self-extract by camming-out under high torque. This was solved by the GKN patented 'Pozidriv' design of 1960 which used a precision forged, straight sided, vertical driving face which not only gripped the screw (a sign of a genuine 'Pozidriv') but also, with its hardened finish, allowed very high torque insertion and minimal risk of camming-out. As with the Bradshaw screw head, the 'Pozidriv' design greatly improved productivity, but sadly the ever decreasing prices of grossly inferior imported screws forced GKN, the leading global producer of high quality fasteners, to pull out of the market in the 1980s - a great and sad loss felt by all true craftsman.

Stereo photography

Bradshaw's marvellous photo-booth invention was followed in the mid 1930s by a design co-patented with Depth-O-Graph Ltd of 9, Mincing Lane, London for stereo photography; he then reformed the business as Stereoscopic Processes Ltd.

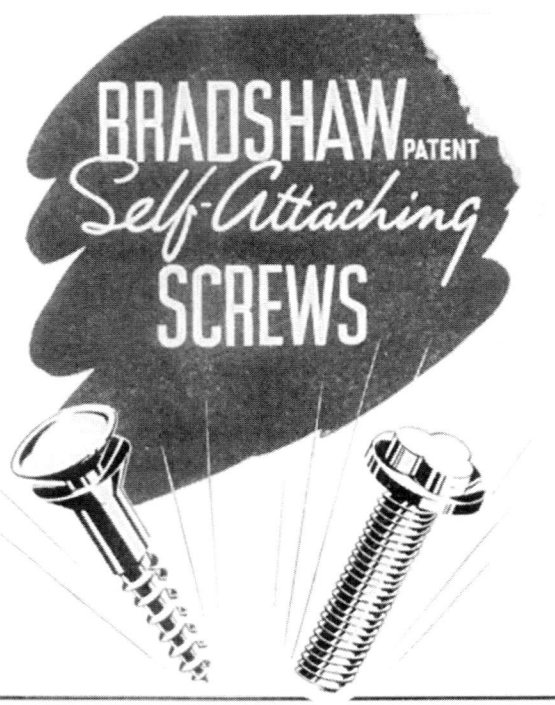

GKN advertisement

Having two eyes, set apart, we have a natural ability to judge distance and depth through 'stereoscopic' vision. This was first described scientifically in 1841 by Sir Charles Wheatstone (of 'Wheatstone bridge' fame), who demonstrated a 'relief' image by having two photographs taken 2 inches apart to represent the eyes' separation. By placing these images in a special, divided, viewing frame they presented a view which the brain merged into a natural, stereoscopic image. In late 1862 Sir Henry Swan patented a compact prism 'cube' stereoscopic viewing device using a pair of miniature pictures and soon stereoscopic photography became the rage of the late 1800s.

In 1903, an American, Frederick Ives, developed flat parallax stereogram images using a series of very narrow alternating strips of an image as seen through the left and right eye. By using a special comb grid laid over the photograph, the left eye saw only its own images and the right eye, likewise, giving an illusion of depth - a 'three-dimensional' image.

Although stereo photography was in the doldrums in the inter-war years, the development of Ives' principles continued, leading to the introduction of a series of finely embossed cylindrical lenses in a transparent film which, when laid over the parallax stereogram images, so described, acted in the same way as the comb. These proved immensely popular in 'stereo' picture postcards in the late 1930s and early post-war years.

This prompted Bradshaw to develop a system of taking such images. It involved six patent applications in 1934 (provisional patents: 12556, 12557, 12558, 15708, 16821, 25903 followed by 24774 of which no details survive). Three patents were granted, Pat Nos: 437517 - 519, all laid on 24th April

1934. His patented process offered an improvement to the whole process of using prism lenses to form narrow strips, at 12 to 15 lines to the inch, of alternating images on Ives' parallax stereo principle. The patent covered the frames and tracking device by which a camera moves completely around an object, photographing it in a series of exposures. On reviewing the photographic image, by using back lighting and prisms in a similar apparatus, an apparent three dimensional stereoscopic image was formed. A further Patent (Pat No: 460655 laid 30th May 1935) refers to an improvement to the camera mounting and movement. The advantage of Bradshaw's system was that large advertising hoardings could be erected which incorporated the back lighting and prism system, making them ideal for night time display.

He sold Stereoscopic Processes Ltd for a £5,000 profit but though it had some early commercial success in the advertising field, the company and his designs soon faded into obscurity.

In 1961, Eastman-Kodak patented a much improved system which they called, Xography. This principle was then improved upon in the 1970s by Jerry Nims and Allan Lo who used a new 4 lens system, producing stereoscopic images on specially prepared lenticular paper at 400-500 lines per inch. Despite much praise, their 'Nimslo' system failed to gain widespread market acceptance, requiring as it did special processing. Although their system is still available, a far more realistic stereo imagery system, using Holograms, has superseded it. Holograms were developed by a Hungarian, Dr Denis Gabor, using lasers, for which he received a Nobel prize in 1971. (Dr Gabor's work on digital sound recording will be found in the author's tape recorder books).

3-D colour television

Bradshaw's fertile mind turned to stereo imagery once again in early 1951 whilst laid up with pneumonia in his temporary home at Noxa, in Audlem, Cheshire while doing work on the Bond Minicar.

Many attempts were being made by engineers to create the lifelike illusion of 3-D in cinema and, now, television. While the brain easily and quite magically melds two images into a single '3 dimensional' image using a still camera, moving pictures present major new challenges.

A phenomenon well known to artists and photographers is to introduce a red coloured object in the foreground of a landscape to give an impression of depth, as the eye tends to focus on and draw forward the warmer, shorter wavelength reds and oranges, while pushing the colder blues and greens into the distance. It had also been discovered that 3-D imagery could be achieved by tricking the brain using 'anaglyph' photography in which an image is captured in two different colours, one in red (left eye), the other in blue-green (right eye). By superimposing the two coloured images, but offset by about a quarter of an inch, and then viewing the combined image through special coloured filter spectacles, they produced an eerie and quite nauseous sensation of a 3-D image. It worked in a fashion and proved a popular novelty in the 1950s wooing audiences back into the cinema and to the new television receivers.

Others attempted splitting images, using the previously described Ives' method, into series of fine vertical lines by prismatic beam splitters alternating between left and right eye, and then viewing them through polarised filters in special viewers. One such polarised filter advocate was Leslie Dudley of Twickenham who had been working on 3-D film technology since 1929, forming a company called Steroptics Ltd. During the war, he worked at the Admiralty Research Laboratory at Teddington on stereographic aerial photography for intelligence. He may well have known Bradshaw through his work at the Admiralty. By 1951 Dudley had developed an entirely new technique, but abandoned it in favour of an improvement to his original system and in April 1953 announced a 3-D television system which, unfortunately, would require the BBC to carry out major work in adapting their existing process, but as their Charter did not cover 3-D technology, they were not prepared at that stage to invest in it.

Like Dudley, Granville's approach was also to use a polarised system but dispense with spectacles. He accordingly improved upon the system he had developed for Stereoscopic Processes in the 1930s, now using lenticular screens to give an illusion of depth to the picture. He replicated that process by using a fine 'comb' made up of a series of fine lines in a plastic or glass plate, pitched at $^1/_{80}$th inch, which made them 'invisible' to the eye. When an image was taken by moving the camera around the object (and photographing it through the fine screen), when viewed through this 'comb', held about $^1/_4$ inch above it, each eye saw a different part of the picture and by moving the head slightly one way, then the other, the brain 'saw' a 3-D image. Family and friends who saw the demonstration were most impressed.

While it was easy to achieve with a static image, cine-film cameras ran at 24 frames per second, and it was found impossible to satisfactorily take and produce those images required at that speed. Both Granville and his electronics expert son, Geoffrey, believed that the new electronic cameras being developed by EMI for television work, which took a single picture element (a pixel) in one five-millionth of a second, was well able to tackle the problem at such film speeds. And so, in 1952, when living at 11, Nevern Square, London SW5, Granville sought out interest from both the BBC and Christopher Grey Tennant (Baron Glenconner) at ITA as well as The Rank Organisation, the major film, cinema and television camera and projector group. Having seen an impressive demonstration, all parties showed considerable interest and an agreement was drawn up in spring 1953 between Granville and Lord Glenconner to form a new business, The Stereo Television Company, based at Glenconner's home, 4 Copthall Avenue, London EC.2. The first board meeting was held on 17th June 1953 and applications for shares were received: Granville Bradshaw, 19,999; Francesca King (Granville's companion), 10,000; Tennents Estates (1928) Ltd 9,998 and just one single share for Lord Glenconner. The board's minutes acknowledged ongoing negotiations with the Rank Organisation,

The formation of the new company involved the purchase of Bradshaw's six provisional patents for the sum of £1,500. These included 4360/53 (laid 14/2/53) 'Improvements in three-dimensional moving pictures' - ie: the use of screens; 5478/53 (laid 27/2/53) 'Improvements in devices for making and showing stereoscopic moving pictures' - ie: the mounting and movement of cameras in a scanning box; 6091/53 (laid 5/3/53) 'Improved means of taking and showing stereoscopic moving pictures' - ie: relating to electronically controlled illumination of the screened object to mimic the 405 lines of current TV screen. This patent was modified in 7143/53 (laid 16/3/53) to 'Improved means of transmitting stereoscopic moving pictures': 8670/53 (laid 30/3/53) 'Improvements in stereoscopic moving pictures' - ie: which added to the electronic imagery by using colour filters in each camera to give the colour spectrum for colour 3-D viewing through a very fine series of coloured screens on the television receiver; 9546/53 (laid 8/4/53) 'Improvements in 3-D moving pictures' - ie: covering a back-projection screening box for electronic scanning; 10903/53 'Improvements in colour television' - ie: a simpler version of the full spectrum television receiver screen using 3 prismatic lined screens in the three primary colours.

The agreement however did not include a novel idea from Geoffrey in patent application 7482/53 (laid 18/3/53) for 'Improvements in stereoscopic moving pictures' which related to interlaced electronic imaging using multiple cameras, running at 2,000 images a second, to be encoded into a mosaic pattern of pixel images brought together as one when unscrambled electronically, to present a 3-D television image. In its prototype form the 'scrambler' box weighed some 10-lbs and cost £250, quite a bit more than the television.

Granville saw great potential for the Queen's Coronation in 1953 (as did the American 'View-Master' company with their 3-D colour transparency images). Meanwhile, research engineers at the

BBC Kingswood Warren research centre reported in early 1953 that Bradshaw's 'Natural eye' 3-D system would certainly "appear in theory to be applicable to the existing TV system" (referring to their existing electronic cameras using the pre-war British 405 line system), but would require much modification to raise the necessary definition (even with interlaced scans) and reduce inherent flickering to match the proposed alignment with the European 625-line standard. To that end influential bodies such as the Post Office and the broadcasting industry would have to consider a radical alteration in 'telecasting' methods. Once again, approval to fund this advanced technology was outside the BBC's Charter and so they abandoned the project in 1953.

In an interview in *Picture Post* of July 1953, Granville reported that "The (3-D television) pictures have been acclaimed by experts both here and abroad as the best ever. No spectacles are needed, they show full and accurate depth, they are clear and precise and restful to the eye and, if one moves to the right or left of the picture one can actually 'look around the back' of a person in the picture just as in real life". He concluded that not only would 3-D work, but it would be technically easier to achieve than colour television.

Although Bradshaw ceased further work on the project, he proposed a simpler version of the scanning box (patent application No. 657/55 - laid 8/1/55) while living at Hill Cottage, Eversley. This new system would decrease the requirement for greater definition under the earlier provisional patents. It retained the four camera set up for filming and a fine, vertical v-shaped lenticular lens screen placed on the receiver to 'unscramble' the image. He submitted it to the Rank Organisation. Their research engineers were quite impressed and an internal report said: "Summing up, the system could be made to work and it is a clever idea, but the difficulty in making the receiver sufficiently accurate, would make it impossible to use a modified production model. The adjustment would have to be of such accuracy that switching on an electric fire in the same house (a power surge) would be sufficient to completely destroy the three-dimensional effect".

Meanwhile, in 1954, a successful colour TV system had been launched in America; it would not be until 1967 that colour TV using the 625 line European PAL (Phase Alteration Line) technology was first broadcast in Britain. But 3-D television remained as elusive as ever, although Philips in Germany persevered with experiments throughout the 1960s, still using colour filter spectacles. The first 3-D television broadcast was made by Philips in Germany in March 1982, followed by one in Britain that November with an ITV broadcast of *The Real World* on a huge TV screen in Covent Garden, but again using the old anaglyph coloured spectacle process. Philips remained committed to the 3-D project and in summer 2007 promised the commercial launch by 2008 of a new multiple camera image system through a lenticular lens LCD television screen. This however requires a quite critcal viewing position, but the commercial launch was abandoned due to low consumer take-up of 3-D television. However, spectacle-free 'auto stereoscopy' lenticular systems, developed in the 1930s, are still regarded as the system of the future but today only Toshiba remains in that field while Samsung has concentrated on the anaglyph spectacle 3-D system.

War against grime

Having completed his work on the Bond Minicar while temporarily living in Audlem, he moved back to London, taking up residence at 11, Nevern Square, London SW.5 and then in June 1953 moved to 1, Glazebury Road, Fulham, London W.14. It was here that in his idle moments, he developed a simple kitchen waste-disposal unit that he had devised in 1952. This used a mains water pressure supply to oscillate, by piston motion, a sliding body which held a pair of cutting blades. These systematically sliced and macerated the food waste as it was flushed through the waste pipe.

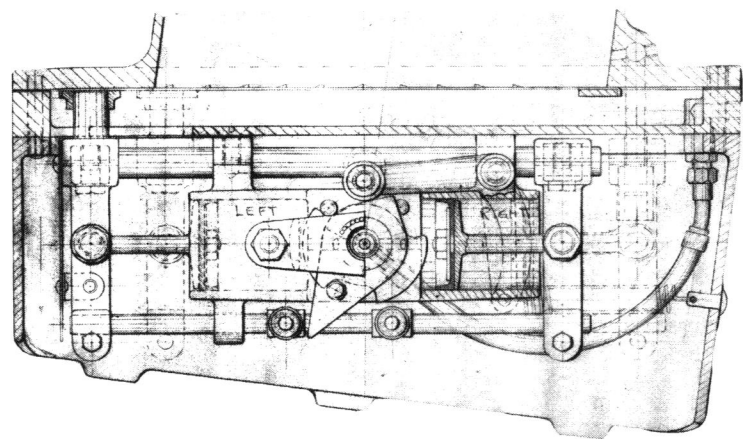

Side elevation of Bradshaw's pump action, reciprocating blade waste disposal unit (GB)

Knowing that these were all the rage in America, a working model (protected by provisional patent) was duly produced with the aim of quickly selling the invention. By chance, he spotted an advertisement in the personal columns of an August 1953 issue of *The Daily Telegraph* by a manufacturer offering spare capacity in their engineering workshops. Bradshaw responded by offering them the rights to his 'hydraulic waste disposal unit'. The mystery company was very keen and invited him down to their works.

J E Shay Ltd of Basingstoke were precision, production engineers. They explained that they produced products for the ironmongery and horticultural trades and that they had also recently established a hydraulics department. They also owned the Lansing-Bagnall and Rotoscythe companies. Intrigued, Bradshaw then offered them his 'Bumblebee' V-twin engine and the waste disposal unit was soon forgotten as discussions turned to rotorvator engines and a new Peoples' car, discussed later.

Motor car lights

Prompted by Parliamentary debate on the need for side lights to be used on parked cars at night, Granville modified a provisional patent (12827/53 of March 1953) for a large reflective system for illuminating rooms in a house by capturing available light from a single bulb, or sun, and then reflecting it into 'local' lamp shades to provide local illumination. The idea was seriously flawed when used in large rooms due to the inverse square law of light intensity over distance, but it could provide short range 'local' lighting. He put this to good effect on a motor car by clipping a lens to a car window frame which, by using a series of mirrors, reflected the light from a street lamp through a coloured lens to act as a side light, thereby preventing drainage of the car's battery. While it met the proposed legislation, the idea was never taken further but it is interesting to note the same principle of light being transmitted from remote locations along fibre-optic filaments to a central point being used today to illuminate a car's instrument panel.

He even proposed a farcically simple device while at Shays, for his Peoples' car, with a semaphore direction indicator which added 'a touch of luxury', but his design was neither solenoid operated nor illuminated and required the driver to merely slide a lever on the dashboard to the right, or left! As the signalling arm emerged, the oncoming air flow passed through a hole in the arm creating a whistle, warning others of the driver's intention. But at what speed the whistle became effective, is not recorded - nor are Shay's comments!

'Telescopic' sail

When living at Wooton Bridge near Ryde on the Isle of Wight in late 1958, from where he had a perfect view of the ships, from yachts to the *Queens Mary* and *Elizabeth*, as they navigated The Solent, Bradshaw reasoned that an aerofoil shaped disc could be used as a 'sail' creating low pressure on one side with the faster moving air over the aerofoil providing 'lift', just as happens over the upper surface of an aeroplane's wing, or Frisbee and indeed in 'tacking' a yacht against a headwind.

By having the 'sail' in segments, much like a bulb flash-gun reflector, they could be fanned out to as many as necessary to catch the wind from which ever direction, just like a windmill. He proposed that the normal spruce main-mast be replaced by a tubular steel mast, rotated by an electric motor which could also alter the angle of attack so the sail could be both turned into the wind, as well as be tilted like an aircraft's wing to give some lift to the bow, reducing resistance as the boat raced through the water. The entire operation could be controlled single handedly from the stern by hydraulics. There is no evidence of a working model having been made.

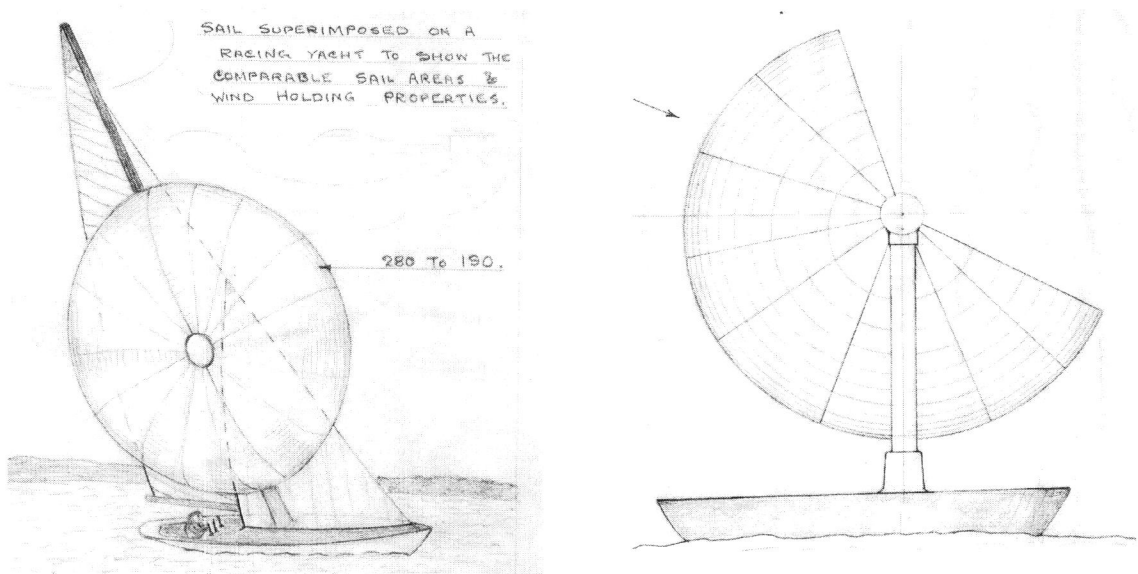

Bradshaw's proposed rotary sail (GB)

Recoverable golf ball

To the entrepreneurial delight of small boys working as golfer's caddies, reasonable pocket money could be had from recovering golf balls. It appears that Bradshaw painted some golf balls with, possibly radio-active luminous paint as used in wristwatches, so they could be detected by Geiger counter in the long grass. It is said that Dunlop bought the idea from him, but this remains unsubstantiated.

16
Bradshaw at War

Florence Nightingale

Bradshaw had an interest in St Thomas' Hospital, London, opposite the Houses of Parliament; his wife, Muriel, was a Life Governor and he, an Honorary Life Governor. Housed within the hospital was Florence Nightingale's historic carriage, which she used on her errands of mercy in the Crimean War and, on her return to England, in her own duties. Of Russian origin, the body was of a canvas covered wicker work frame on a light wooden chassis with rear leaf spring suspension. It was presented in 1930 to the Nightingale Training School, which Florence had founded at St. Thomas' but by the late 1930s, the carriage was in quite poor condition suffering from woodworm, general decay and the scars of many a student rag week.

During the Blitz of September 1940 to April 1941, the hospital and carriage suffered considerable damage and in February 1942, Granville personally arranged and paid for the carriage to be restored by Mr T Welland and his 81 year old uncle, Mr A Welland, at their 200-year-old family firm of coach builders on the Portsmouth Road, Esher, Surrey. The Wellands had already done much restoration of Science Museum exhibits. Two months later, it was presented to Sir Arthur Irving, Clerk to the Governors by Mrs Bradshaw and was thereafter much used by the hospital to raise funds before being retired to the Queen Alexandra's Army Nursing Corps museum at Aldershot; the carriage is now at Claydon House, near Buckingham.

Muriel Bradshaw presenting the restored carriage to Sir Arthur Irving in 1942. Granville looks on (GB)

Drag-anchor barrage

At outbreak of war, Bradshaw was contracted by the Admiralty's Royal Navy Scientific Services on £2/10/0d a day, to work at a 'secret location somewhere in Surrey'. He duly moved from Hanover Square, London to be closer to work, renting Lovelace Lodge, Woodland Drive, East Horsley while employing Miss Betty Harries as his personal secretary.

It was suggested that he should go into uniform and be commissioned at a high rank to reflect his technical standing, but he would have none of it for he preferred to remain a mere civilian, allowing him to follow his own profession as a consulting engineer and preserve his independence and freedom to write, lecture and carry out design work for others. Indeed, he regularly wrote under both his own name and a *nom de plume* (often 'Designer') in learned, if often highly controversial, articles for various technical, flying, motorcar and motorcycling magazines.

By the late 1930s, Britain was well advanced in the detection of airborne invaders through acoustic listening posts and the fledgling, coastal Chain Home and later Chain Low radar detection network. But attacking the enemy aircraft by conventional anti-aircraft gunnery was, literally, a futile hit and miss affair, even with primitive proximity fused shells, until the development of radio-proximity fuses of 1944.

However, during the Great War, a system of aerial mines suspended by cable from barrage balloons proved very effective at catching out the unwary enemy pilot, or an unlucky allied pilot at night or in cloud. But should a mine fall, civilian fatalities below were inevitable. High altitude balloon barrages, with or without suspended aerial mines, were again seriously proposed in 1938 by Frederick Lindemann (Winston Churchill's senior advisor) and by 1940, Churchill was demanding rocket propelled aerial mines on cables to freefall by parachute.

These were first deployed by the Royal Navy in late 1940. Although they proved quite ineffective, such 'Fast Aerial Mines', or PAC (Parachute and Cable) rockets, remained in use by Royal Navy and British Merchant marine vessels throughout the war. A newspaper cutting from around 1942, reports a similar idea from South Africa which reported it had been "adopted by the Admiralty"; it went on to say that the Italians had reported its use against them in March 1941 over the Mediterranean and that, later that year, the Luftwaffe's Focke-Wulf Condor long range maritime reconnaissance bombers, hounding the Atlantic convoys from a safe distance by signalling U-boat packs, had reported parachute retarded cables being used against them. They were even tested against enemy tanks, immobilising them in 600 feet of entangled wire.

One of Bradshaw's first contributions to the war effort came in October 1940 when he designed an anti-aircraft barrage which could be launched by rockets or dropped by bombers from 40-50,000 feet. It comprised a 2lbs plumb-bob suspended on a 500ft line of quarter inch steel wire. Its descent to earth was retarded by a small parachute and as the wires floated silently to the ground they gave 30 minutes of almost invisible defence against enemy aircraft. The device could then be collected and recycled. He suggested that a single aeroplane could deploy these over a 20 mile radius of London in around five minutes protecting the city at a moment's notice.

The theory behind Bradshaw's concept was that a fast approaching aeroplane's wing would snag the wire; the drag from the parachute would embed the wire into the wing and cause the aeroplane to spiral out of control. A variation to this used an anchor instead of a plumb-bob. The anchor's arms were sharpened to a knife edge and along the wire were spliced a series of sharpened barbs which sawed through the aluminium wing while the anchor's sharpened edges severed the main spar leaving the flukes to rip the wing to pieces - well, that was the idea.

Fairmile Marine Company

That 'secret location somewhere in Surrey' turned out to be a land-locked Fairmile Marine Company on the A3 Portsmouth Road, Fairmile Common, near Cobham, Surrey. Formed by Noel Campbell Macklin, his works comprised a large single storey workshop in the back garden of his home, Fairmile Cottage. Noel had a keen interest in boats and was a naval reservist before the Great War, being commissioned Lieutenant in 1916, rising to the rank of Captain.

He also had a keen interest in motor racing and after the war, co-founded with Hugh Eric Orr-Ewing, the Eric-Campbell Company of Cricklewood to produce a sporty 10hp two-seater car with a Coventry-Simplex engine. The car was built for them at the nearby Handley-Page aircraft factory, but so few orders were received that they sold the business to the Vulcan Iron Works of Southall who persevered until 1926. Dissatisfied with the Eric-Campbell, Macklin set up on his own in 1920 as Silver Hawk Motors Ltd at his Fairmile home and produced a superior sports car based on the Eric-Campbell, but now with a tuned 35hp, 1,370cc engine and lightweight aluminium body. That project fared even worse, collapsing within a year.

It was not until 1925, when he teamed up with his neighbour, Oliver Lyle (of the Tate & Lyle sugar family), to produce the 'Invicta' car, modelled on the American philosophy of a heavy chassis with a powerful if lazy engine, that he found success. With its Meadows 6-cylinder engine, the Invicta was an expensive but much admired car. However, Macklin's involvement ended in 1933 when he sold the Invicta to Lord Fitzwilliam who moved production to Chelsea. This allowed Macklin to produce a better car, based on the American Hudson Terraplane chassis, for which he sought advice from Reid Railton who, with John Cobb, had had much record breaking success with Napier-Railton racing cars gaining the World 24hrs at 150mph. (Their 1938 twin Napier Lion W-12 engined, four wheel drive Napier-Railton 'Mobil' special driven by John Cobb at Bonneville, took the land speed record at 333mph followed in 1939 by a modified Railton at Utah at 369.7 mph).

Their new Railton motorcar was built at Fairmile and had a powerful 4-ltr straight eight Hudson engine fitted into a heavy chassis with a sporty, British coachwork body. They proved very popular with the Police drivers at Scotland Yard, but with a decline in sales and realisation that war was imminent, the Railton business was sold to the Hudson Motor Company of London and by 1940, Macklin's Fairmile factory was ready for war work.

Coastal patrol boats

As a Royal Naval Volunteer Reserve Lieutenant on the Dover Patrol between 1916 and 1918, Macklin realised the Royal Navy lacked a crucial fleet of fast coastal craft to protect Britain's merchant shipping from submarine attack. He was not alone in his observations for Vice Admiral Cecil Vivian Usborne, a retired naval ordnance expert in the First War, was also pressing the Admiralty to build a fleet of fast submarine chasers.

An Irishman, Usborne had joined the Royal Navy as a cadet in 1894 at the age of 14 and became a gunnery expert, inventing a quick firing 'pom-pom' gun in 1902 followed by 'fall of shot indicator' for naval gunnery in 1912. Soon after that he was commissioned Commander of *HMS Colossus*, a two-funnel Dreadnought. It is interesting to note that both she and her sister ship, *HMS Hercules*, rarely fired a broadside with their ten 12" guns, as such practice badly stressed the deck; even Naval architects can get it wrong! In 1916, he patented a 'paravane' anti-mine device which comprised a pair of steel wires fitted with hydrovanes, which created an underwater device to deflect and cut the mooring cables of tethered sea mines, bringing them to the surface for destruction by gun fire. So successful

Below, the Fairmile 'B' type RML used by King George VI while reviewing the D-Day landing fleet in June 1944. Opposite, the 'C' type engine room showing the three Hall Scott engines

was the paravane device that Usborne was awarded £6,000 by the Royal Commission of Awards to Inventors in 1920. He also patented lifting gear and tackle for ships' life boats and launches. In 1928 he was elevated to Rear Admiral and in 1930 became head of Naval Intelligence; he retired from the Navy in 1935.

Drawing then on their combined naval expertise, Usborne resigned from his Home Office job to join Macklin as joint Managing Director of their newly formed Fairmile Marine Company and, with Norman Hart (a leading naval architect) as designer, developed a kit-built, marine-plywood motor torpedo boat which could be assembled anywhere; this principle became known, generically, as the 'Fairmile type'. The prototype 110ft Fairmile 'A' was built in July 1939, at Macklin's own expense, by the Woodnutt boatyard on the Isle of Wight and was powered by three American Hall-Scott 1,200 hp V-12 Defender engines.

Despite pressure from Usborne and Hart, the Admiralty at first showed little interest in the Fairmile 'A', for they already had experience of wooden hulled fast motor launch submarine chasers in the First World War. However, they were impressed enough by Macklin's work and determination that in September 1939 they contracted him to produce 12 boats for sea trials. By early 1940 the Fairmile works had been commandeered by the Admiralty who then recalled Usborne from retirement to the Admiralty in 1941 for whom he remained on 'special service' until 1945, probably still based at Fairmile Marine, for his work there included his patented design of improved engine room ventilator cowls which deterred entry of water. He also patented a means of operating sets of fluid control valves in one operation for rapid response in emergencies.

Though this first batch of Fairmile 'A' boats proved underpowered and too flimsy, with the help of Sydney Graham, and now working to an improved Admiralty design, they developed the 112ft Fairmile 'B' fast motor launch, powered by two Hall-Scott V-12 engines. Being a coastal submarine chaser, it was fitted with 'ASDIC' detection, depth charges, a 3-pounder gun and a pair of .303" Lewis machine guns. Some 568 'B' types were built. The B type RML 513 and 518 feature in the war-time classic air-sea rescue film, *For those in peril*.

These were soon followed by the 23 knot Fairmile 'C' Motor Gun Boat of 1941 with three supercharged Hall-Scott Defenders laid out in V-formation, but the 'C' type was not a success and was replaced in 1942 by the larger, William Holt designed, planing type hull, 115ft Fairmile 'D' Motor Torpedo Boat powered by four Packard 4-M, 5,000bhp engines, giving 31 knots. Some 229 'D' types were ordered and in late 1944 about 40 modified designs with copper-sheathed hulls, to combat toredo worm attack, were destined to serve with the RAF as air-sea rescue launches in the Far East, but the war ended whilst in passage!

Plywood boats

The beauty of the 'Fairmile' launch was its versatility from its simple block construction. Each boat was built from a kit contained within six large crates; each crate was designed to fit the load bay of a standard 5 ton Royal Navy lorry to allow delivery to boat yards across Britain on a just-in-time basis.

The kits assembled into several watertight compartments, bolted together. For greater strength, their bulkheads were built up of 18" x 30" scarph jointed, bias-laid 9-ply marine grade plywood. The mahogany boarded deck had special steel runners laid in, which allowed the rapid fitting of a wide range of armaments making them an incredibly versatile boat. The 'Fairmile' design was produced by all manner of companies throughout Britain and the Commonwealth during the war with several boats serving in the US Navy.

Chief Admiralty Boffin

Describing himself as a 'Chief Admiralty Boffin', Bradshaw initially worked on engine room design and propulsion for the Admiralty and is known to have worked on torpedo propulsion. Titch Allen recalled of Bradshaw's 3-cylinder motorcycle engine, that this was originally developed for a small, remote controlled torpedo to attack the German U-boats sinking Britain's vital Atlantic shipping. This was probably developed from an earlier compact 3-cyl radial torpedo engine designed by Peter Brotherhood Ltd. Such confined torpedo engines were usually aspirated by compressed air/oxygen and fuelled by kerosene held in tanks within the torpedo body. The engine is said to have had a minimum of steel, to minimise magnetic or sound detection, and used bronze disc, or rotary valves. Bradshaw's war work was, by necessity, not openly reported, nor is it recorded in his archives, so whether he was ever involved with the Motorised Submersible Canoes, or other submersibles for clandestine Royal Marine Commando operations built under Admiralty orders at Fairmile in 1944, is not known; for these used electric motors.

He also soon found himself designing all manner of components at Fairmile. In June 1941 he patented a telescopic stanchion (Pat No: 547279 laid 9th June 1941). The contract to produce these was most probably that awarded to Phelon & Moore Ltd who were known to have produced ship's stanchions as part of their war-work.

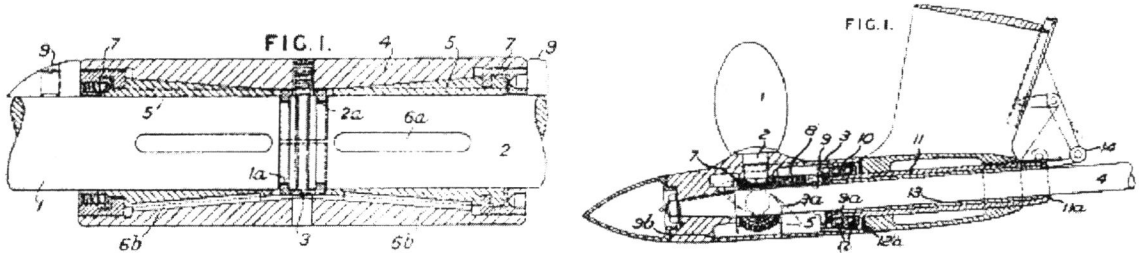

Left, patented shaft coupling. Right, variable pitch propeller

His work on propulsion schemes for the Packard engined Fairmile 'D' motor torpedo boats is not recorded in his archives in detail, however, on 23rd June 1941 he submitted a patent (Pat No: 581760) for an epicyclic reverse gearbox for ships and launches with a clutch linked to the throttle control to enable a rapid and smooth transition between forward or reverse by cutting and then restoring engine power as the clutch control was operated.

He also patented an adjustable pitch, marine propeller whose mechanism was lubricated by circulating water around the rust-proof, bronze ball, or roller bearings set in stainless steel races (Pat No: 549585 laid 27th June 1941). The boat's speed was controlled by lever through a simple rack and pinion which altered the blade's pitch. In 1942, he patented a propeller shaft coupling with a wedge lock device, to securely join two long shafts together (Pat No: 578515 laid 7th August 1942).

Radial engines

With the constant threat of U-boat attacks on convoys and the consequential loss of imported Hall-Scott engines from San Francisco Bay, Bradshaw proposed fitting British made engines into the motor launches and, being a keen advocate of air cooled radial engines, persuaded the Admiralty to equip one Fairmile 'D' type ML with four 1,600hp Bristol Hercules XI 14-cylinder two row, sleeve valve aero-engines giving it incredible performance.

The scheme was not developed further by Fairmile and it appears that Bristol themselves then stepped in, during 1944, as a private venture, taking up the challenge with Fred Cooper, who had designed the British Power Boat Company's fast launches at Hythe and who was at the time living close to Bradshaw in East Horsley. Cooper also developed an adjustable hydroplane system for Fairmile. Using a long range rescue variant of the Fairmile 'F', christened *Celerity*, four Bristol Hercules engines were duly fitted and produced 36 knots in sea trials, but the air-cooled engines soon overheated. They then fitted large cooling fans, drawing air from huge rearward facing cowls with Usborne's ventilator flaps, but these fans merely sapped the engine of its power. In 1945 *Celerity* was reequipped with the improved Mk.XVII Hercules engines and was duly despatched to the Far East for air sea rescue operations. In the end the Royal Navy adopted high speed diesels.

Although it is not clear if Bradshaw was involved with *Celerity*, his knowledge of oil cooled engines resolved one major problem with fast motor launch gearboxes which, after several hours of high speed running, ran unacceptably hot. Engineers had stripped down one gearbox suspecting bearing failure, but Granville believed the problem lay in the gears themselves which, in tests, showed they transmitted power with only 97% efficiency. He reasoned the 3% loss was through friction. Thus a 3bhp loss through friction in a gearbox on a 100bhp engine was the equivalent of 60bhp loss on the 2,000bhp fast motor launch engines, generating the equivalent heat of several electric fires! The solution, he believed, lay in a pumped oil cooling system via an external oil-cooling radiator.

At war's end, Noel Macklin was awarded a knighthood and in recognition of Fairmile's work, the Admiralty presented the people of Cobham with a ship's bell from *HMS Cobham*, an inshore minesweeper. The Fairmile Marine Company then moved up the road to Green Lane, Cobham where they remained as an independent company until 1984, using a shipyard in Berwick-on-Tweed to construct their boats. Their old Fairmile Common engineering works, now known as Cottage Laboratories Ltd, remained under Admiralty control until 1949, after which they were leased to Plessey for Sonar development until 1963. Between 1968 and 1995, these old works became part of the Ministry of Defence torpedo research establishment. The site was demolished in 1998.

Crippled for many years by arthritis, Sir Noel Macklin died in November 1946. Vice Admiral Cecil Vivian Usborne died in January 1951.

Bradshaw air pump

While at Fairmile, Bradshaw worked on three unique engines: the previously mentioned 3-cylinder radial, an 'air-pump' (circa 1939) and a toroidal (annular ring) chamber, reciprocating piston rotary engine (circa 1942). It is understood from Bradshaw's letters that these last two were specifically developed to Admiralty requirements for Fairmile's MGBs (motor gun boats) and MTB (motor torpedo boats).

But what exactly this 'air pump' was remains a mystery (pre-war parlance suggests this was a super-charger). Judging by P&M's involvement in late autumn 1939, around the time Bradshaw joined Fairmile, it is most probable that its design had originated with the Turner Manufacturing Company while Bradshaw was almost certainly brought in at the behest of the Admiralty to improve upon its design.

How much development work on the 'air pump' was actually undertaken at Fairmile is not known and, unfortunately, the only eye-witness description given me in the 1980s was later contradicted. It appears that only a prototype was made and was described thus: "a simple design based on a Roots type blower, with 'sliding' blades on a stationary core within a rotating chamber", not dissimilar, it would appear at first guess, to the Wankel principle, but reversed. In a conventional Roots positive displacement compressor, designed in the 1850s by Philander H and Francis M Roots of Indiana, a stationary 'figure 8' chamber has two inter meshed impellers; as air enters one chamber, it is passed to the second chamber by gears, whose lobed 'teeth' simultaneously compress and expel the gas at a higher pressure.

The prototype 'air-pump' was developed and built by the Turner Manufacturing Company of Wulfruna Works, Moorfield Road, Wolverhampton, (P&M archives say Birmingham) in the late autumn of 1939. Turners were a long established engineering company who had produced cars at various times, but during both world wars they had turned over entirely to government contracts which now included machine tools and aircraft undercarriage components. They also developed a rotary, multi-piston pump - but this does not correspond with the description given of the air pump.

Having worked closely with Phelon & Moore, Bradshaw now asked them to consider manufacture of the air-pump, so Richard Moore, P&M's co-founder, their Works Manager and Works Foreman all visited Turners to assess manufacturing requirements. They considered the project would cost between £20-25,000 (a considerable strain on their limited resources) and would require much new machinery. Besides which, the 4-5 months required to set up production would jeopardise potentially more lucrative contracts which they hoped to secure for exclusive motorcycle orders for the RAF, just as they had in the First World War. In declining the offer, Richard Moore told Bradshaw that if Turner were to produce the jigs for them, P&M could be persuaded, but they didn't - and neither did P&M secure that RAF motorcycle contract.

Free-piston engines

As a senior Admiralty 'Boffin', it is entirely possible he was aware of the use by the German Navy of 'free-piston engines' in their submarines to expel torpedoes by compressed air. By definition, a free-piston engine is one where the piston is free to move in a cylinder, unhindered by connecting rods and cranks. This type of engine usually employs two double headed pistons which meet at the centre of the cylinder, forming a combustion chamber between the two heads, as in an opposed piston engine. As the volatile charge explodes by spark plug or compression ignition, the two pistons are propelled apart compressing the gases behind them, forcing these gases to turn a turbine and output shaft. The pistons are returned to the centre by some of the compressed gases diverted to a 'bounce chamber' - and so the cycle continues with the absolute minimum of mechanical components.

The free-piston design was perfected in the 1920s by Raoul Pateras de Pescara, an Argentinean engineer living in France and working for Alstrom at their works in Belfort, close to the Swiss border. It was here that he developed a free-piston gas turbine engine but with France falling to the Germans in 1940, production ceased and some of the new technology passed to the German Navy who already had a diesel powered free-piston air compressors in their U-boats for discharging torpedoes under high pressure. Having captured several U-boats in 1943, the technology was now shared between the US Navy and the General Machinery Corporation of America who made ships' engines. The potential for maritime use was huge - from out-board motors to aircraft carrier catapult launchers and by the mid 1950s, General Machinery had merged their interests with the Lima Corporation, who made railway locomotives, and as Lima-Hamilton, they began work on a 4,000hp free-piston locomotive engine running on coal-gas. Meanwhile, the General Motors car company began work on a smaller (though still massive) 250hp engine for lorries using a turbine drive. Bradshaw later incorporated the free piston principle into a twin toroidal chamber 'gas generator turbine' engine in 1958, when at Cowes on the Isle of Wight.

Oscillating piston toroidal engine

In a letter dated March 1960 to his nephew, Col Reg Gray at Bonds, Bradshaw reveals that the overhead valve gear on the Packard aero-engines used in the Fairmile 'D' launches, were always needing attention. Admiral Usborne accordingly asked him if he could design a less problematical engine. One presumes Usborne was thinking of a rotary or sleeve valve machine as fitted to the Hercules radial aero-engine, but Bradshaw had other ideas and around 1942, he formed a new company, Bradshaw Prime Movers Ltd, and set to work on a weight saving single crank, valve-less toroidal rotary engine.

The Admiralty was most interested and, as Bradshaw later said, 'several engineering companies already on Admiralty contracts', were at various times instructed to build prototypes - but who these were, to what extent and when they became involved, is not stated.

This unconventional engine design would, in theory, produce considerably more power, and at a faster rate, than a conventional internal combustion engine. While his later 'Omega' engine used scissor action pistons moving around a stationary toroidal chamber, his earliest engine had the scissor action pistons constrained and through cranks, rotated the toroidal chamber which drove the output shaft, in the same way as earlier patents by Frederic Beck and others, before the Great War.

The idea dates back to at least the turn of 20th century and is first properly described by a Frenchman, Frederic Beck, who presented two patented designs in 1909 and 1910 which overcame the

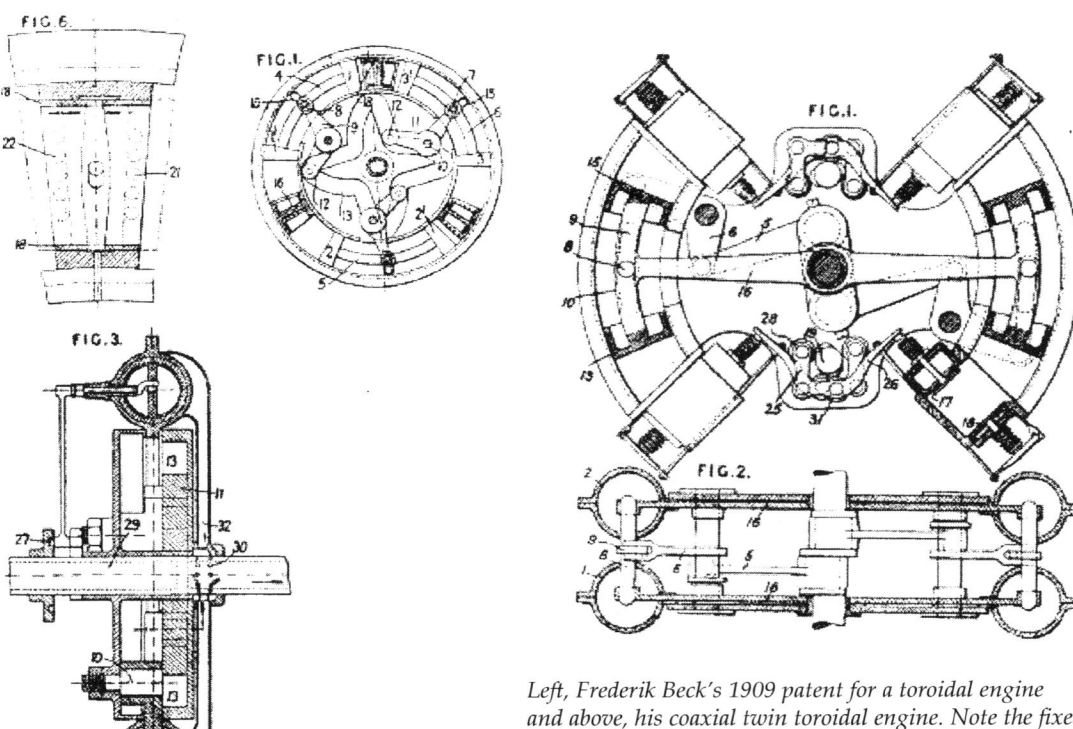

Left, Frederik Beck's 1909 patent for a toroidal engine and above, his coaxial twin toroidal engine. Note the fixed quadrant heads causing both piston and chamber to rotate together

inherent inertial forces in the rotary aero-engines then being developed. Beck's 'rotary explosion motor', (UK Pat No. 1909926914) comprised three interconnected double ended curved pistons, oscillating within and rotating with, a quadrant shaped annular chamber. Each piston head compressed gases at the quadrant's fixed combustion heads where there was an automatic inlet valve fed through the hollow, stationary shaft and a cam operated exhaust valve. Induction was through ports exposed by the pistons. Note from the illustration (above) that, where the connecting rod enters the chamber, there is no loss of combustion into the crankcase.

Beck offered an alternative in 1910, (UK Pat No. 191122798), in which two pairs of coaxial cylinders and conrods now operated in a scissor fashion, using mechanically operated valves. It was a design which inspired another Frenchman, aero-engineer, Bauart Esselbé in 1913, to use a pair of oscillating pistons within a single, 65mm diameter continuous toroidal, annular chamber, producing a theoretical 60bhp. In this case, as one piston was thrust backwards by the exploding gas mixture, it caused the opposite piston to compress the induced gases against a fixed head, at each quarter.

Esselbé also produced a conventional 7-cylinder rotary aero-engine in late 1913. The use of double acting pistons was also used in the conventional, Joseph Day double action two-stroke engine motorcar and radial aero-engine of 1911.

The great advantage of toroidal engines is the lack of inertial forces and loads on the pistons and connecting rods. By also using two-stroke porting arrangements, as proposed in 1931 by another Frenchman, Alfred Aumeteyer, the earlier restriction on rotational speeds of 10-12,000rpm, imposed by valve springs was eliminated. Two-stroke engines also make for a lighter unit, running at far higher speeds, firing at twice the firing rate per revolution than a conventional four-stroke engine, producing a very high power to weight ratio. Furthermore, the two-stroke cycle also ideally lends itself to the simpler diesel (compression ignition) system which will run happily on most hydrocarbon fuels. Indeed, multi-fuel engines are now standard in military vehicles in theatres of war and this proved a deciding factor in Rover abandoning their kerosene fuelled gas turbine motor-car, as 'free-piston' engines would run on cheaper, unrefined fuels.

Esselbé rotary engine of 1912

P&M toroidal engine

It very much appears that Bradshaw built a prototype toroidal engine - but in which year is not known, for there is, unfortunately, conflicting evidence without supporting documents.

It appears he took this along to Phelon & Moore and set up a demonstration with Admiralty officials, Richard Moore, Ben Hey (P&M's chief draughtsman) and Percy Shaw (P&M's works foreman). The top secret engine, which had '3 or 4 carburettors' had to be started by hauling on a belt; it fired up and then ran at a tremendous speed, but only for a very short time before stopping.

It then appears that towards the end of the war, Bradshaw again approached Phelon & Moore to develop what seems to be a new prototype toroidal engine to which Bradshaw later refers (confusingly) as the 'New Action' toroidal engine. This used an oscillating piston action which, through sun and planet gears and twin cranks, operated the piston connecting rods by 'scissor action' through 30°. As the pistons were forced apart by the exploding gases, the cranks rotated the toroidal cylinder by a gear train, translating it into a rotational force through the output shaft. It is not clear if P&M actually built this from scratch, but it was reported that the 'P&M' prototype eventually fired up and 'spluttered'; many more abortive attempts were said to have been made to get it to run for any length of time, but with the war now over, P&M was keen to return to motorcycles and the project was happily abandoned with the engine relegated to a dusty corner in the stores. As the engine had been developed in secret, and as its workings had not been publicly disclosed, there was no need to lodge patent applications and indeed after the war, all the admiralty rights were handed back to Bradshaw. In August 1948 he drew up plans for his new 'Humming-Bird' $2^1/_2$ hp toroidal motorcycle engine, described below, but nothing more was done on his toroidal engines until around 1952 when he was working at Bond Cars where he realised its potential in the Bond Minicar. The original P&M-built engine was duly recovered from Phelon & Moore, who gladly and unceremoniously dumped it in the back of Bradshaw's Ford V-8 to be rid of it!

Humming Bird toroidal engine

Bradshaw proposed a miniscule $2^1/_2$ hp motorcycle version, the Humming Bird, measuring only 5" in length; whether this was intended for Phelon & Moore is not known but as it weighed only 3lbs, it is almost certainly not the original P&M built toroidal engine and, as little else is known, it is doubtful the engine was ever built. The full sized drawing (No. 48128) shows the piston to be $^{15}/_{16}$" diameter with a $3^3/_4$" overall diameter, rotating annular chamber. It also shows the use of disc type connecting rods with seals around the inner circumference of the chamber.

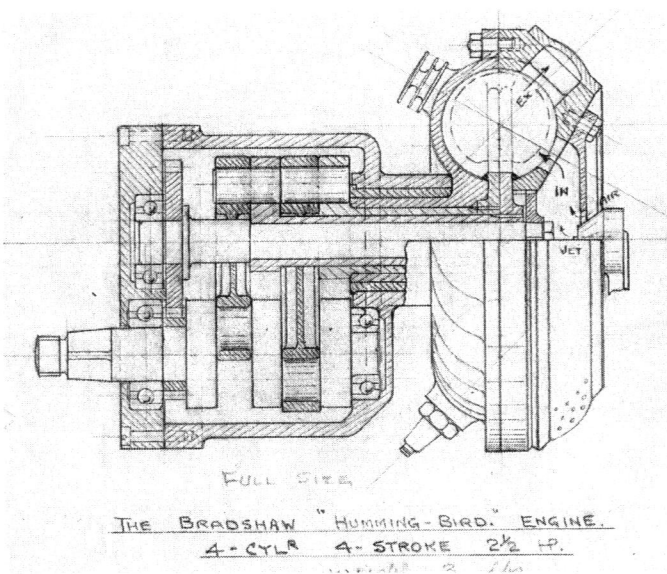

Bradshaw's Hummingbird rotating toroidal engine of 1948 (GB)

In June 1960, Bradshaw actively revived the idea and presented it to Edward Turner at Triumph for potential use in a lightweight motorcycle or scooter. Turner showed much interest in the toroidal engine concept as they were already closely following progress in the then very troublesome, Wankel rotary engine; but in October 1960 further talks with Bradshaw ended and Triumph thereafter concentrated on the Wankel before the project passed to BSA's Research and Development team, in conjunction with Fichel-Sachs of Germany and thence to Norton-Villiers-Triumph.

Whereas those Beck, Esselbé and Aumeteyer toroidal engines proved dismal failures, Bradshaw's determination continued over many years, working almost single handedly on a shoe string budget, until he finally produced a working, if unrefined, engine in the early 1960s. Any hope of its commercial success was killed by the Wankel engine which had received huge financial backing from German, American and Japanese industries. Engineers around the world continued working on experimental toroidal reciprocating piston engines, similar to Bradshaw's, such as the Kauertz and the Tschudi engines both from 1967; the Rocha-Cano of 1971; the Taurozzi of 1972; the Farrokhzad of 1991 and the Kaloustian of 2003, all using free or gear driven pistons, or vanes, as internal combustion engines or compressors.

17
Peoples' Car

With his work done at Fairmile, Bradshaw moved to Lake Cottage, North Chapel, Petworth, West Sussex and continued writing his *Motor Cycle* articles under the 'Technicus' *nom de plume* in the 'Jottings from a Designer's Notebook' series. Cleverly written, these gave little immediate clue to the writer's true identity and as always he and his its editors welcomed all critical letters.

Sharps Commercials

In late summer 1950, Granville was called upon by his brother, Ewart, to do some design work at Sharps Commercials at Preston, on the new Bond Minicar. Ewart had established himself in a small motor garage while in his early 20s, buying the business of J H R Loxham & Co Ltd (Motor and Mechanical Engineers) of Charnley Street, adjacent to his father's opticians shop on the corner of Fishergate and Corporation Street, Preston. He went on to establish Bradshaw's Motor House Ltd of Marsh Lane, Preston as one of the first Ford dealers. His Motor House and Loxhams garage businesses rapidly grew to around 25 garages and soon became one of the largest chains of motor-car dealers in the North; Loxhams later included a large portfolio of local commercial properties. Ewart later founded a finance house, Astley Industrial Trust and along with Sharps Commercials, ran his empire until his death on 5th June 1959, just before his 75th birthday.

Henry Ford's Model T was first seen in Britain at the 1908 Olympia Show; it cost £225. In 1911, Ford opened a British factory at Trafford Park, Manchester and, wisely, Ewart Bradshaw became one of Henry Ford's first seven dealers in Britain. By 1919, the Model T accounted for 41% of Britain's car production making Ford the UK market leaders. However, with the introduction of the new RAC hp tax rating in 1921, the big bore Ford T was now rated at 22.5 RAC hp and Ford saw their position plummet to 4th, behind Austin and Morris, who like Bean, had ironically adopted Ford's production methods to reduce costs, exacerbating Ford's woes.

Ford dealers were forbidden from selling any other make of car but under a personal agreement with Henry Ford, Bradshaw's Motor House was the Ford main dealer while Ewart was permitted to sell any make through his Loxhams business, which included Austin, Morris, Rolls-Royce as well as several motorcycle marques. Following a disagreement with Herbert Austin, he famously returned all the Austins in his showrooms to the factory! He was forbidden to handle Ford's American competitor, General Motors and their new British subsidiary, Vauxhall who had begun life in 1857 as the Vauxhall Iron Works in Vauxhall, London and had progressed to marine engines, diversifying into motorcars in 1903; they moved to Luton in 1905 and appointed the 21 year old Lawrence Pomeroy as their Chief Engineer in 1912.

In 1910, General Motors of America began exporting their Buick car chassis to London and, fitted with British coach work, were marketed as Bedford-Buicks. Then in 1920 they began exporting their Buick lorry chassis to Britain with final assembly at GM's works at the redundant Government aircraft and munitions factories along the Edgware Road, Colindale, Hendon, opposite the tram depot.

In 1923, they began importing their new Chevrolet bus and lorry chassis in 'Complete Knock Down' (CKD) kit form. Fitted with British made bodies, these Chevrolets soon proved increasingly popular with the smaller owner-driver, particularly bus operators.

Keen also to produce large engined, 'American type' motorcars and establish a British manufacturing base with access to the vast Commonwealth market (under the Imperial Preference Scheme), they bought Vauxhall Motors Ltd in 1925. This very British luxury car maker, much favoured by royalty, was in deep financial trouble, but GM's action caused a furore over American ownership of British companies.

Despite the collapse in the motor industry following the Wall Street crash of 1929, GM voluntarily imposed an embargo on all further Chevrolet imports to allow their UK assembly operation to establish itself while Vauxhall enlarged their works in Luton to build the new, British 'Bedford' chassis. Production began on 20th March 1930, at which point Hendon closed. The Bedford soon became an established, firm favourite at home and abroad in both civilian and military guise. With General Motors now seen in a more favourable light, they resumed imports of the Chevrolet chassis whose UK distribution rights were granted to Pass & Joyce Ltd.

Ewart was most impressed by these Chevrolets, but was forbidden by Henry Ford from selling any of GM's products. However, Pass & Joyce's rights were due to expire in 1937, and as Ewart believed he could secure the dealership, he acquired, privately, Paul Sharp's 'Sharp & Company' of the Lea Garage, Long Lane, Ashton, near Preston, renaming it Sharps Commercials Ltd. His gamble paid off and in 1938 Sharps began assembly of Chevrolet lorries from CKD kits, continuing right through the war under the Lend Lease scheme. Sharps also rebuilt many war-damaged lorries for the Ministry of Supply in a recently acquired former rope-works off Ribbleton Street in Preston. Fortuitously for Ewart, his eldest daughter, Anne, had recently married Major (Hon. Lt. Colonel) C R Gray MBE, formerly of the Royal Artillery, whom Ewart duly installed as managing director at Sharps Commercials in 1946. Gray was bestowed with the Territorial Efficiency Decoration (TD) in February 1952. They later lived near Longridge, 3 miles north of Preston.

In 1946 the Chevrolet contract to supply the Army ended, followed in 1948 by cessation of the Ministry of Supply refurbishment contract, leaving the Ribbleton works idle.

Bond Minicar

In the 1920s, Lawrie Bond was apprenticed as an engineer at Atkinson & Co of Preston, famous for their steam and, later, diesel lorries built at nearby Bamber Bridge. He then moved to the famous engine makers, Meadows Engineering of Wolverhampton, before joining the Blackburn Aircraft Company at Brough, but in 1944, Bond left to establish his own business, the Bond Aircraft & Engineering Company, outside Blackpool, where he worked on government engineering contracts. When these contracts ended, he moved to the Towneley Works, Longridge and kept his staff occupied by making small 500cc JAP single cylinder engined racing cars, with which he had some success. Then came his idea for a small and very basic, three-wheeler 125cc 'Shopping Car' and tradesman's truck which evolved into the Bond 'Minicar' of 1948 - a name coined long before Alex Issigonis's transverse, front-wheel-drive Austin XC9001 'Mini' prototype of 1956 which first ran in October 1957 as the ADO 15 and evolved into the Austin Seven and Morris Mini-Minor launched to great public applaud in August 1959.

Encouraged by good press reports and with an advanced order for 25 cars, Lawrie Bond tried to persuade a very sceptical Ewart to build these for him at Sharps Commercial's redundant Ribbleton works. Ewart's caution came from having no experience of manufacturing. It appears that Bond knew Col. Gray and Granville in passing (as Bond had briefly been at Dick Kerr's in the 1920s) and managed to get Granville to intervene and persuade Ewart to proceed. Even Granville was at first sceptical but, as he had long maintained in his magazine articles, there was a huge potential market for cheap, everyday transport.

Although it was a very clever design, using a lightweight aluminium stressed skin monocoque body, it was soon realised there were major production and design problems which needed resolving. But nevertheless by late spring 1949, using Bond's original jigs, production slowly got under way with much help from Autoys Ltd, who had an engineering workshop next door, at a rate barely able to meet the ever increasing demand. Incredibly, despite the Minicar being a very basic form of transport, it was soon mimicked by a very short lived near copy, the 'Lamb', produced by an optimistic competitor.

The Minicar's resounding success prompted Ewart to acquire the sole rights, at which Granville was aghast, for he entirely disliked the cheapness of the Villiers two-stroke engine and the car's mechanical design but, regardless of his brother's highly regarded feelings, Ewart pressed ahead and for a consideration of £8,000 gained the rights along with Bond's crucial staff, while Lawrie Bond helped with many of the initial production problems in consultation with Col. Reg Gray, who had been given free reign over the project. It was not long before Granville was engaged to improve the car's mechanical design and suspension. Over the following years, Lawrie and Granville were regularly called upon for ad hoc consultancy work.

The original Minicar's strength, literally, lay in its simple monocoque body and absence of doors which gave it an incredibly low weight of 308lbs. The 122cc Villiers Mk.10D two-stroke motorcycle engine developed 5hp at 4,000rpm, giving it a top speed of just under 40mph, and a 104mpg economy. The production Bond Mk.A however, mostly used the 197cc Villiers 6E engines. Of great importance was the fact it could be driven on a motorcycle licence, but although a modifed Albion 4-speed gearbox with 3-speed and reverse gear was offered, as drivers with only a motorcycle licence were not permitted to use reverse few such 'boxes were ordered.

Unfortunately, as the Minicar had no front brake, it was often driven into the back of another car! It was incredibly basic and spartan, with its crude steel cable and bobbin steering, steel lined cast alloy brake drums and a rigid rear axle which, like Bradshaw's Skootamota, relied on air pressure in the 4.00 x 16" balloon tyres for its suspension. The wheels were of the split rim type, bolted directly to the hubs. The engine, steering and front wheel were all built into a simple, cast alloy fork arrangement with a friction disc shock absorber.

Though the Bond Mk.A was greatly improved over time through Sharps' own in-house work, they encountered major reliability problems even on cars with less than 1,000 miles on the clock. These included body fractures, broken front wheel spindles, fractured cast alloy front fork assemblies, and worse, cases of engines falling out. Ewart and Reg Gray once again called upon Granville who, suffering severe respiratory problems, then moved from London to the drier, warmer climate of Cheshire rather than the damper, Preston and, along with Francesca, duly rented 'The Elms', in Audlem, near Crewe before moving down the road to 'Noxa', commuting to Preston when necessary.

Granville's first task was to set up a proper drawing office at Preston engaging two junior draughtsmen to create production drawings. These young lads' antics and inexperience unwittingly presented Granville with many a comical observation on life for his learned 'Technicus' magazine articles. In one such, he entertained the readers with an earnest discussion by the lads on the much heated debate over first pouring milk or tea into a cup; the consensus was that by putting milk in first and then adding the hot tea, it produced a more fulsome flavour by maintaining a uniform temperature of milk and tea as the tea was poured, whereas adding milk to the hot tea, simply scalded the milk and tainted its flavour. Useful knowledge indeed when picnicking quietly at the roadside admiring your new Bond Minicar.

Having established a set of working drawings - and drunk his tea - Granville then systematically eradicated the problems in the Minicar's design. The biggest of these was traced to a lack of adequate suspension which caused the cast alloy fork assembly to absorb all the road shocks, causing its failure. Granville duly designed a new, fully sprung front fork assembly complete with front brake which later appeared in the Mk.C. He also added a simple rear suspension using an aluminium casting, fixed to each side of the body, in which were a pair of sliding pillars with a stub axle for the wheel. A heavy, adjustable coil spring was fitted each side, supporting the body weight, while a pair of lighter springs damped the rebound. Primitive it may have been, but it provided independent suspension and, as any Morgan sports car driver will attest, sliding pillar suspension really does work; it does however require meticulous regular maintenance, which any proud Morgan owner neglects at his peril!

This improved Villiers 197cc powered Bond Minicar Mk.B was duly launched in July 1951 and was soon followed by the greatly improved Mk.C of 1952 which benefited from better engine mounting and a new worm and sector steering which gave it an incredible 180° arc; this had been developed by Granville for the 'Commercial 3cwt'. The Mk.C also adopted a new, but less expensive 6" trailing arm rear suspension using 'Flexitor' rubber bonded bushes to provide a totally maintenance free suspension. (See also Appendix 5 on Walter Adams). The 1953 Mk.C got a Dynastart electric start - a great improvement over the earlier and rather crude starter motor atop the cylinder head, driving the combined flywheel-magneto. The standard starter was a through-dash cable hand pull start; this was fine provided the front wheel was pointing straight ahead for direct pull - otherwise the cable tended to snap forcing the driver to open the bonnet and operate the kick-start!

In 1953, Pashley of bicycle and motorised rick-shaw fame, also proposed a similar design to the Bond Minicar, but their project went no further,

Commercial and Unicycle

Encouraged by progress on the original Minicar, in 1951 Granville set about building a simple three-wheeler, 3cwt truck which, in the absence of a suitable Villiers engine, employed a Brockhouse built, Indian 'Brave' 250cc ohv motorcycle engine and three speed gearbox driving the front wheel, Bond fashion. The squared off body had an open rear cargo area. Like the Land Rover prototype, the driver sat in the centre, but the 3cwt 'Commercial' looked more like a fair-ground dodgem car than a work-horse. Bradshaw even developed a clever 'worm and sector' steering system which allowed the front wheel to turn through 180° within its own diagonal length, which made it ideal for use within small factories and market gardens. In essence a light duty version of the ubiquitous Lister 3-wheel

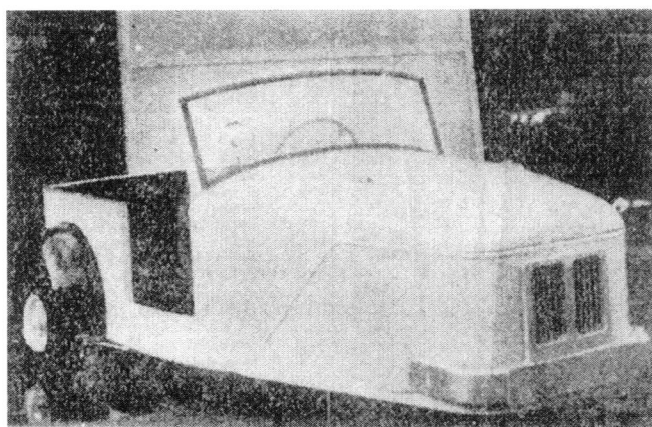

A very early Bond Commercial

platform trucks, and being capable of 40-50 mph and 70+mpg, it was ideally suited for scurrying around airfields and industrial sites. This steering was then fitted to the Mk.C Minicar.

It was a surprise entry at the November 1951 Cycle and Motorcycle Show, yet despite enthusiastic plans for a van version and wide-spread pre-production publicity, it appears only one prototype 'Commercial' was built along with four or five motorised chassis before the project was abandoned, probably in favour of the modified Bond Mk.B Minitruck (with short, flat platform body) and the very boxy Minivan, both of which sold reasonably well.

Doubtless influenced by his earlier bolt-on Bumblebee engine, Granville saw great potential in the Commercial's Brockhouse driven wheel, gearbox and associated steering gear as a self contained power unit which, using a simple four bolt attachment, could be fitted to any hand-truck or invalid car. Alas, the 'Unicycle' power unit, as it became known, failed to attract much interest and only a handful were built.

Reliant Motors

Though remarkably crude vehicles, the Bond Minicar sold in large numbers throughout the 1950s, at home and abroad, to an eager public desperate for independent travel on the cheap - and at 45mph and 90+mpg it was more family friendly for the motorcyclist, who could legally drive one, than the alternative motorcycle and sidecar. With competition from Reliant's Regal, three wheelers became ever more luxurious and acceptable, particularly during the 1956 Suez Crisis, but alas, the Bond Minicar could never match the car-like engineering standards and comfort of the Reliant, for it must be borne in mind that Thomas Lawrence Williams had designed the original Raleigh Safety Seven of 1933, which he continued to produce as the 'Reliant' after Raleigh ceased production. Tom also produced an ohv aluminium alloy version of the Austin Seven engine which gave the Reliant its car like qualities and high power to weight ratio. Furthermore, Reliant produced their own sturdy gearbox and axle and developed extensive skills in glass-fibre technology which, together with their high standard of engineering, allowed them to set up indigenous car companies in Israel, Turkey and Greece, as well as produce the truly superb, trend-setting, Reliant Scimitar GTE sports-estate car of the 1970s.

Reliants were of course keen to see off the Bond Minicar and its not very succesful replacement, the short lived Hillman Imp engined Bond 875, but they particularly saw the new Triumph Herald based Bond Equipe as an ideal companion to their Scimitar and, more importantly, the trading agreement between Bonds and Triumphs would give Reliant a much larger distribution network and parts backup. With Bond Cars now in a perilous financial situation, unable to pay either Hillman for the Imp engines in their 875, or Triumph for their chassis, and with Reg Gray now nearing retirement, it was agreed to sell Bond for a 'nominal figure' to Julian Hodge's Reliant Motor Company in 1969.

The Bond 875's major failing was that, as with all glass-fibre bodied cars, small scale production does not yield attractive economies of scale. Against this, its inability to fall the right side of the 8-cwt maximum weight saw much imaginative skimping in trim and finish which did little to endear it to the public. Reliant always used the lightest bodies in the de-luxe saloons to ensure that trim levels remained at the highest standard - the heaviest and strongest bodies ended up in their vans. The 875 was soon axed, leaving Bond's almost redundant Preston factory available to produce the unique, Reliant designed 'Bond Bug' from kits shipped to them from Reliant's Tamworth factory, but in the event only a few Bugs were assembled at Preston. The Bug certainly offered an exhilarating, if noisy, driving experience beyond belief!

Unfortunately, the merger between Leyland (which owned Triumph) and BMC in 1968 brought the unforeseen axing of the popular Triumph Herald in 1970 and with that, the intended Equipe replacement failed to materialise along with the benefits of Triumph's distribution network. Rationalisation by Reliants brought the inevitable closure of Bond's Preston factories in late 1970.

The full story of Lawrie Bond (1907-1974) and the Bond Minicar is told in *Lawrie Bond - the man and the marque* by Nick Wotherspoon. The reader is also drawn to the Reliant story, *Rebel without applause*, by Daniel Lockton.

Meanwhile, Ewart's garage group had become part of Evans-Halshaw after his death, passing to Dutton Forshaw Group and by the mid 1970s to LCP Holdings. Ewart's hire-purchase and finance arm, the Astley Industrial Trust, run in association with the District Bank and National Provincial Bank, passed to the Mercantile Credit Co.

Villiers acquired J A Prestwich Industries in 1957 and soon after, BSA's industrial engines business. By 1963 Villiers had become part of Manganese Bronze, who in 1966 had bought Associated Motor Cycles (Matchless, Norton), but the combined engine businesses were unable to compete on quality against Briggs & Stratton of the USA and various Japanese makes. Out of the collapse of AMC in 1966 emerged Norton-Villiers-Triumph, but the famous Villiers engine business survived independently, briefly producing industrial engines before selling the manufacturing rights to the SAEPL Company of Madras, India in the 1980s. In 2000, the Villiers Group pulled out of engineering altogether and transformed itself into a lucrative health care business as Ultrasis Ltd.

Toroidal Minicar engine

While at Bonds, Granville continued his experimental work on the toroidal rotary engine. It seems he called upon the engineering skills of Autoys and it appears up to three further examples were built. He certainly did try to interest Ewart and a drawing dated 14th August 1950 shows a 300cc rotating chamber toroidal engine for the Minicar measuring 7" diameter and 7" in length. This had a conventional car type flywheel housing containing a flywheel, V-belt drive to a dynamo, a single plate clutch and very short drive shaft with bevel gear to a transverse gearbox bolted to the flywheel housing. The 3-speed and reverse gearbox had through gate selection like a motorcycle gearbox (R-1-N-2-3) and was fitted with cable start from a hand lever. This was to be mounted with the centre line of the engine in line with the centre line of the driven front wheel. The chain final drive was via a transmission brake.

The assembled engine, gearbox and final drive measured 16" long, 11" wide and 8" high. Sadly, this innovative design was not to be.

In-line twin

It is also clear in a letter dated 1st January 1953 that Granville had implored Ewart to fit a cleaner, smoother running, lightweight four-stroke vertical twin to improve not only starting and passenger comfort, but also to raise the Bond's image from a cheap £300 runabout, to a more appealing car-type design for the general public to compete against the Reliant.

In early spring of 1952, Granville had designed at Audlem a conventional four-stroke, 5 RAC hp, in-line air-cooled twin cylinder engine especially for the Bond. The 402cc (63.5mm x 63.5mm) side valve twin would develop 12hp at 4,500rpm. Unusually, it used a single throw crankshaft design in which both pistons rose and fell at the same time, but within separate barrels each with their own combustion chamber and head. But as good as Granville's toroidal and in-line twin might have proven to be, Ewart was not an engine maker. Neither was this the time to experiment, for crucial to Bond's cash flow was

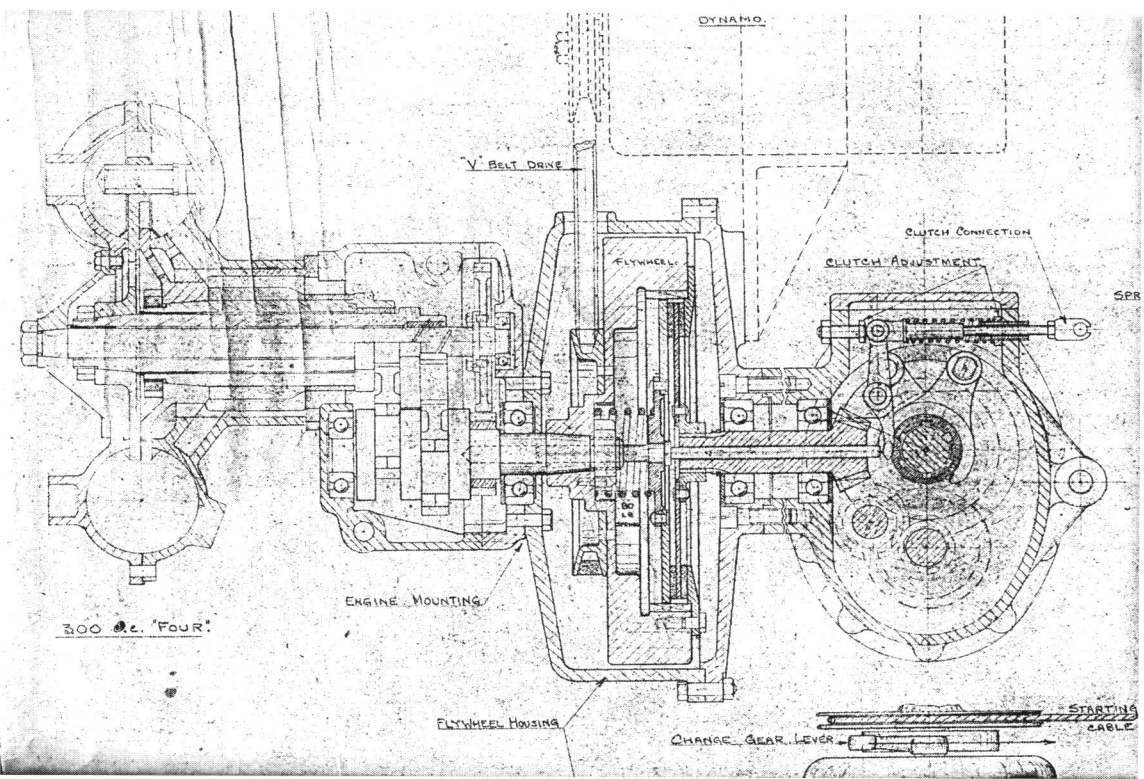

Part of a 1950 drawing of a rotating toroidal chamber engine, with transmission, almost certainly for the Bond Minicar (GB)

a commercially successful car powered by a proven and reliable engine with an established nationwide spares and servicing facility - which the Villiers offered. More importantly, the utilitarian Minicar was yielding a healthy profit and required only a largely unskilled, if haphazard and lackadaisical, work force to churn out some 50 cars a week. They also had a healthy order book, despite delivery delays of four months, against stiff competition from the AC Petite (two months delay), the more up market Reliant Regal (three weeks delay) and an ever increasing competition from Continental Europe. Furthermore Ewart was quite content with the Minicar and the Villiers engine as it was - after all, Villiers were more than happy to build or modify any engine to meet Bond's requirements and even succumbed to Ewart's pressure (possibly using the threat of Granville's in-line twin as an incentive) to build the much praised Villiers 250cc unit.

Unperturbed at Ewart's disinterest, Granville redesigned his twin as a 500cc ohv motorcycle unit and in an effort to increase fuel economy to approach that of the two-stroke Villiers engines, he proposed partial recycling of the exhaust gases via ports drilled into the cylinder just above the piston at bottom dead centre which were exposed, two-stroke fashion, as the piston descended (described below). He carried out experiments on his Ford Consul car; the results greatly impressed both Ewart and Fords. Confident of its commercial success, in late 1952 he proposed a small four-wheeler Bond car powered by this new fuel efficient twin cylinder engine, fitted with an easy, but undefined, 'hand starting device' (Provisional Patent 17019). This appealed to Ewart, but it was conditional upon a complete car being presented to him for inspection. However, Granville was personally unable to fund the project and now set out to sell outright his fuel economy design to Fords, Standard-Triumph and even Harry Ferguson. He confided to Ewart in a letter of 4th January, 1953 that, "...for the first time I have not the money to pay last month's rent (£18) and grocers (£13)... if Harry Ferguson doesn't give me a fresh start I just do not know what I shall do." He signed off: "It is a pity I was not born with less technical vision and with more commercial acumen. You were the lucky one!" It appears that he finally got Ewart's backing and the 500cc twin engine was built, possibly by specialists in Blackburn (probably, Projects & Developments Ltd).

Though the engine failed to enter production, in 1956 Col Reg Gray approached Granville to develop a new 'motorcycle' engine to power a range of future Bond Minicars. Bradshaw's immediate response was to redesign the ohv version of the 500cc in-line vertical twin, still using a single throw crankshaft, which he had developed for Shay in 1953 (described later). However, he soon realised that the existing Bond Minicar chassis could simply not handle the extra power and performance safely, so he abandoned further work. As an aside, in 1967, Reliant engineers had fitted a Ford 1600GT engine in their truly superb 4-wheel, 700cc Rebel saloon - at 200hp per ton, the Ford engine gave the 'Rebel GT' phenomenal performance and while it predated the 'original hot hatchback' Golf GT concept and, even though the Rebel chassis had very sporty handling, such potential performance would have been too dangerous for the British roads at that time. If only Reliants had persevered!

Fuel economy device

One major problem pre-war (and especially with Pool grade war-time petrol which varied between 67 and 80 octane!), was pre-ignition (pinking) which prevented the development of more powerful and fuel efficient high compression four-stroke ohv engines. With petrol rationing ending on 26th May 1950 and the abolition of Government price controls in April 1953, branded octane graded petrol gradually became more widely available, but with a 2/6d tax levy.

Engineers have long dreamt of higher octane petrol allowing them to design more powerful, smaller under one litre engines, but it would not be until the mid 1960s when 100 octane (5-Star) petrol arrived (the same grade as war time Avgas 100 for Spitfires). Alas, by this time, much smaller, more fuel efficient and ever more powerful engines were being developed in continental Europe using their standard, low-grade, 95 octane (2-star) 'Regular' petrol.

Engineers had also long dreamt of designing a variable compression engine which would ensure maximum temperature and thermal efficiency, with minimum pre-ignition, at all throttle settings. Just before the war, an American engineer had proposed such a high compression engine, offering greater power at low throttle settings, by fitting additional combustion chamber capacity accessible only by mechanical valves at higher speeds; this would reduce the effective compression ratio and prevent pinking.

Pre-ignition had long been controlled by tetra-ethyl lead additives which helped delay burn rates, but it was also known that recycling some of the 'inert' burnt exhaust gases back into the combustion chamber delayed pre-ignition, allowing the freshly induced volatile gases to explode nearer to top dead centre, providing a much greater downward pressure on the power stroke and increasing the engine's efficiency and economy. However experiments had shown that if these exhaust gases were induced with the fresh mixture, it both reduced the oxygen intake and further dampened the flash-rate of the burning fuel mixture.

What Bradshaw now proposed was injecting some inert exhaust gases just ahead of the rising piston. This would not dilute the freshly induced volatile mixture but would only mix with a very small portion of this fresh mixture at a lower strata, preventing pre-ignition. The hot, inert recycled exhaust gases were already under partial compression and thus helped raise gas pressure in the cylinder, improving volumetric efficiency, power, performance and economy. In later experiments it was found that on each successive cycle, the pressure from these recycled gases gradually raised that in the combustion chamber by around 30psi. He duly submitted four Provisional Patents Nos: 6930, 10582, 23927 and 25956, of which there are, alas, no specific details.

Clearly for it to work, the engine had to have an even number of cylinders such that as one was on its exhaust stroke, it bled a metered volume of inert exhaust gases through a control valve linked to the throttle, and thence to the next cylinder starting its compression stroke, via ports drilled into the cylinder at bottom dead centre.

Granville now approached Sir Rowland Smith at Ford's, who promised him 'every assistance' and supplied a specially sleeved cylinder block which Bradshaw duly modified and fitted to his 4-cylinder, 1,508cc Ford Consul car (reg no ECK 738). By late 1952 Bradshaw had completed his own tests on the Consul and reported average fuel economy increased from 28mpg to 38mpg together with increased top speed from 68mph to 76mph and a noticeable boost in acceleration from 40mph, just as one gets from a modern turbocharged car. In a head to head comparative test by his son,

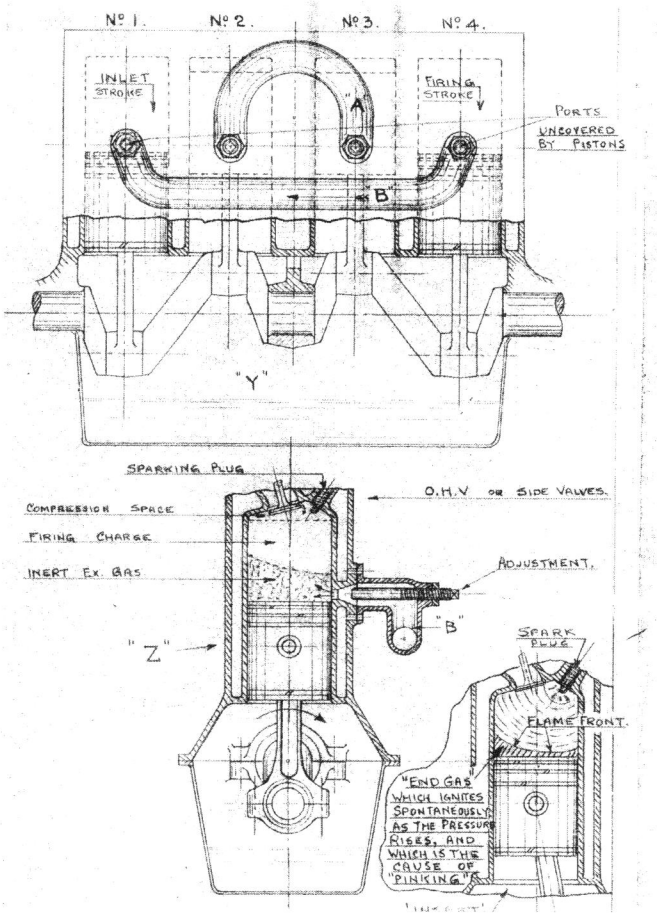

Bradshaw's fuel economy system of 1952 (GB)

Geoffrey, who had a standard 2,263cc 6-cylinder Ford Zephyr, no discernible difference was noticed between the two engines other than the Zephyr's better top speed of 83mph. Granville estimated that his system had raised the 'effective octane rating' of the low grade petrols then available to 85 octane and, accordingly, engine designers could now raise the standard 6.8:1 compression ratio to 7.5 or 8.0:1 with further benefits of fuel economy and power.

Remarkably, the modification gave the 4-cylinder engine near perfect, vibration-free running which prompted Granville to mischievously suggest a most beneficial application in Edward Turner's notoriously rough Triumph twin motorcycle engine! Alas Granville's archives do not record whether Turner was approached, or, if he was, what his comments were.

After 3,000 miles the Ford Consul engine was stripped and it was discovered that the pistons ran very much cooler than normal, but there was evidence of greater than normal carbon build up from engine oil, which he put down to the induced exhaust gases acting as a protective thermal buffer at the piston crown. Granville was confident he could resolve that problem.

When he had an example of his 500cc in line twin so modified, he reported 75mpg. Keen now to get Ewart to back him in the construction of two Bond prototype cars which could compete favourably with the new baby Austin and, knowing that Ford were greatly impressed by the fuel economy device, he offered Ewart 15% of any deal he secured from Ford. Others in the motor industry were now also expressing keen interest, so Granville began seeking licence agreements, including from Harry Ferguson.

Harry Ferguson

Born in 1884 on the family farm in Dromore, County Down, Harry George Ferguson and Granville Bradshaw had a similar start in life. Harry had joined his brother, Joe, in Belfast as a cycle and car repairer. Excited by flying, Harry had built his own Blériot type monoplane in 1909 powered by a 35hp V-8 air-cooled JAP engine; he became airborne in December 1909 and is said to be the first Irishman to fly. He had a narrow escape in 1910 when his engine faltered over water, just managing to coax enough power from it as the Blériot kissed the water's surface to make a safe landing on *terra firma*.

Opening his own business as the Austin motorcar dealer for Belfast, he then added American made tractors. He built his first tractor in 1933 and entered negotiations for its construction with David Brown, at the Huddersfield Gear Works, followed by a licence with Henry Ford as the Ford-Ferguson. Finding the standard tractor drawn ploughs heavy going, he also invented in 1936 a controllable, 3-point linkage with which he would revolutionise farming.

Both Ford and David Brown produced limited numbers of the Ferguson design, but by 1945 negotiations to renew the contracts had floundered, at which point Ferguson secured a contract with Standard cars in Coventry to build a new lightweight Ferguson TC-20 tractor using a Standard 4-cylinder engine. This contract maintained employment at Standard's war-time Banner Lane factory while the much loved 'Grey Fergie', as it became known, secured the future of both companies.

After the war, Harry Ferguson began work in secret with Fred Dixon and Anthony Peter Roylance 'Tony' Rolt on a revolutionary four-wheel drive car chassis, whose origins lay with Dixon and Rolt's pre-war, four-wheel drive land-speed record project. Dixon was a well known tuner of Riley cars while Rolt was a racing driver and serving officer in the Rifle Brigade. As a prolific escapee, Rolt served time as a POW at Colditz and became instrumental in the famous Colditz glider episode. In 1950, they teamed up with Ferguson to form Harry Ferguson Research and with Claude Hill (an experienced Brooklands racer and ex-Aston Martin engineer) developed a Formula 1 racing car. Driving for Jaguar, Rolt won the 1953 le Mans 24-hour race.

While these projects were secretly undergoing development, Harry Ferguson revealed in January 1953 that he was looking for a small, 1-litre air cooled engine for a proposed 6-seater 'World car'.

At this point, Bradshaw offered him the chance to invest in his 500cc twin, a proposed high performance racing version (with a centrally mounted $1^1/_2$lb vibration damper piston set between the two connecting rods), the fuel economy device and his own proposal for a Peoples' car which he enthusiastically described at great length in an introductory letter. Bradshaw's technically sound design was aimed at the lower income bracket, on the lines of the 500cc Fiat Topolino, but with a roomy and properly constructed, weatherproof body with proper glass windows, sturdy fittings and an interior heater as standard. Designed to suit the young housewife, burdened with shopping and children, he proposed it should have two adult seats at the front with a single bench seat behind for two children or, in 'an emergency', an adult. Given that his war-time prophecy was that post-war cars would be of an air-cooled rear engined design, he conceded that front engined cars would be easier to sell especially as a rear boot lid could be left open to accommodate an excess of luggage - or children.

The car should also be able to withstand unmade roads and the trauma of novice drivers; it should be half the weight and cost of the new 8hp Austin A40 but should avoid miniature wheels as fitted to the Bond as their lower brake drums were prone to being flooded in puddles and brake linkages damaged in ruts. He much preferred the simplicity and reliability of leaf spring suspension and accurately predicted that cars of this class would retain a light but compliant chassis and use, he proposed, modern glass reinforced plastic panels. For economy in running and repairs, the cars would be built from proprietary parts readily available at any garage. His new vertical twin was, he proffered, the perfect choice for such a model.

Ferguson, Major Tony Rolt and Bradshaw duly met at Coventry on 8th January, 1953 - letters reveal that Granville's fears of snow that day proved unfounded! Though the meeting went well, Bradshaw increasingly felt ill at ease with Ferguson's long term policy of a thorough testing programme; he was also somewhat dismayed at their dismissal of his prediction that his car would, by using modern plastic panels, weigh only $6^{1}/_{4}$ cwt (700lbs). In his defence, he pointed out that the Bond three wheeler only weighed 4cwt.

Unbeknown to Bradshaw, Harry Ferguson was a this point only really interested in testing the modified Ford Consul engine which Bradshaw had currently sent for examination to Mr Grinham, Technical Director at Sir John Black's Standard car company. Granville even confided in Harry that he had given Sir John his first job at ABC and that whenever Sir John saw him at Standard's factory, he tended to keep well away! Nevertheless, pleased with their interest in the fuel economy device, Bradshaw sought a mutually attractive contract to act as a consultant engineer to Ferguson for a six month's trial at £1,000, and as a matter of faith, sent his five relevant provisional patents to Ferguson for scrutiny, together with general arrangement drawings of the engine and his proposed crankshaft damper for vibration free running. But by the end of January, it was clear Ferguson had no interest in Bradshaw's car, the revolutionary use of plastics, the 2-cylinder engine nor retaining Bradshaw as a consultant engineer. Yet while his interest in the economy device remained strong, Ferguson was unhappy at the device being scrutinised at Standard's and demanded that the Consul engine be removed from the car and be fitted with a suitable adaptor for bench testing at Coventry with and without the economy device active. Bradshaw however was adamant the engine should be road tested to better observe its benefits of improved economy and performance, but by now the relationship had begun to sour and negotiations were abruptly terminated.

In September 1953, Harry Ferguson merged his tractor business with the Canadian Massey-Harris company, and a new 12 year contract with Standard was signed; but in 1957 Standard sold their tractor interests to Massey-Harris.

In 1955, through his Ferguson Research team, Harry Ferguson finally developed his unique R5 4-wheel drive estate car, followed by the Coventry-Climax engined P99 racing car successfully driven by Sterling Moss in 1960. Harry Ferguson died suddenly on 25th October 1960. The Ferguson formula four-wheel drive system was adopted by Jensen cars for their revolutionary Jensen FF of 1966. Harry Ferguson's biography was later written by Colin Fraser. Tony Rolt died on 6th February, 2008.

J E Shay Ltd

Undeterred by Ferguson's rejection, Bradshaw had high hopes for his Peoples' car, but with his work at Bonds finished, he retuned to London and rented 1, Glazebury Road, Fulham, London W.14 where he turned his attention to selling his other patents. These included a waste disposal unit which led to a meeting with Messers J E Shay Ltd of Basingstoke. Formed by John Reginald Sharpe and Emmanuel Kaye (later Sir Emmanuel), 'Shay' came from their surnames.

During the war, as precision engineers, Shay had produced many aircraft components but post-war, they sought general engineering contracts and were, thus, very keen on adding Bradshaw's hydraulic waste disposal unit to the range of products they already manufactured for the ironmongery and horticultural trades; this would also take up spare capacity in their recently established hydraulics department. They had also very recently bought Power Specialities Ltd, who produced the 'Rotoscythe' grass-cutter; they also owned the Lansing-Bagnall company, which they had bought out of bankruptcy in 1943 for £3,000. Formed before the war at Isleworth, Lansing-Bagnall had pioneered battery and petrol driven platform and fork-lift trucks; they had also produced a small aircraft tractor-tug in 1940.

While being shown around Shay's Basingstoke factories, Bradshaw was introduced to one of the works foremen who, by chance, was a keen motorcyclist and knew Bradshaw's work on the ABC cars, motorcycles and engines. On now being made aware of who they had in their presence, John Sharpe and Emmanuel Kaye confided in Bradshaw that they had, for some time, been considering producing a small economy car. Granville duly seized the opportunity and offered them both his new vertical twin engine and his Peoples' car, but for now, Shay's more pressing need was refining their Rotoscythe.

Rotoscythe engine

The unique Rotoscythe concept of a rotating blade, or disc, was pioneered and patented in 1933 by Power Specialities Ltd of Maidenhead, established by M Cobourn; they later moved to the Bath Road, Slough. It was a vast improvement over the traditional agricultural reciprocating cutter (typically found today in hedge trimmers) which was very prone to damage from stones and required constant oiling to prevent seizure. A further patent by Harold Inderwick, for Power Specialities, came in 1939 for a moss-scarifying tine attachment to the rotor cutter disc to which were also added vanes to 'vacuum' the debris into a rear sack (the opposite of a 'Flymo') and as a result, these early Rotoscythes were often referred to as 'vacuum cutters'. The rotary-scythe system is now the standard form of grass cutter.

Originally powered by a 98cc two-stroke Villiers 'Midget' engine, with massive flywheel magneto, the engines soon suffered from bent crankshafts under heavy loads when tackling thick grass. Villiers were unwilling, or unable, to design and build a more robust engine causing Shay to urgently seek an alternative unit and with Bradshaw now on hand, they specified a single cylinder $1^3/_4$hp to 3hp petrol engine for the Rotoscythe, preferably fuel injection rather than carburettor, with a vertical output shaft and 'silent' running. Granville immediately proposed an 89cc version of his patented oil and air cooled Bumblebee V-twin engine, fitted with a simple and efficient air-flow, vane-damped engine governor to allow for heavier loads on the engine when cutting dense tufts of grass. He also proposed fitting a cam on the output shaft to operate a reciprocating hydraulic pump driven power take-off (PTO) as well as a splined PTO shaft at the top of the engine for rotary attachments, such as the 'Tarpen' or 'Heli-strand' flexible drive cutters and hedge trimmers. In addition, he proposed a PTO driven rotary cultivator/tiller attachment.

A suitable specification was soon approved with Alfred Arnot, their chief engineer, and so it was, that on 21st September 1953, Bradshaw was contracted to design the new Rotoscythe engine for a fee of £200. He further proposed a small air-compressor attachment to power all manner of garage and industrial tools, but while a very attractive proposition, Shay deferred their decision preferring to get a basic machine onto the market as soon as possible. By late October 1953, Bradshaw had presented them with his general arrangement drawings and at that point, further correspondence in his files on the Rotoscythe ends! Of the waste disposal machine, nothing more was ever heard.

The outcome was a unique multi-purpose 'BUX' two-stroke engine of around 150cc; but whether this was designed by Bradshaw is not known. Shay duly produced a BUX powered 18" Rotoscythe motorised pedestrian grass cutter, a 20" self propelled rotary grass cutter and an 18" 'Rotogardener' rotary cultivator, patented in 1954. The BUX was largely replaced in the late 1950s by the 'Aspera' two-stroke engine. Shay later offered a 14" battery powered motor mower, but around 1960-62, these products were sold to Webbs, of mower fame. Around the same time, Dennis Selby, who had been with Power Specialities from the start as their salesman, left to form the Mountfield horticultural/mower business. While at Shay, he was instrumental in pioneering mower service centres at retailers who sold the Rotoscythe products - other manufacturers required the machine to be returned to their factory. Selby also introduced the revolutionary Swedish 'Flymo' hover mower to Britain - what a shame Power Specialities hadn't reversed the 'vacuum' to form a hover-mower back in the 1930s!

Bradshaw's Peoples' Car

With the Rotoscythe project under way and Shay still showing no immediate interest in his Peoples' car, Bradshaw resumed his search for buyers of his fuel economy device and offered *The Motor Cycle* an exclusive review in September 1953. While they turned this down, Temple Press did raise his fee to 6gns per 1,000 words for his regular 'Jottings' articles! He then reported his fuel economy findings in one of his 'Technicus' articles of late 1953, disguising his identity by saying, "I understand that over 5,000 miles have been covered on the road and the device shows from 30% to 40% (improved) fuel economy.. (and that) really steep hills can be climbed on top gear with no signs of pinking".

In a surprise move in early September 1953, Sir John Black announced his new, compact monocoque construction Standard 8 (£339 plus Purchase Tax) proclaiming it as Britain's first post-war 'Peoples' car'. It was a more modern, direct competitor to Alec Issigonis' Morris Minor Series II (£573 plus PT) and the new 2-door version of the Austin A30 (£475 plus PT), which since Austin and Morris merged in 1952 to form BMC, now used the same drive train. These were joined by Ford's new 100E, which was set to replace their much loved 'sit up and beg' Popular E93. The new 2-door Anglia 100E was offered at £360 (plus PT) and the 4-door Prefect at £395 (plus PT).

Confident of being able to undercut the Standard 8, not from mass production methods but by careful and efficient design - the forerunner of 'value engineering' - Bradshaw earnestly proposed that Shay sponsor the construction of his two prototypes using his existing 500cc vertical twin engines. They

People's car chassis and engine (GB)

Left, road testing the prototype chassis. Right, engine in Utility (GB)

approved and in mid October 1953, a formal agreement with Emmanuel Kaye secured the exclusive rights to Bradshaw's cars, the 500cc engine and the outright purchase of his fuel economy device with which Shay's hoped to earn a quick return through licensing agreements. For this, Bradshaw received a salary of £3,000 p/a as a consulting engineer plus a quarterly royalty of 2% nett for each car up to the first 1,000, decreasing pro-rata to 1% after 2,000 cars. The agreement was, however, conditional upon the satisfactory testing and acceptance of the agreed prototypes. Bradshaw duly had his experimental 500cc engines, then stored at Crewe, delivered to Shay along with the jigs and tools held by Bonds at Preston. He then moved himself and Francesca King from Fulham to a former vicarage, Little Heckfield House in Heckfield, four miles north of Basingstoke, which he rented from a Colonel White who was temporarily in America on business. A former vicar living at Little Heckfield House was the father of Anthony Trollop, the famous novelist.

In keeping with his war-time design philosophies, his modern Peoples' car would be simple to manufacture, easy to repair, cheap to run and maintain. He also proposed that the car should be offered in a wide variety of body styles to cater for different markets and customer needs, using a common platform chassis, just as did Rolls-Royce and others for various coach builders. Bradshaw rightly maintained that in these post war years, the customer no longer wanted an austere 'cheap' car, but one which was luxurious, spacious, sporty but above all, inexpensive - times haven't changed! He also maintained that, as most post-war cars were now being bought on hire purchase, buyers could afford to go 'up market' for 'a few bob more'.

Left, Bradshaw's original car proposal with cast alloy bumber. Right, pre-production Utility with rear door open (GB)

To that end, he proposed a simple, fabricated open channel, or box section, parallel ladder type steel chassis, on which was a flat floor pan. By having front wheel drive, independent rear suspension and with the absence of a rear driven axle, the spare wheel and tool box could be kept in a well under the floor between the wheels allowing a very low loading height. He initially proposed the 6-gallon fuel tank and battery were located under the bonnet, but in the final event they were mounted low in the chassis, ahead of the rear axle and under the passenger seat. By having front wheel drive it was a very simple matter then of producing a three wheeler chassis, enlarging Shay's potential market share. The simple chassis had an underslung,

Granville with the Bradshaw Utility (GB)

front and rear transverse leaf spring and upper wishbone suspension with simple telescopic shock absorbers (a design later employed on the Triumph Herald). 'Silentbloc' bushes were used throughout. Steering was by worm and nut.

The four-wheel chassis had a wheel-base of 6'8", front track of 45", rear track 44" and an overall length of 10'5". Its intended unladen weight was 800lbs for the Utility estate car.

A chassis was quickly built-up and a rudimentary body rigged up for road testing. Once happy with the basic chassis design, work began on the bodies for which Bradshaw proposed lightweight, but immensely strong, hand-laid glass-fibre construction with lightweight Perspex sliding and fixed windows. There were to be three body styles; a 4-seat Utility estate (not dissimilar to the 1954 Hillman Husky) with sliding doors, full height lift up rear boot-lid and round windows; a 4-seat saloon with a windowless, fabric rear body/convertible; a two seater sports car with a dickey seat for luggage or the occasional passenger and a shapely 2-seat, rear engined sports car for export markets, not dissimilar from the Volkswagen Beetle convertible. As with the Citroen 2CV, the interior was incredibly spartan with bench seats of a simple foam rubber matting slung over a bent, tubular steel frame. In all versions a simple kick start was proposed; this comprised a pedal in the driver's compartment which was 'prodded' briskly by the foot, pushing a long rod connected to a quadrant kick start, as often found on large motor mowers.

In the event, only three prototypes were built within Shay's large Nissen huts: a 4-seater fabric bodied saloon, 4-seat Utility estate and a smart sports car. The glass-fibre bodied saloon was truest to Bradshaw's original design with, in plan view, curved sides between the wheel arches in a uniform 9'0" radius for which he proposed a large hinged door. The car's nose also had a 9'0" radiused bow to which was fitted a slatted, 30" wide cast aluminium grille and a pair of front bumpers (complete with over riders) cast in aluminium alloy and bolted to short, deformable steel brackets. The bonnet was a massive one piece clam shell design which included the upper wheel arch wings.

The Utility was redesigned with the assistance of Shay's design engineer, Edward Wright (who designed the Shay Rotocultivator) and Mr Bowden, their prototype engineer. This model now had a parallel sided aluminium panelled body with a pair of narrow sliding doors instead of the large bowed door. It had a smaller conventional bonnet and a one piece rolled steel bumper.

The sports car had small hinged doors and rear dickey seat/boot with a very small bonnet aperture, much like a Reliant 3-wheeler. The body closely followed Bradshaw's proposed rear-engined export

Left, Bradshaw at the wheel of the Shay Sports, showing the elegant, sporty lines. Right, nearside view of the Shay Sports with Francesca at the wheel (GB)

model, but this now more closely resembled the later 1955 Kieft style sports car with larger rear and front wings. It was almost certainly built for Shay elsewhere and, most probably, in glass-fibre rather than aluminium. Surprisingly there was no provision for a hood and, in fact, as the one piece windscreen was totally frame-less it would be incapable of taking one. Thankfully, the electrics were entirely conventional and Shay wisely avoided Granville's proposed audible semaphore direction indicators!

Most remarkable was the transverse, twin-cylinder 488cc (70mm x 63.5 mm) ohv engine which developed 12hp at 4,500rpm. This was developed from the 402cc side valve unit he had proposed for Bond, but Shay were keen to have an overhead valve engine despite Bradshaw warning them against it on the grounds of ohv clatter and that motor dealers were far from enamoured with ohv engines, much preferring the more reliable L-head side inlet, overhead exhaust layout. Bradshaw even brought Jack Emerson (his former ABC test rider - now Jaguar's Chief Tester for their racing and LeMans cars) down to back him up, but Shay had their way in the end and the tall, ohv engine was duly built.

The ohv engine retained the single throw, $1^{3}/_{8}$" diameter crank pin with both pistons rising and falling simultaneously in their own air cooled barrels. It used $1^{1}/_{8}$" exhaust and $1\ ^{3}/_{16}$" inlet valves hidden under cast alloy rocker box covers. It was mounted well ahead of the front wheels, with chain drive from the cone clutch to the transverse, three-speed gearbox mounted between the engine and the chain driven final drive and differential encased in an alloy housing. Drive to the front wheels was by the usual half-shaft to Bradshaw's own design of rubber cush-drive, spring loaded, constant velocity joints. Following the launch of Alec Issigonis' Mini in 1959, Bradshaw approached Mr Baston at Hardy Spicer in November 1959 offering them his constant velocity universal joint design, but as Bradshaw had not taken out a provisional patent, they declined to take the matter further. Col Reg Gray then persuaded Bradshaw to approach Birfield's, who had built the Mini's spherical constant velocity joints, but Bradshaw declined. However, in October 1961, having seen a new BRD Metalastic universal joint at the motor show, Reg Gray persuaded Granville to allow details to be sent to Metalastic's (a subsidiary of the John Bull Tyre company) major competitors, the Avon Rubber Company who, having expanded from tyres into suspension units, believed their rubber technology would be of benefit to Bradshaw's design; on 19th October 1961, he duly submitted a Provisional Patent No. 037448 and at the same time also sent drawings to Wellworthy (then part of the major Associated Engineering automotive components group) who had produced most of his piston and piston ring designs, but neither showed any interest and nothing more was heard.

The proposed kick-start gave way to a proper Lucas electric starter motor with Bendix drive to the exposed flywheel. A belt driven dynamo had a cooling fan attached. A column gear selector was employed while the brake and clutch pedals were fitted to a chassis mounted pedal box. The cable operated throttle pedal to the forward mounted carburettor was fitted to the bulkhead footwell.

Maintenance was made intentionally easy by using just two sizes of spanner, just like the Royal Enfield Bullet motorcycle, allowing the engine to be completely removed and stripped by any garage or competent mechanic in the least possible time, but such time saved was offset by the use of hidden, inboard, front drum brakes also used by the Citroen 2CV.

Bradshaw's records on these cars stop in January 1954 with a letter to Harry Louis, editor of *The Motor Cycle*, dated 5th January, inviting him to test his new Peoples' car, which he hoped to have on the road in February. Certainly by early spring 1954, both the sports car and the Utility estate, finished in a mid-blue livery, were being road tested and receiving favourable reports of fuel consumption around 70mpg.

Priced from £340, it would have been the ideal Peoples' car, but alas they never entered production and in later years Bradshaw felt that Shay had 'robbed him of his design' - but in what context is not clear. Quite why Shay stopped is also not known. Geoffrey believed 'big business' had put pressure to bear on Shay not to proceed and there are certainly subtle hints in Bradshaw's correspondence with Shay that Emmanuel Kaye was becoming increasingly cautious with the project as more and more of the major motor manufacturers released ever better budget cars with more luxurious appointments. It is entirely feasible that they now saw their very spartan car as no longer viable and thus, a potential drain on their resources which were better directed at their Lansing Bagnall operations.

Much to Bradshaw's utter dismay, the People's car project was finally abandoned in early 1954 and, his work done, he moved to Hill Cottage, Eversley renting from General Lindsey, who was stationed in Burma. Keen to develop the Utility further, by now registered OAA 438 (a Southampton CC registration of spring 1955) as a 'Bradshaw', he approached the local garage proprietor, Mr Hopkinson of Finchampstead to take the project up. Hopkinson had done much trials driving and had also supplied Rotoscythe products, but soon after this approach, Bradshaw moved to near Chichester and all contact was lost. With the Suez crisis of late 1956 making fuel economy uppermost in everyone's minds, the Bradshaw Utility may well have stood a good chance of success against the short lived Anzani 'Astra' Utility, Gill saloon and the imported Goggomobil, but no further development work took place, and the car remained outdoors, undercover, for many years before being sold in 2006 for restoration by Hopkinson's son John, after his father's death. John was most impressed with the very advanced design and sound construction; he even got the engine to run after 40 years 'hibernation'!

Of the fuel economy device, this too was abandoned primarily it appears from the Government relaxing petrol distribution and pricing controls in 1953, allowing the reintroduction of improved petrol mixtures with higher octane grades, making Bradshaw's system redundant.

Under Emmanuel Kaye, J E Shay survived, just, within his Kaye Organisation which included Lansing Bagnall, and now Henley Forklifts; they became world leaders in materials handling equipment but the group was sold to the German Linde materials handling company in 1989.

Car heater

Granville's final motor car related design was an 'instant' car heater for Bonds in January 1959, in which Geoffrey had an input. This was in essence a simple perforated jacket around the car's exhaust pipe. Ambient fresh air was drawn into the jacket and propelled into the car's body by an exhaust driven turbine to which was attached, by a dog clutch, the heater's vane blower. The engine speed alone determined the blower's rate. A provisional patent was submitted.

18
Bradshaw Pulsation Motor

In the post war years, engineers such as Bradshaw became almost obsessed with the potential of free-piston rotary and gas turbine technology as the next stage in the development of the internal combustion engine. Unfortunately, although many of his drawings survive and are referred to in his letters by number, there are many instances where he refers to a specific engine in such a vague manner as a generic 'gas cycle', 'gas generator' or 'pulsation motor', that is not always easy to determine whether this referred to his 'Bradshaw Pulsation Motor' or his quite different toroidal rotary engine.

This, and the following chapters on toroidal engines, are therefore the best interpretation of, at times, quite confusing records.

Bradshaw pulsation motor

Keen to get the maximum horsepower from his engines, Bradshaw often considered supercharging which had successfully been used in pre-war racing cars (so successfully that superchargers were at one time banned). They were also used to great effect in the Spitfire's Rolls Royce Merlin engine giving it superior performance over the Luftwaffe's 405mph Focke-Wulf FW.190A-3 with its Methanol-water injector boosted 14cyl BMW 1,700hp radial; the later 436mph FW.190A-8 used a Nitrous-oxide booster in the supercharger raising output to 2,050hp. Nitrous-oxide introduced extra oxygen and inhibited pre-ignition.

However, conventional axial and centrifugal supercharger blowers were constant velocity devices which, due to internal resonance, affected the speed of the expelled air. What was really needed was a high pressure, instantaneous pulse of air which could inject, say, 30% more oxygenated compressed air than an axial compressor as the inlet valve opened.

Indeed it was the pulsating 'ram jet' which gave Hitler's FZG-76 'Doodlebug' V-1 flying bomb its, then, phenomenal speed of 390mph from its Argus 014 'Propulsive duct' engine. This incredibly simple design had a bank of spring loaded vanes at the front, opened by the oncoming air flow admitting air into the combustion chamber where fuel was injected and auto-ignited; the exhaust gases firmly closed the sprung vanes and as they were expelled, created a drop in pressure in the chamber below atmospheric which allowed the vanes to reopen; this occurred 40-45 times per second giving the 'Doodlebug' its characteristic 'burbling' noise. The engine developed 750hp, but required preheating and a catapult launch to get the ram jet cycle started. However, once started, it was impossible to control engine speed making it unsuited to aeroplanes. The principle is still being developed for hypersonic flight with experimental aeroplanes already reaching Mach 3 and above.

By autumn 1957, now living at Fishbourne Green on the Isle of Wight, Bradshaw formulated ideas for a double acting two-stroke piston engine using a double ended piston operating in a similar fashion to a conventional two-stroke engine. As the piston descended, it drew a petrol mixture into the sump, forcing it out through a port in the cylinder wall and thence to the combustion chamber, at the same

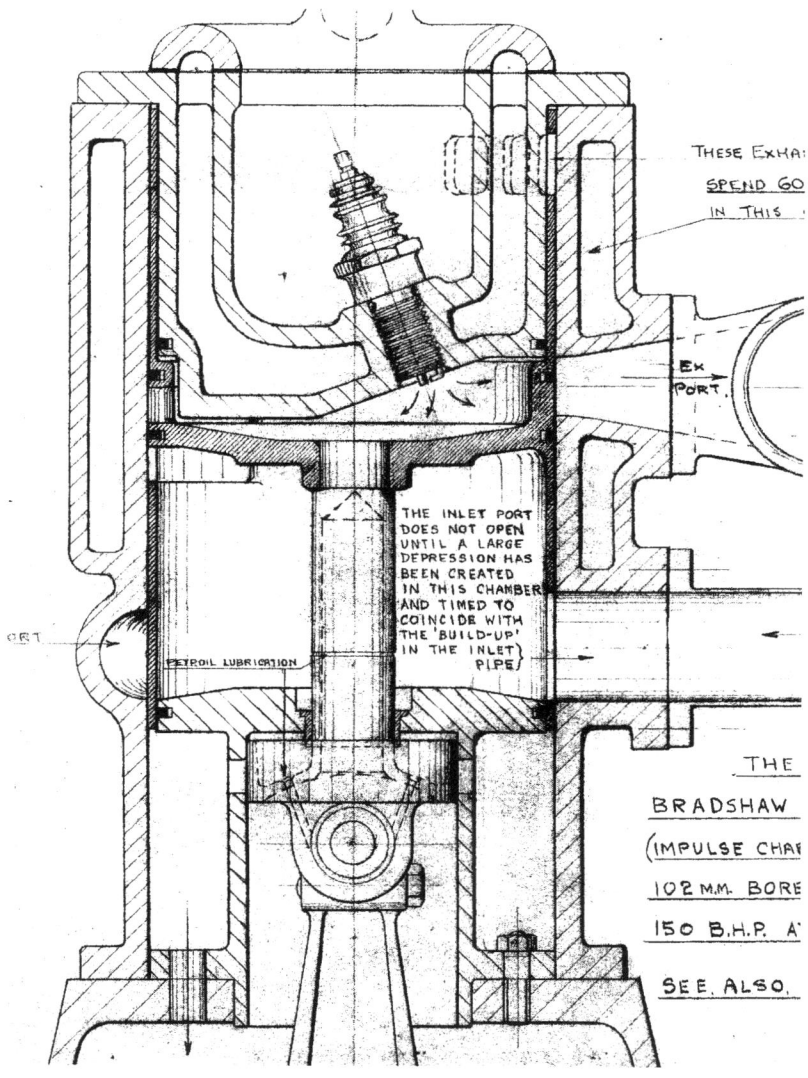

500cc Bradshaw Pulsation Motor. Note inverted ported piston upper skirt is cooled by a water jacket, while the lower ported skirt controls the inflow of volatile gases (GB)

time forcing out the exhaust gases through a further port exposed by the descending piston. In this way the two-stroke piston engine provided a firing stroke every cycle compared to every second cycle on mechanically valve operated four-stroke engine and in that regard two-strokes were potentially capable of faster running with greater power output.

On the conventional two-stroke, the induced petrol mix is under atmospheric pressure and only then compressed by the ascending piston. What Bradshaw proposed was to partially compress the incoming petroil mix, as in a free-piston engine, by the descending skirted piston before it was transferred to the combustion chamber. Furthermore the incoming petroil mix would be controlled by a port in the piston which, as it ascended, caused a partial vacuum some 40% lower than atmospheric

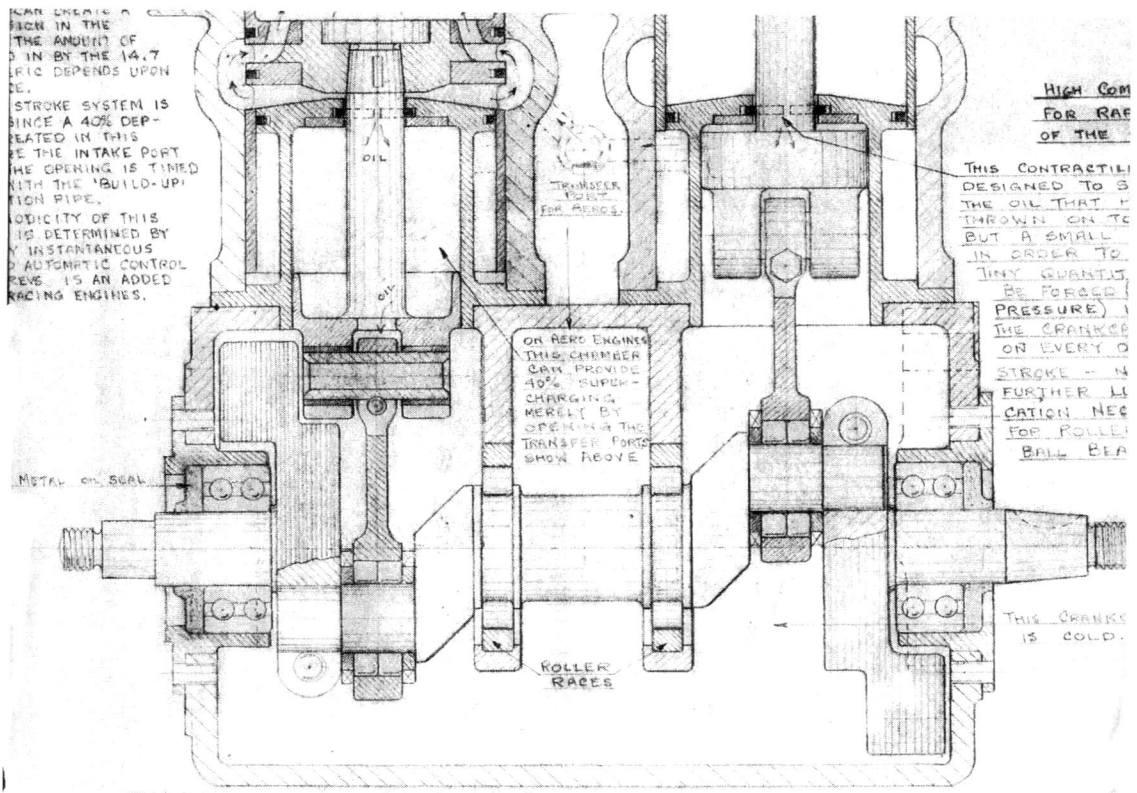

Cross section of 250cc twin BPM unit. Note the new secondary piston on the lower con-rod for supercharging or as an 'air-brake' in aero engines. The transfer port is just visible in the centre (GB)

pressure, drawing in large volumes of petroil mix in sudden pulses. This he terms variously as 'impulse charging' or 'gas cycles'. He also provided for fuel injection at the induction port if so required.

As seen in the diagram, the piston's inverted upper ported skirt, which allowed exhausts to be expelled, was cooled by its own water jacket at the top of the cylinder head, and by having a double skirt, it avoided creation of hot spots. Furthermore the incoming volatile petroil mix was at ambient temperature which helped cool the piston crown on combustion.

The engine would, in theory, have a minimum of inertial loadings on the bearings as the usual loads imposed on the power stroke were counterbalanced by the loads imposed by the piston as it compressed the incoming gases. He calculated an overall reduction in frictional losses incurred from these normal inertial loadings would yield around an 8bhp boost in the engine's power.

Arnfield

General arrangement drawings for a 500cc (102mm x 60mm) two-stroke Pulsation Motor, with a potential 155bhp output at 8,000rpm, were duly presented to R J Thomas at J & E Arnfield Ltd of Audenshaw, Manchester in late 1957. The prototype would use machined, forged steel barrels and pistons. Over the following months, drawings for a 250cc twin appeared, now fitted with improved roller bearings on the crankshaft to reduce the need for excess oil in the petroil mix and improve the engine's power.

Arnfields were an engineering subsidiary of the large Mono Pumps Company and would have been well versed in such prototype work, however they were soon having major problems with Bradshaw's new engine which, rather oddly, he put down to it being a unique engineering project of which Arnfield had little previous experience. However, by March 1958, Arnfield's chief engineer, Mr Birchenough, was reporting satisfactory running at a tick over of 400rpm and top speeds of 3,500rpm although the engine did at times run erratically. Bradshaw referred back to his earlier experimental notes and soon recognised the problem. In a four-stroke engine, the piston expels the exhaust gases before inducing the fresh mixture whereas in a two-stroke, the fresh intake from the crankcase purges the exhaust in the same cycle. In February 1959, he duly set to revising the design by introducing an internal 'cross flow' porting system, via a series of induction ports which fed the chamber through five ports in the piston crown.

Despite these modifications, they made little improvement as the piston rings partially obstructed the ports and the full charge was never received at the right time, causing starting problems. By May 1959, a further improved design was reported to be starting first time and running well.

In February 1959 he submitted details of a 300bhp, V-4 1500cc (87mm x 62mm) racing car engine and simultaneously submitted a Provisional Patent, No. 6836/59 on 27th February.

He then optimistically proposed that the sleeve, below the main piston on the short connecting rod, be converted into a booster compressor to add 40% supercharging especially suited for the rarefied atmosphere of high-altitude aeroplanes or, reversed, as a rudimentary 'air-brake' to retard the engine; but when so modified, blow back was being reported in the induction system for which Bradshaw proposed, in September 1959, fitting rotary or sleeve valves. Drawings for the first 350cc engines with a rotary disc valve were presented in November 1959. A further modification employing a 'cross-head tunnel' for aeroplane engines was duly carried out on a twin cylinder engine and in March 1960 the engine was reported to be running well at 5,000rpm. He duly submitted a Provisional Patent (No. 30957) for this two-stroke porting system.

In March 1960, Mr Birchenough offered Bradshaw a radical new design for the double skirted piston and cylinder head which would make the engine more compact, more efficient and far easier to manufacture, but Bradshaw was not impressed and soon afterwards, J & E Arnfield Ltd declared they were unable to expend further time, energy or money on the project and returned the prototype and incomplete engines to Bradshaw in the late spring 1960.

Knowing that the engine did indeed work, Col Reg Gray proposed in March 1960 that Bradshaw submit the racing car engine design to BRM as well as approach Jack Emmerson at Jaguars for further development, but no further reference to this is found and it must presumed they showed no interest.

Bradshaw had also experimented with a similar high pressure pulsation induction cycle on a toroidal engine in February 1959, but this caused endless problems and, in the event, neither the piston pulsation motor nor the toroidal rotary pulsation motor went any further as he concentrated his time and effort on his twin toroidal gas turbine version of the promising toroidal engine.

19
Toroidal Engines - Project Omega

The toroidal engine's development is complicated in the extreme for, like the pulsation motor, the many letters, drawings and provisional patent applications are not always clear to whom, to what application or to which type of toroidal engine (single chamber, pulsation or twin-chamber 'gas generator') they refer. Added to which, there is rarely a description of the provisional patents.

Over a ten year period, he drew up so many variations on this simple idea, some with single cranks others with twin or more cranks, that it would be impossible to describe every modification made or proposed. It is therefore necessary to concentrate only on the essential features. Fortunately, Granville made several wooden demonstration models to make it easier to see (rather than describe!) how it works and how it fires at up to four times per cycle. Alas, due to an imperfection in the construction of the surviving $4\frac{1}{2}$" diameter model, pistons 2 & 1 and 4 & 3 do not fully close on their firing cycles, nevertheless, the sequence of photographs (page 233) show the general principle of what is, incredibly, a full-size, 350cc engine developing 40bhp.

It should be noted that on a toroidal engine, just as with the Wankel, the unique cycle and nature of the chamber's design do not allow a direct comparison by swept volume to that of a conventional internal combustion engine; hence, quoted cubic capacities are theoretical equivalents derived by complex formula. What is not difficult to understand though, is the incredible compactness and weight saving, which gives the toroidal engine a much higher power to weight ratio over a conventional internal combustion engine. I hope I have correctly untangled this confusing 'mess'!

Project Omega

As previously related, Granville Bradshaw's first involvement in the toroidal engine came during the war when asked to overcome inherent flaws in the mechanical valve closure on the Packard aero-engines used in Fairmile's motor torpedo boats. With work at Phelon & Moore aborted at war's end, nothing more was done on the engine until 1950 when, it appears from Granville's archives, his interest was renewed by the Rover Car Company's development of a gas turbine car engine as a replacement for the internal combustion piston engine - this is discussed in the next chapter.

Now working for his brother Ewart, he realised the toroidal engine had great potential in the Bond Minicar and duly recovered the prototype from P&M. Ewart's initial reaction was one of rejection, probably because it wasn't properly working, but he was sufficiently interested to financially support Granville's efforts in developing the engine further. It appears that Granville may have got Autoys at

Preston involved in producing (or improving upon) one or two prototypes of his 'New Action' toroidal engine. These were then stored in the basement of Ewart's 'Greyfriars' home in Preston along with the aborted 500cc in-line twin. Later prototypes were completed at Sharps and it is evident that Lawrie Bond, who was also retained as a consulting engineer, did quite a bit of precision engineering work on the engine in their machine shop. There are also several references in Bradshaw's letters and diaries in the early 1960s which refer to an unidentified 'Mr Fisher at Preston' doing prototype work on both the 'Omega' and the later twin toroidal gas generator engines.

In 1954, now living in Eversley, Hampshire, Granville Bradshaw worked earnestly to redevelop his earlier spinning chamber prototype. Though we have no idea how large his Admiralty/P&M engine was, surviving drawings show his original designs were of incredibly small units, some 250cc 12bhp engines barely weighing 10lbs. But to power a medium-sized car, a drone aeroplane or collapsible boat would require in excess of 40bhp. His constant efforts to perfect the design led early on to a more efficient, leakproof, static toroidal chamber with rotating, reciprocating pistons. This design became known in the motorcar and motorcycle press as the 'Omega' engine.

How it (should) work

His first experiments with a spinning toroidal chamber were very unsatisfactory mainly through an inability to run continuously due to excessive gas leakage. What all pioneers of toroidal engines had also overlooked was Newton's Third Law of Motion: 'To every action there is an equal and opposite reaction'. In theory, as the fuel enriched gases explode, the forward piston (No.4 in the photograph) would be propelled forward, dragging the rear most (No.1) with it through scissor action of the connecting rods via the 2:1 gear. In practice, the exploding gases caused both pistons Nos 1 and 4 to try to move in opposing directions, tending to stall the engine! It was soon found necessary to build up considerably the initial inertia by electric motor before firing but even the addition of a heavy flywheel, to maintain that inertia, failed to keep the engine running for more than a few, very raucous, seconds - applying a load would almost certainly have killed the engine. Added to which, a major problem also existed in trying to find an effective flexible seal which would allow the oscillating 'connecting rods' to move as the cylinder rotated and, at the same time, maintain a good gas tight seal around the inner circumference between the two halves of the doughnut shaped toroidal cylinder.

Bradshaw's solution lay in a static chamber with two co-axial 'connecting rod discs', running with 0.002" clearance, each with two 'ears' acting as connecting rods to curved, cylindrical pistons within the toroidal chamber. As these discs were coaxial and flat, a series of concentric bronze (or later, lead-indium coated bronze) O-rings between the discs provided a satisfactory, if imperfect, seal against the compressed gases escaping. Bradshaw later improved its efficiency by bleeding off some of the compressed exhaust gases from the combustion chamber, to pressurise these outer seals for maximum sealing effect. Tests showed that the specially shaped piston rings proved only 95% gas tight but it was felt that a 5% loss was an acceptable irritant. Though Bradshaw remained confident it could be improved, it eventually proved virtually impossible to achieve despite considerable help from Wellworthy Ltd of Lymington, who had produced both the seals and the unique, curved pistons.

The static toroidal combustion chamber was of cast aluminium formed in two precision machined halves, bolted together around their outer circumference. The inner diameter of, say a 1,500c car engine, was 6" with a cross section of 2" making for a nominal 8" outer diameter. The chamber's inner surface was later hard-chrome plated - a similar process later adopted by Moto-Guzzi for their hard wearing 'Nikilsil' cylinders.

Running around inside the toroidal chamber were two pairs of double ended, heavy cast-iron curved pistons (later aluminium-alloy for petrol engines) with, variously, plain or spring loaded bronze piston rings which helped dissipate heat uniformly around the air cooled outer wall, eliminating the usual hot spots in conventional or opposed piston internal combustion engines. An allowance of 0.004" running clearance was made.

The pairs of pistons were driven by way of co-axial shafts through a 2:1 sun and planet gear which oscillated them, scissor fashion, within 30°, causing one piston to move at full engine speed and the other at half speed so that they played 'cat and mouse' chasing and catching up with each other as they spun around the cylinder. Ports in the cylinder wall were exposed and closed by the pistons, inducing fuel enriched gases which were then compressed as the pistons caught up. On ignition the pistons were forced apart, driving, via cranks, an output shaft through a conventional clutch to a gearbox, propeller or final drive axle.

The great advantage of a toroidal engine is the lack of inertial forces on pistons, connecting rods and bearings at top and bottom dead centre and, by using porting arrangements like those of a two-stroke engine, it eliminates spring assisted mechanical valve gear whose very springs severely limit an engine's maximum speed to around 10-12,000rpm. Devoid of these masses and dynamic components, a much lighter, more powerful faster spinning engine is possible.

In a toroidal engine, it is the heavy pistons which act as the flywheel. As the engine contained only nine moving parts, this resulted in a minimum amount of friction and energy to overcome inertial loads which, as Bradshaw reasoned, increased available power output by at least 25% over a conventional piston engine. Furthermore, as it operated on a two stroke principle, each piston pair fired twice per revolution, doubling the effective output and power - hence four firings per cycle which, he suggested is equivalent to a 4-cylinder, four-stroke engine. Hence, in proposals drawn up in 1955 in which Bradshaw illustrated how two or more (for example 1.25-ltr engines) could be coupled side by side, driving by cranks a common output shaft, it became what he termed a 2.5-ltr 'V-8'.

By close coupling engines co-axially, it was possible to have a 360bhp, double row (180hp), diesel toroidal aero-engine measuring 15" diameter and weighing only 180lbs - half the size and one third the weight of his ABC Dragonfly radial aero-engine.

Manufacturing problems

When Harry Mundy of *The Autocar* saw drawings of the proposed 'Omega' toroidal engine in 1955, he showed considerable scepticism over its success. However, when he saw a wooden demonstration model of the toroidal engine in 1960, he was now most impressed. Indeed, everyone bar none was amazed at the engine's simplicity, even if they were unable to fully comprehend the complex sun and planet gear and 2:1 crank ratios. However, Mundy remained concerned at the seemingly impossible

Toroidal engine cycle (opposite page)
Spark plug 12 o'clock; exhaust port 7 0'clock; inlet 5 o'clock

Anti-clockwise from top left

A	No 1 and No 4 fire
B	No 4 pushed forward and exhausts 4:3
C	No 1 exhausts 4:1
D	No 4 induces 4:3
E	No 1 induces 1:4
F	No1 compresses 1:4 ready for firing

Meanwhile 2:1, 3:2 and 4:3 have also fired!

Toroidal Engine Cycle

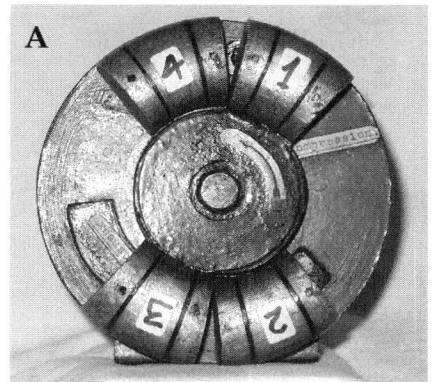

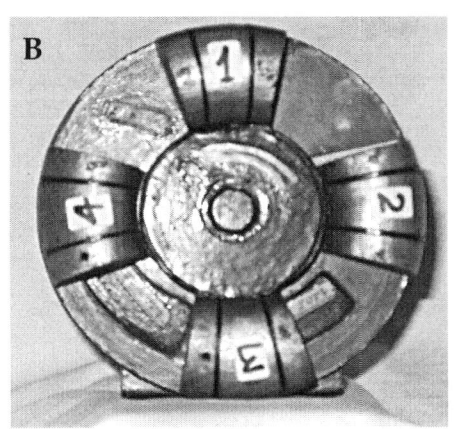

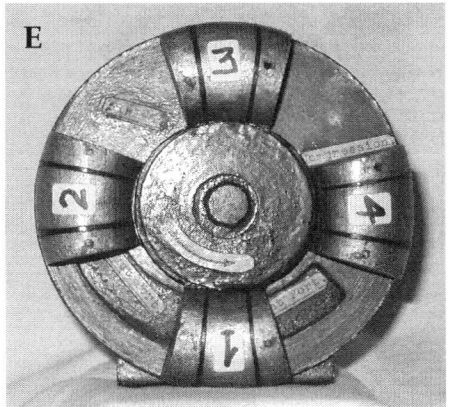

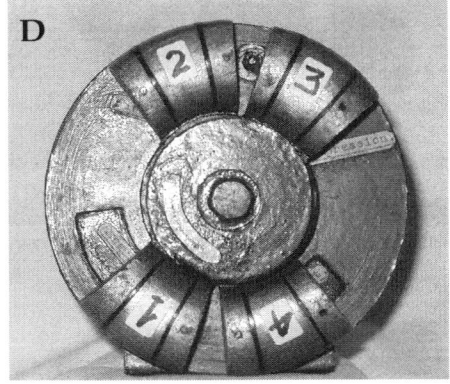

accuracy required in machining the matching split halves of the toroidal chamber. He was also concerned over the asymmetric heating of the chamber by the exploding gases and saw problems with piston rings seizing in their grooves from an inability to rotate about the piston - yet, two-stroke engines had spigotted piston rings to intentionally prevent their rotation - they rarely seized.

As indeed it so proved, manufacturing the engine was fraught with problems not least in the precision turning of the toroidal chamber, but also added complications of the intense heat from combustion, being conducted through the connecting rod 'ears' making them expand and in so doing, attempt to push the pistons against the outer wall causing their seizure. This was resolved initially by looser fitting pistons, placing greater reliance on the spring loaded bronze piston rings, then from 1961, by clever use of an eccentric little-end bush allowing the radial expansion of the connecting rod 'ears' within the piston.

Cylinder wall lubrication was by oil-enriched fuel (such as petroil or diesel fuel oil) and only the reduction gears and oscillating con-rod and crank system, encased in a cast iron housing, had positive or oil bath lubrication but in 1961 he proposed a conventional sump and skew gear driven pumped oil system - one letter mentions this in a marine engine application.

He soon discovered however that the engine's failure to run properly was down to misfiring from the normal spark plug's inability to cope with continuous high speed operation; alas there is no record of how that problem was overcome other than the fitting on some engines of twin spark plugs fed from a high tension supply via a slip ring to ensure correct timing, as shown in some drawings. Furthermore, at low throttle speeds with low piston inertia, there was insufficient force to propel the pistons (which simultaneously compressed the induced gases) causing their failure to fully expose the inlet port, which brought the engine instantly to a stop. This in turn placed a tremendous stress upon the oscillating con-rod crank mechanism. While Bradshaw acknowledged that although a heavy external flywheel would help, as well as provide a means for an electric starter, it would reintroduce heavy crankshaft loads back into the engine, negating his objective. However, he remained as confident as ever that these issues would be resolved and in early 1961, he proposed a heavy gun-metal 'cage' enveloping the 2:1 crank assembly which would serve as a flywheel.

Fortunately, experience with compression ignition free-piston engines, running on diesel fuel oil, meant that the toroidal engine would also work with greater reliability on most types of hydrocarbon fuel - and indeed, multi-fuel engines are standard in military vehicles in theatres of war and was a deciding factor in Rover abandoning their kerosene fuelled motor-car gas turbine car engines as the less complicated 'free-piston' engines can run on cheaper, unrefined fuels.

His earliest drawing of a diesel toroidal engine dates from March 1956 for a 60bhp unit of under 12" diameter, weighing under 50lbs. His final projects in the early 1960s also preferred the more reliable and less complex, if more heavily built, diesel versions for motor car use at a time when UK restrictions on untaxed diesel fuel only allowed the occasional diesel Land Rover, Austin FX3 or Morris Oxford taxi; yet export markets thrived on diesel fueled cars!

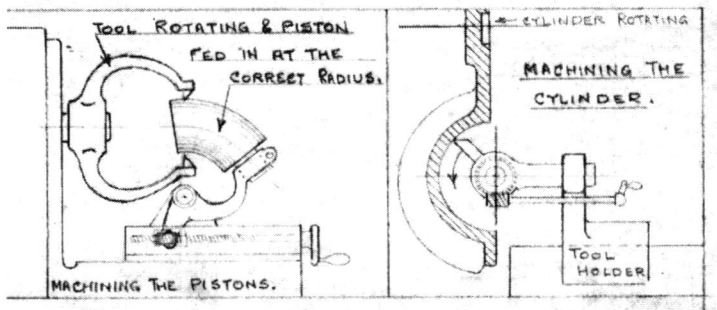

Simplified machining process for piston and toroidal chamber (GB)

Bromega Ltd

At his companion, Francesca Beaven King's suggestion, in late 1955 Bradshaw set up a new company, Bromega Ltd (from **Br**adshaw and **Omega**), at 2 Holmes Road, London NW.5. They then approached the National Research & Development Corporation for government funding for the toroidal engine, but were unsuccessful.

Incorporated on 27th January 1956, Bromega was capitalised at £10,000 through 200,000 one shilling shares. Bromega's other directors included a solicitor, Matthew Gibb (Chairman), another solicitor, Mr Gill, an unidentified Mr R.F.L (introduced to him by Francesca and to whom Bradshaw took an instant dislike and distrust!) and a C Ruth Moore. Major shareholders included Granville and Francesca each with 80,000 shares and Ewart Bradshaw. In Ewart's case this was through Sharps Commercials in repayment for work already done for Bradshaw on the toroidal engine.

As his companion, Granville had already granted Francesca joint rights to all his new patents, including his 3-D television and toroidal engine designs. On 30th January 1956, Granville and Francesca jointly sold four provisional patents to Bromega for £5,000: Provisional Patent No. 17046/55 - Improvements to internal combustion engine; Prov Pat 2619/55 - Improvements to internal combustion engine; Prov Pat 31283/55 - Improvements in prime movers and Prov Pat 1735/56 - Improvements to internal combustion engine. Here again, as provisional patents there are no further details. The agreement ensured that any improvements to these designs would be granted to Bromega; in addition, Granville transferred all benefits from the sale of the patent rights relating to the previously described Lemery chain saw engine.

In a surprise move, Granville then resigned as a Bromega director in May 1956, probably to concentrate on his design work. Col Reg Gray, became a director on 14th April 1958 and went on to control Sharps' investment in Bromega, following Ewart's death in June 1959, as his executor, and at the same time took over as chairman of the Loxhams Garages holding group as well as maintaining chairmanship of Sharps Commercials, with Tom Gratix now as their Managing Director.

Prototype engine on left and with front half of toroidal chamber removed on the right (GB)

'Omega' engine

Unbeknown to Bradshaw, and much to his and Col Gray's anger, unnamed directors at Bromega had prematurely released information on the 'Omega' toroidal engine to the motoring press forcing Bradshaw to announce its details before he was ready. Both Bradshaw and Col Gray were mystified at the source of Jim Bennett's famous artistic impression, which appeared in *MotorCycling*, showing a rotating vane induction system and extra, superfluous exhaust ports, of which there is no record in Bradshaw's archives! Luckily, the conceptual drawing is otherwise faithful. Despite this most unwelcome premature publicity, Bradshaw eagerly emphasised the engine's benefits to car designers: the air cooled engine required no energy sapping water cooling system and its compactness allowed the engine to be set well forward to increase passenger space and greatly reduce the car's overall size, much as Alec Issigonis achieved with his revolutionary Mini. He confidently predicted the 'Omega' toroidal engine would make the conventional internal combustion engine obsolete - and that it would be cheaper to build!

The 'Omega' naturally received considerable interest and in a Pathé News report from 1955, filmed at Granville's Eversley home, he was shown explaining the theory of the Omega engine. Unedited clips have survived and can seen and heard on the Pathé, web-site. *The Daily Telegraph* eagerly announced on Saturday 17th December 1955: 'The wonder car engine for Britain' declaring it to be friction free and running on any fuel, giving 20-30% more power than an equivalent conventional engine with a 200,000 mile life and capable of running at 12,000rpm through having no valve gear or friction from moving parts. They also reported Bradshaw as saying that with this design he could build a 1,000bhp engine (he had in fact

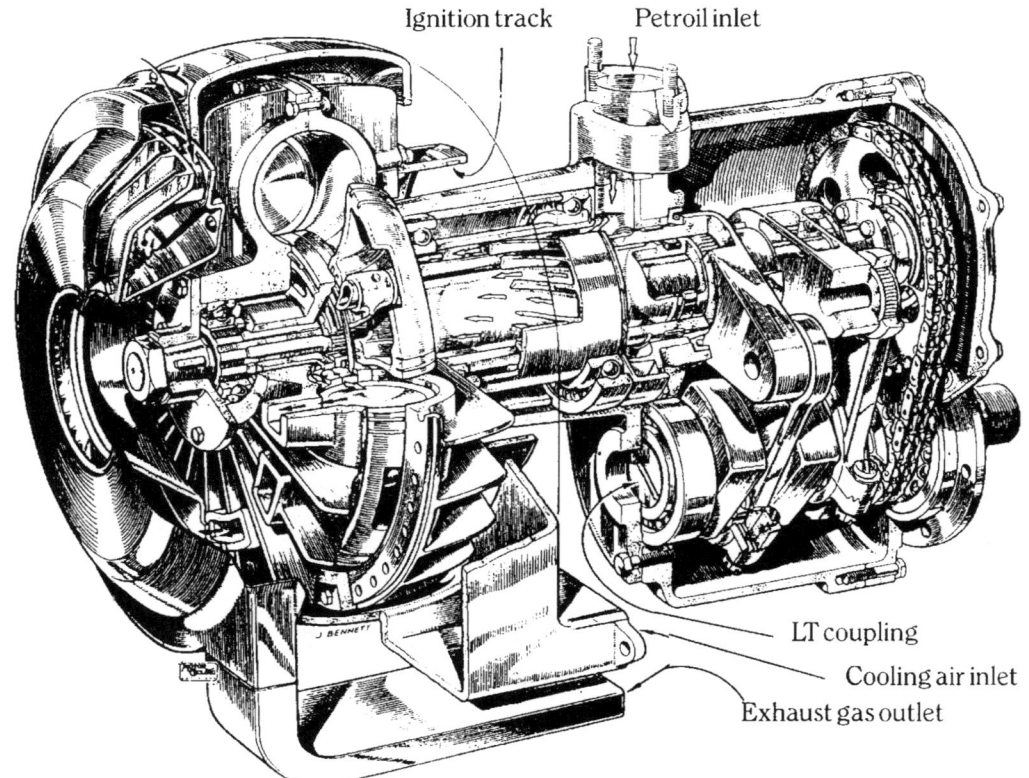

MotorCycling *magazine's flawed artist's impression of the Omega engine*

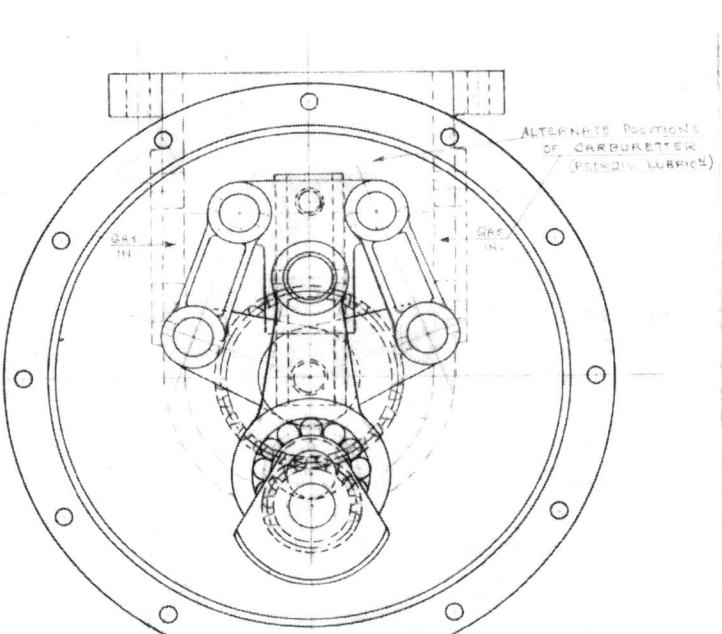

Crank details of the Omega engine

calculated a potential 3,000bhp aero-engine of only 29" diameter - a 380hp Formula One Grand Prix racing car engine, for example, would weigh only 110lbs compared to 336lbs for a then current BRM racing car engine; Bradshaw then declared his dream 'to see Britain triumphant and sweeping the racing honours of the world'. He went on to say that three engines were already built and were being held at the Bond car company in Preston and that their managing director, Col Reginald Gray, had declared that 'this engine will revolutionise motoring as we know it today'. Bradshaw advised *The Daily Telegraph* that he had three or four (unidentified) 'big people' in the motor industry keen to develop it, especially the Grand Prix engine.

The Daily Sketch of 17th December proclaimed; '1,000hp Wonder car engine will go in a biscuit tin' and fit under the bonnet of a Morris Minor, while in *The Star* of 21st December 1955, Laurence Cade announced the Omega engine alongside Harry Ferguson's revolutionary new 4-wheel drive, independently sprung car chassis and held his hopes high for both. He also hinted at a new Peoples' Car from Bradshaw (discussed earlier).

Courtney Edwards' report in *The Daily Mail* of December 22nd, 1955 was equally effusive, but John Kinsey of the Automobile Association was not totally convinced, saying:

"The engine is an interesting departure from orthodox principles. It holds possibilities of interesting developments. It is a power unit which may well fulfil the inventor's claim for improved power to weight ratio, high permissible speeds and fuel economy.

"The use of opposed pistons with a common combustion chamber - the basic feature of this design - is not unique. The method employed would appear to permit high compression ratios without detonation or knocking.

"It also substantially reduces one of the limiting factors in power output and maximum engine speeds and inertia losses in the reciprocating parts.

"Features in Mr Bradshaw's design which give promise of an engine of outstanding performance are: the ingenuity with which the piston to crankshaft linkages are devised; retention of the four

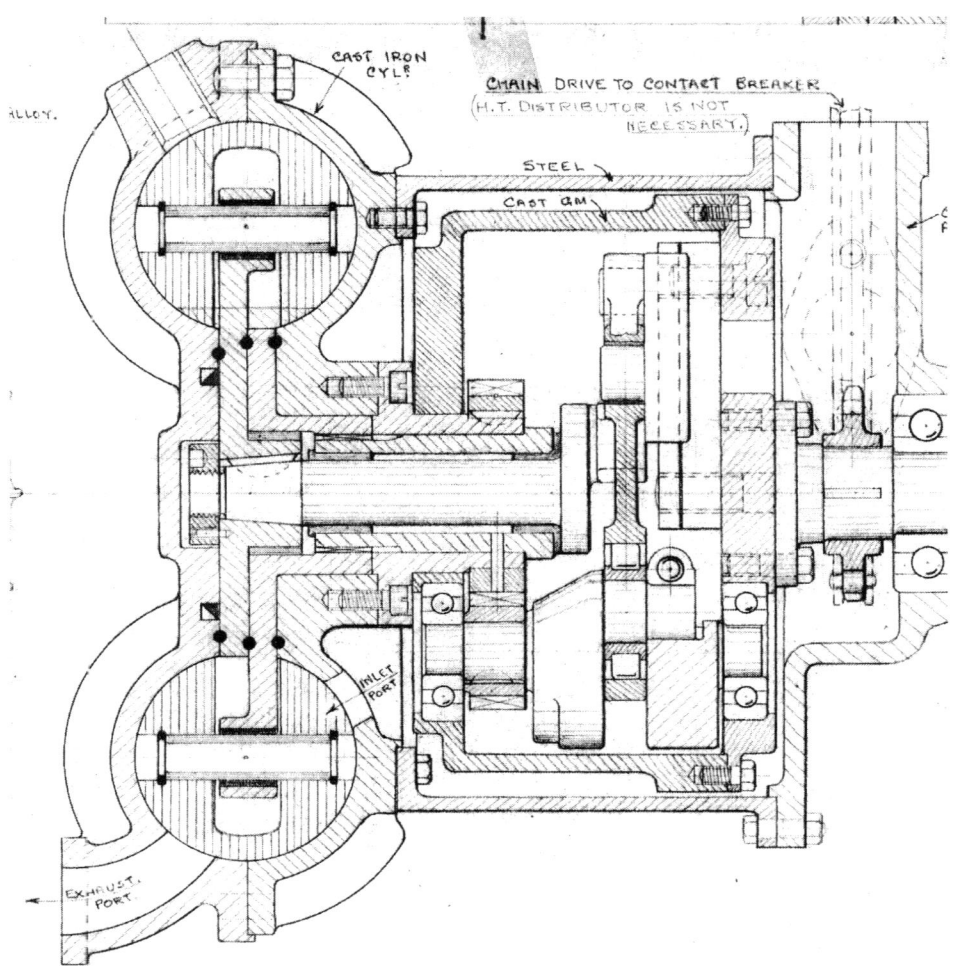

9 inch diameter, 1,500cc single crank design of 1960 (GB)

stroke principle without the need for valves or valve mechanisms and the balance of reciprocating and rotating parts.

"The virtues of Mr Bradshaw's principles remain to be established when a prototype is subjected to extensive tests for performance, cooling efficiency and durability."

In response, Granville announced that full technical details of the engine were to be held back until early January 1956, while he negotiated terms with interested buyers. Certainly, the automotive world had high hopes for it, as it would allow them to revolutionise their designs with a range of compact, powerful engines, but against them were the engine makers, whose expertise was needed to develop the engine, but for whom this revolutionary design conflicted with their vested interest in maintaining development of their core market in piston engines!

Bradshaw's letters reveal considerable interest from Canada, Holland, Germany and France and a $1 million offer from an American motor manufacturer who sent over a team of engineers. It is believed this was General Motors who, at the time, were developing a free-piston engine. Bradshaw, however, was most keen to sell his design to a British company, yet only one 'major' (unidentified) British car maker had shown any interest and that was conditional upon exclusive rights, on which Bradshaw was not keen, for he firmly, if optimistically, believed that his engine would become 'the standard in all cars'.

The publicity brought correspondence from fellow inventors around the world. One such was Fr. Beyersdorf in Germany who had submitted a provisional patent in 1922 for a revolving piston design for naval applications; he invited Bradshaw to co-operate in a joint patent, but there is no record of Bradshaw's reply. Further interest came in a letter in June 1960 from a Russian engineer, I V Korolkov, who had also been working on a similar engine and was most interested in collaborating with Bradshaw, but Granville was very cautious of any such dealings for he was unable to protect his patents in Russia. Interestingly the letter from Russia had been sent to Bradshaw's old address stated in the patent - it took several months to find him!

Cash flow

All patentees tend to get caught up in a vicious 'Catch 22': without a working model, no one will invest in the project and without that investment, the project would not yield a return with which Bromega could fund further research into perfecting a working engine and thereby pay Bradshaw his due income. A letter from Bradshaw's solicitor, Matthew Gibb, in June 1956, resolving a second petition from the Inland Revenue for Bradshaw's bankruptcy against unpaid taxes (Bradshaw was virtually claiming anything and everything as tax deductible research costs - but the Taxman would not allow it!), referred to negotiations between a Fritz Clarence-Langford and a Government sponsored organisation for rights to the engine (possibly the NRDC), which Mr Gibb had every confidence would prove fruitful with, at the very least, a consultant's fee able to provide Bradshaw with a reasonable income. But these proposals never came to fruition.

The engine's potential and eager reviews encouraged all involved; both Bradshaw and Bromega's directors were certain it would yield 'millions' and rather foolishly, in retrospect, Bradshaw decided to distribute his Bromega shares among his family so they could all share financially in the engine's huge potential. Even Ewart now expressed a keen interest in the rotary engine by paying Granville a personal deed of covenant of £50 a month and, on his death, an annuity of £500 per annum for 7 years towards the upkeep of his design office in exchange for 30,000 Bromega shares.

Although Bradshaw enjoyed some lighter moments during 1959, such as preparing his autobiography, his financial position remained dire and in May 1959 he cashed in his Post War Credits in full, to the value of £345. Following Ewart's death in June 1959, most of his executors (which included Messers John Whitehead, Marden & Huck, solicitors and Col Reg Gray of Sharps Commercials), wished to discharge the covenant giving Granville a £1,000 lump sum. However Col Gray, acting in the best interest of Ewart's wishes, and knowing that development costs were financially crippling Granville, preferred a controlled regular payment; he won the others over and became Bradshaw's principle contact and ally over the toroidal engine's development.

Bradshaw's despair at the way Bromega was being run and the fact the twin toroidal 'gas generator' engine was finding no commercial outlet, led to a constant battle with Mr Gibb (as the company's solicitor) which came to a head during late summer 1959 when Bradshaw decided to keep all future patents to himself. By September he had regained control of both the proposed scooter engine for Edward Turner at Triumphs, and the 'Rotoscythe' engine rights. Although this boosted his confidence, his financial problems remained as dire as ever and even Francesca had now been forced to borrow money from her father's estate to buy 'Grey Cottage' in Spencer Road, Pelham Fields, Ryde into which she and Granville settled in the summer of 1959, with Col Gray acting as guarantor over the mortgage gained, it appears, from Bromega Ltd. Thankfully, they were saved further expense when a meteorite struck their neighbour's roof on the night of 10th October, 1959!

But in 1960, he was once again in the taxman's bad books, yet he was still working tirelessly for Bromega without income from them, living almost entirely off the generosity of Ewart, family and friends.

His newly patented 'gas generator' twin toroidal engine of January 1960 and progress on Geoffrey Wade's offer (see later) for the existing toroidal engine, further boosted Granville's confidence, but his failing health, much of it from shingles, required Francesca to tend more to him, leaving her unable to find gainful employment. Still tormented by lack of funds from Bromega, things became more acrimonious between him and Mr Gibb regarding the way in which the company had been set up to provide for Bradshaw's work yet, increasingly, it appeared to Bradshaw that he was being steadily forced out of the company and denied future benefits. Even the relationship between him and Francesca (a Bromega director) began to sour as it appears other directors at Bromega were now attempting to prevent Bradshaw selling any further rights to 'their' joint patent!

Bearing in mind his experience in the Clarence Hatry case, on more than one occasion he threatened to report Gibb to the Law Society and in his desperation, he was now keen to buy out Bromega's interests and secure total control over the 'gas generator' engine jointly with Sharps Commercials (represented by Col Gray) and Francesa King, each of whom now held 85,000 Bromega shares. But unable to fund such a buy-out, due to his perilous financial state, the only salvation would, in his eyes, be resolved in Ewart's annuity being discharged as a lump sum, together with the outright sale of his original Admiralty pattern (P&M built) rotating toroidal engine, which was not controlled by Bromega. But to temporarily relieve the financial situation, Bradshaw had little option other than to steadily sell off his own shares (and potential profits) in Bromega to friends and business acquaintances, often in lieu of work done, such as in February 1960 when he transferred 16,000 shares to Rex Norman Lowin, on the Isle of Wight, who had been machining toroidal chambers and who was more than happy to return his confidence in both Bradshaw and the engine, by a £10 per week 'personal loan' towards running Bradshaw's office.

By early 1960 the first real hopes of a financial return on his original toroidal engine came from Geoffrey Wade and by late 1960, Mr Gibb was now more than happy to advise Bromega shareholders to sell, in exchange for monies invested and be rid of the company, as it was still not making any returns on the twin toroidal 'gas generator'.

Geoffrey Wade - G E Motors

During the late 1950s, Dowty Marine at Hamble, Hampshire had developed a water-jet propulsion system for small boats which, by not having a propeller, allowed them minimal draught and the ability to navigate freely in marshes, foul and weed strewn waters. Those who followed *The Prisoner* ITV television series will have seen them in action. Several boatyards on the Isle of Wight, including Fred Cooper at Attrill & Sons, had been involved in this work for Dowty Marine.

By November 1959, keen interest was being shown by a local man, Geoffrey Wade, for an outboard motor for his proposed water-jet propelled boat. Wade, then aged 34, had worked at his father's boat yard and now ran G E Motors, a petrol filling station and garage off Swains Road, Bembridge. With the acrimony between Bromega worsening, Bradshaw was more than agreeable to Wade carrying out development work on his original Admiralty/P&M toroidal rotary engine as this was not related to the Bromega patents so, in December 1959 detailed negotiations for Wade to acquire the rights began. In early 1960, Bradshaw recovered the engines from his, now deceased, brother's home in Preston, resubmitting his original design for the engine as a complete patent.

Wade knew of a small precision engineering workshop, H A Wills Ltd of Pyle Street, Newport, Isle of Wight, who had recently diversified into making less rewarding go-karts to maintain cash flow, but who were well versed in aviation work and more than capable of developing the toroidal engine. Bradshaw duly proposed that Geoffrey Wade set up a new company, the Isle of Wight Marine Engine Company to market it and, following talks in early spring 1960 with their local MP, Mark Woodnutt, they were now hopeful of a £10,000 Government Development Grant for the engine to which, provided the MP saw a working example, he would 'move heaven and earth' to secure.

Satisfied with the engine's workings, the contract was signed in July 1960 to rights for Wade's use in small boats, a £1,200 p/a salary for Bradshaw as consulting engineer, plus 5% commission on all engines sold. Wade also sought the services of Mr R Howe at Hobourn Aero Components Ltd, Rochester, Kent to produce the engine. They were part of Hobourn-Eaton oil pump company (a subsidiary of the huge Thomas Tilling Group) with whom Bradshaw had earlier discussed a 180hp 3-ltr version in the belief they were interested in buying a license.

However, in late autumn of 1960, Bradshaw received news that his patent applications on his original, Admiralty/P&M rotating toroidal engine had been rejected as much of its content originated from an existing French patent from the 1920s - which of course, Bradshaw knew. Being unpatentable, he was now unable to secure the £10,000 Government Development Grant, so withdrew the contract with Wade in November 1960. But that was not the end of the story.

Light aeroplane engine

Meanwhile, from a chance interview with Tony Bruce over Bradshaw's earlier proposals for man-powered flight, David Scott, an old journalist friend, suggested Bradshaw talked to a friend of his, Arthur W J G Ord-Hume, who owned the recently formed Phoenix Aircraft Ltd at Cranleigh, Surrey, established to produce light aeroplane designs. This he duly did in August 1960, but while Ord-Hume's advice on man powered flight was limited, it led to an interest in Bradshaw's proposal for a light aeroplane toroidal engine and later that month, Bradshaw was introduced to Harry Wills at H A Wills Ltd, one of several small businesses owned by Ord-Hume. They were a long established local business as both the island's local electrical goods wholesaler and distributor. They were also experienced Ministry of Defence precision engineering contractors to the RAF and Royal Navy and had also produced the first special, precision, rectangular-section curved tubing for wave-guide work in early radar detectors. During lulls in these contracts, H A Wills Ltd produced the 'Watkins Converter' allowing Aga cooker ranges to run on gas, built motor boat trailers and, more recently, go-carts!

As an ex-RAF pilot, Arthur Ord-Hume saw Bradshaw's engine as a potential alternative to the American Lycoming flat four and the promising, but still troublesome Wankel. It ideally met the specifications for his new solo and 2-seater Phoenix light aircraft, so he was quite keen to see the engine developed. At long last, Bradshaw now had a very real chance of making the engine a success for, being lighter than most, it was quite possible to modify a second engine with pivot points for fitting within a frame to provide vertical thrust for short take off.

With the expertise of their toolmaker/engineer, Vic Newton and works director, Harold Oatley, work began at Wills in September 1960 on a 500cc static toroidal engine which had both marine applications for Geoffrey Wade and aeroplane applications for Ord-Hume. By October 1960 Newton had got the engine running, albeit seizing at times (through internal con-rod expansion), but by November 1960 with the modified eccentric small end bush to allow for con-rod expansion, they reported it was running well. By December 1960 they had incorporated Bradshaw's modified spring, and gas assisted, O-ring sealing system, described earlier.

The eccentric little end bush seen on the 1500cc car engine shown on page 238

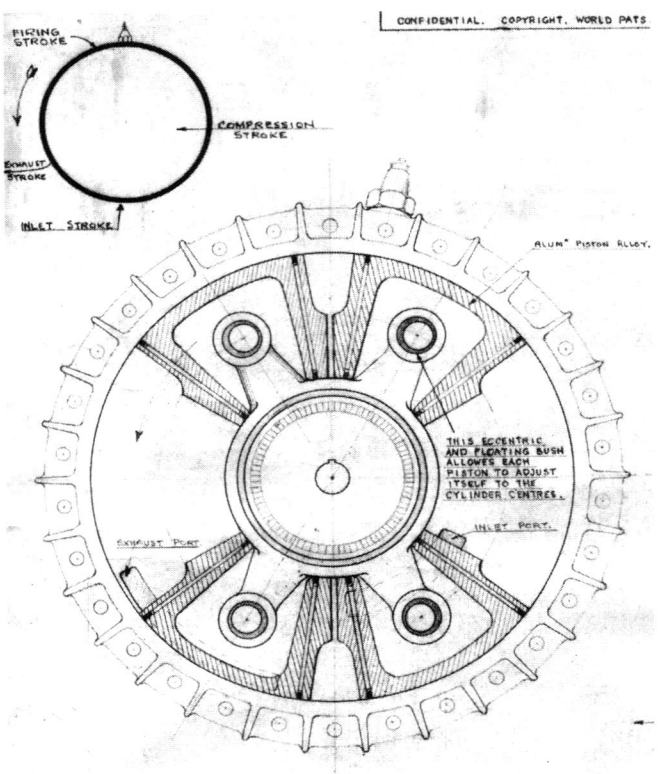

Marine engines

Meanwhile, encouraged by the engine's progress, and armed with several demonstration models (both working and not) from both Wills and Bonds in Preston, Bradshaw abandoned further discussions with R H Howe of Hoburn Aero Components for they were only interested in helping build and not buying the design. In September 1960, he approached Mr Russell at Dowty Marine expressing interest in developing a jet propulsion engine for their boats. Dowty showed great interest, as it was an ideal alternative to the heavy Ford Zodiac 6-cylinder motorcar engine which they were currently using, but in the event, they rejected the design in favour of their own axial turbine water jet propulsion system, which was later successfully fitted to the Army's Alvis Stalwart amphibious vehicles.

In October 1960 Bradshaw approached Samuel White, the local boat builders, and proposed a 500cc marine engine which was followed in January 1961 with a further marine engine design. The old established J Samuel White & Co, shipbuilders, of Medina Road, West Cowes, had been contracted to build Short flying boats during the Great war. Following Bradshaw's approach, their technical expert, Dr Terence Fursdon Crang responded that they were also developing a gas-turbine engine and asked for the fullest of details, advising that they would be happy to take on production of Bradshaw's engine, subject to inspection.

With such promising progress, and by now well conversant with Bradshaw's problems with Bromega, Col Reg Gray and Harry Wills both agreed in October 1960 to support Bradshaw's proposal to form a new company to supercede Bromega and be known as Toroidal Engines Ltd, working solely on development of these, now seemingly, viable toroidal engines.

Success

Meanwhile development work on the aero-engine was continuing at H A Wills, but in January 1961 carburettor problems were reported which prevented the engine running properly; thankfully it was discovered that there was no air vent in the petrol tank which had merely caused a vacuum, starving it of fuel! By February, the exhaust gas and spring loaded sealing rings, designed in conjunction with D A Law at Wellworthy, were reported to be working well and Bradshaw now submitted a further Provisional Patent, No. 007394, on 1st March 1961.

With the toroidal engine now at long last starting to run first time, Bradshaw probably felt his work was done and he could safely end his direct involvement in the project for in March 1961 he now seriously considered offering the toroidal engine patents and designs to others, including Dr G S von Heyderkampf at NSU in Germany, who headed the Wankel project. Whether he made contact or not, is not recorded, but it is doubtful at this stage in the Wankel's development that NSU would have been at all interested.

Unfortunately, Harry Wills was most unhappy at Samuel Whites having been invited to take part in the engine project and asked Bradshaw to keep them out of the picture. By April 1961, and now often working all day without food in his desperate drive to bring these projects to a viable commercial proposition, Bradshaw began work on a water cooled version, but unfortunately in May he became incapacitated with shingles and remained so until August and though by then sufficiently recovered to resume talks with Samuel Whites, shingles debilitated him again in October and December.

Meanwhile, Ord-Hume had made available an airframe for ground testing at nearby Sandown airfield, but the engine's troublesome development had by now taken a very heavy toll on the company's, Harry Wills' and indeed Bradshaw's, funds and by March 1962, despite the engine being 'almost ready', Harry Wills and Ord-Hume no longer felt as convinced as Bradshaw that success was 'just around the corner' so at this very late stage, they reluctantly took the decision to abandon the project in its entirety. It is believed some of the Wills built engines survive. With the Wills project at an end, in July 1962 Bradshaw once again approached Samuel Whites to interest them in a more powerful marine engine for larger boats, but to no avail.

Arthur W J G Ord-Hume later restored a 1938 Luton Minor 4A (G-AFIR) and formed the Light Aircraft Association before joining Britten-Norman as a designer.

Toroidal Pulsation Motor

When living at Fishbourne Green, in April 1958, Bradshaw had hoped to use a high pressure pulsed induction cycle in a toroidal engine, as developed in his piston pulsation motor, using an unregulated transfer porting system. A 750cc 200bhp toroidal pulsation motor was duly built, but he soon discovered endless problems with gases under pressure reacting unpredictably. Like vibrating air in an organ pipe, these gases set up varying wavelengths of pulsating gas flows which meant the engine worked efficiently, without surging, only at certain speeds. He found a cure by introducing a rotary disc valve to the induction port to control inflow into a fully purged combustion chamber. He now enlisted the help of the Royal Aircraft Establishment at Farnborough to check the design. They thoroughly approved of it despite their concerns over sufficient cooling of the valves as they were experiencing similar problems on the experimental supercharged Rolls Royce Crecy two-stroke sleeve-valve engine then under development, but Bradshaw was confident he could overcome the problem through the toroidal engine's unique induction design.

Farnborough suggested the toroidal pulsation motor may have an application in helicopters so, in 1961, Bradshaw sent drawings of his 400bhp 'Toroidal Impulse Resonance' helicopter engine (with capabilities of up to 5,000bhp) to R B C Elliott of Saunders-Roe, whom he already knew (was he the same Mr Elliott who designed the ABC Scorpion Mk.II aero-engine?), but while their engineers were impressed by demonstrations on his early toroidal engine, nothing more came of it. (Saunders-Roe had become part of Westlands at Yeovilton on 15th May 1959).

Toroidal Engines Limited

Though clearly disappointed with these setbacks, Bradshaw was now more confident than ever of his twin-toroidal 'gas generator' engines and saw no difficulty in now selling the design. Unfortunately, the acrimony between him and Bromega was getting worse for he had barely received a penny, to which he was entitled, from the provisional patents he was handing over under the agreement. Bromega simply lacked funds to pay, causing him to threaten to withhold further patent rights and, once again, threaten to refer the matter to the Law Society. In a bizarre, yet very sensible precaution, Bradshaw committed his concerns and arguments in letters to Francesca (as a Bromega director), for these would provide the crucial evidence of Bradshaw's belief in Bromega's mismanagement.

Col Gray was also deeply concerned. He however was adamant that, being so close to a finished working example developed by Harry Wills and in view of the fact Wills had been offered a seat on the board of Toroidal Engines Ltd, that Bradshaw and Gibb should work together and in harmony. While Col Gray and others were extremely supportive of Granville and were as equally keen to see the engine come to fruition, they were now trying desperately hard to keep Granville at arms length from prospective buyers lest he worried them about his business and financial allegations against Bromega - and also put them off by 'blinding them with science'!

All concerned fully realised that only the major motor vehicle engine manufacturers, such as Petters and Rolls Royce, had any real chance of making the engine a success. To that end, Col Gray implored Bradshaw to allow him, alone, to try and sell the rights to Bill Lyons at Jaguar, Harry Harriman at BMC and Pat Hennesey at Ford, all of whom he knew personally. Bradshaw meanwhile held out hopes that his old friend, Mr Everden at Rolls Royce, would keep alive his interest in the project.

Granville Bradshaw with a 750cc 50bhp toroidal engine (GB)

Granville finally backed off his assault on Bromega and Mr Gibb and in a letter of apology, he wrote that he was now under doctor's orders and was far from well; he had lost two stone in weight and had barely been out of the house for some months (except for the odd trip to his favourite pub, The Fleming Arms for a medicinal drink and game of snooker). In December 1961, Mr Gibb finally agreed to transform Bromega into Toroidal Engines Limited, registered at Holmes Road, London NW.5. Fearing the worst, Granville had already expressed a wish that should his ill health take its final toll, that both Col Gray and Harry Wills should manage the new company.

Poverty

Although in April 1961, Bradshaw eventually received the £1,000 lump sum for Ewart's annuity from Mr. Huck, his constant lack of funds and Bromega's constant inability to pay Bradshaw for work done, forced him to pay even more bills for engineering work by giving away his shares; this at least had the advantage of their recipients maintaining their interest in the project, but by early 1962, he and Francesca were falling behind on mortgage repayments for Grey Cottage and in March 1962 Mr Gibbs, on behalf of Bromega, was having to threaten eviction if Col Gray, as guarantor, did not make up the shortfall! On 30th March, Bradshaw submitted his long threatened report to the Law Society... and there, annoyingly, his archives end.

Even Sharps Commercials was now suffering with limited funds to maintain Bradshaw's work for, with sales of the Minicar already in decline, a fatal blow came in the 1961 and 1962 Budgets which destroyed the very favourable discount in Purchase Tax between three-wheelers and motor-cycle sidecars over the more highly taxed four-wheeled motorcars. This severely dampened demand for three-wheelers and heavyweight motorcycle and sidecars from companies such as Phelon & Moore, but though P&M went into receivership in November 1961, Bond Cars survived by diversifying at the end of 1961 into lightweight trailer caravans, motorboats and powered water skis.

Final developments

The major set-back over Wills and Ord-Hume abandoning the aero-engine in March 1962, was offset by good news from Col Gray that Mr Gibb has 'practically sold' the Bromega gas turbine engine design to one of America's largest firms and, accordingly, Granville's financial worries 'should soon be over' - but there is no further mention of this development.

Then in May 1963, with renewed, deepening acrimony towards Mr Gibb, Bradshaw received confirmation through Francesca, that a major engineering concern was seriously interested in developing the engine; Mr Gibb's letter states: 'I trust Mr Bradshaw will appreciate that it is not necessary for the engine to be in perfect working order before it is sent to London, and I think you will agree with my view that the persons interested in exploiting the engine are fully capable of developing it to a proper state'. Once again nothing more is heard.

By mid-1963 Granville and Francesca had moved to 2 Holbury Hill, Holbury St. Mary, Ewhurst, Surrey where he produced drawings for a series of toroidal engines each reporting the 'culmination of 20 experimental engines, three of which were now said to be running satisfactorily'. These included a 60hp 1,500cc petrol engine weighing 20lbs (using alloy pistons); a 1,000cc racing car engine; and a 100bhp 1,500cc diesel motor car engine (measuring $9^1/_2$" diameter, 10" long and weighing only 80lbs complete with a cooling fan which drew air into and around a cowled toroidal chamber). There was also a compact 650cc motorcycle engine on the drawing board.

A 'silent' aero-engine

Before his bankruptcy in the late 1930s, Bradshaw lived in the old manor house at Lowfield Park, Lowfield Heath, West Sussex. By cruel irony the house was doomed by the very aeroplanes that he had helped create half a century earlier, for with the transformation between 1956 and 1958 of the pre-war civilian Gatwick airport, Lowfield Park manor house lay in the way of the new 7,000ft concrete runway which would turn Gatwick into London's second airport. Built in 1646, the house is said to have been dismantled brick by brick, beam by beam, and shipped across the Atlantic to be rebuilt in Pennsylvania. The gate keeper's Primrose Cottage, was not so lucky and all that remains of the park today is the lodge cottage and the park's entrance gates. Convinced he could indeed develop a 'silent' engine and solve the problem faced by thousands of people living under the flight paths of major airports, he came up with a design in the summer of 1962. Measuring 18" diameter, the 500hp, single crank toroidal engine could be coupled in-line with a second engine to form a 1,000hp power plant offering 0.65lbs/bhp. But 'silent' was a misnomer, for these cacophonous toroidal engines were as far from being silent as could be imagined; 'silent' only meant the engine lacked the usual mechanical valve clatter! But it kept external interest in the project alive.

Having temporarily moved to his son, Geoffrey's, home in Leatherhead in 1964, he came under the flight path to Heathrow and Gatwick airports and once again dreamt of a truly silent aero-engine! Returning to his lodgings at the Southdowns Hotel near Rogate, Petersfield, on the evening of Friday 27th, August, he drew up specifications overnight for 'silencing gas turbine aircraft engines'. He posted the provisional patent form on the Saturday morning, but sadly, there are no further details - the patent was not granted.

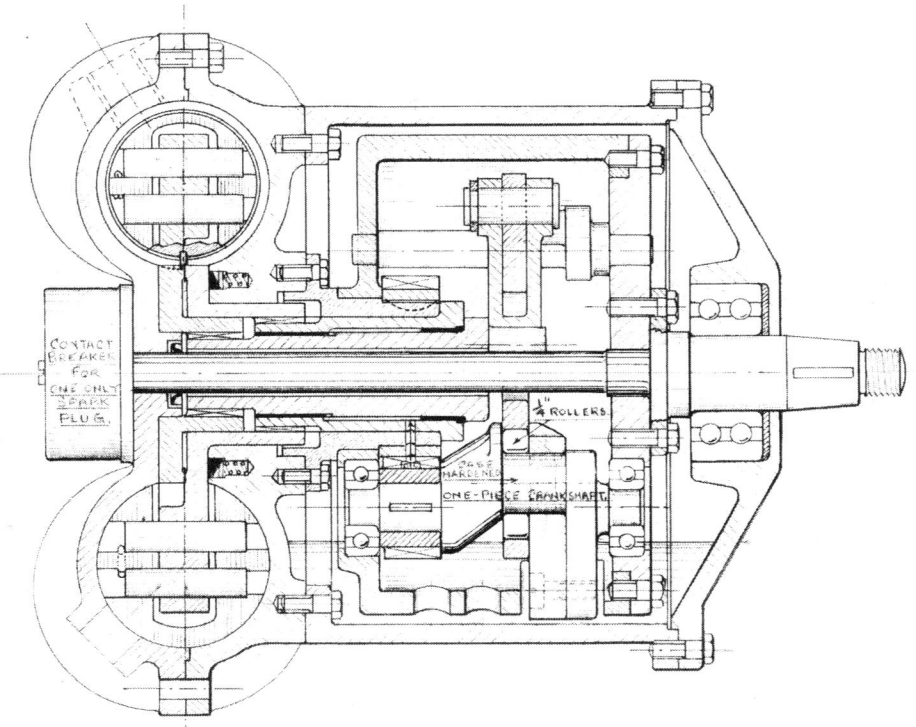

Proposed 650cc motorcycle engine from 1963 (GB)

20
Wankel Opposition Engine

In several of Bradshaw's letters and drawings he refers to his toroidal engine as the 'Wankel Opposition motor', for both his and the Wankel engine (first seen in 1956) were competing against free-piston and gas turbine engines in the race for the engine of the future.

Yet while Bradshaw and Wellworthy Ltd had virtually solved the gas sealing problem on his toroidal engines by late 1960, the Wankel engine was still having major problems with its spring loaded, oscillating tip (or apex) seals which, as the angle of contact varied slightly as the rotor revolved in the epitrochoidal chamber, caused hammering and incomplete sealing. This proved an almost incurable problem right into the 1970s despite a partially successful 'cure' by Daimler-Benz in 1961. Fortunately the three sets of corner seals and rotor end-seals proved less of problem. Even the sceptical Harry Mundy of *The Autocar* conceded that Bradshaw's toroidal combustion chamber design and sealing system was far superior to Wankel's!

Wankel rotary engine

The conventional reciprocating internal combustion piston engine has all but reached its limits of development, power and speed due to the forces of the internal reciprocating masses: pistons, connecting rods, valves. These forces vary as the square of the speed. Limitations also exist in controlling spring closure of valves at high speed. While a two-stroke engine largely overcomes these valve problems, their limitation in speed and power is governed by port exposure in the relatively short time cycle of the reciprocating piston. While Bradshaw's toroidal engines solved the problem of reciprocating masses, it still suffered from limitations in port exposure, but worse, just like free-piston engines and gas-turbines, it was unsuited to low engine speeds and part load conditions!.

Engineers around the world are still striving to find the ideal engine. Dr. Felix Wankel believed he too had found the perfect solution, but even his design can be traced back to ideas at the turn of the 20th Century!

In early 1924, Felix Wankel was working on high pressure lubrication systems in his technical research laboratory at Heidelberg, Germany. While developing new rotary pumps, he saw great potential in a new type of pump seal and in the early 1930s wrote up a purely theoretical design. He realised that, just as in rotary aero-engines of the Great War, their Achilles' heel was distortion through asymmetrical heating which caused improper sealing but which was temporarily overcome by obturator rings. The outbreak of war in 1939 brought his work to an end and he became involved in research and testing of aero-engine rotary valves at research laboratories in Lindau. With his own laboratories destroyed by the RAF, it was not until 1947 that he was able to resume his work on rotary compressors and engines. By 1951 this had attracted the attention of NSU who were seeking a small engine for their post-war motorcycle designs.

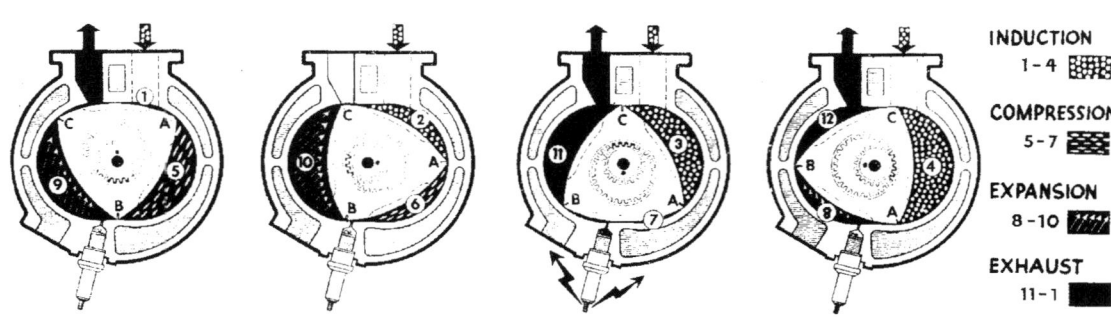
Wankel engine design and cycle

By 1953, Dr Froede of NSU and Dr Felix Wankel had begun their work on a twin rotor, epitrochoidal 'rotary' engine which used an offset, inner triangular shaped rotor, rotating inside an outer rotor travelling at one and a half times the inner rotor's speed. The triangular inner rotor maintained contact at its apex at all times with the 'epitrochoidal loci' figure '8' shaped outer rotor's combustion chamber by virtue of spring loaded tip seals and, in so doing, formed three ever varying volume chambers which allowed for induction, compression and, on firing, expanding exhaust gases pushed against the rotor seals, turning the engine over.

The first prototype ran in February 1957 and though, it was said to have 'worked first time', it soon underwent major redesign using a now static outer rotor in which an inner rotor both rotated and followed an eccentric path along the geometric centre of the figure-8 outer rotor's combustion chamber.

In 1958 a joint development programme was signed between NSU-Wankel and the Curtiss-Wright aero-engine company in America for industrial use with further licenses being granted between 1959 and 1962 to Perkins Diesels in the UK, Daimler-Benz, Fitchel ünd Sachs, MAN, Krupp, Glockner-Humbold-Deutz in Germany and both Yanmar Diesel and Toyo-Kogyo (Mazda) in Japan.

As with Bradshaw's toroidal engine, there is no directly equivalent cubic capacity between a conventional internal combustion engine and the Wankel engine. This caused problems with the German motor-car tax system which favoured small engines, and many a headache ensued as Dr Wankel and the tax-men tried to reach a compromise! Similar problems were encountered in racing circles over the Wankel's true capacity. However, unlike the Bradshaw toroidal engine, the Wankel's power and economy is no better than its equivalent piston engine, but where the Wankel engine excelled was in smoothness, balance and, due to its compact design and lower weight, a greater power to weight ratio.

Extensive testing revealed inconsistent seal-tip wear due to varying contact angles with the figure '8' wall, causing it to chatter and, over time, impress minute indentations in the cylinder wall. Added to this, the gas ports and spark plug apertures formed corrugations on the apex seal. Curiously, not all of these defects were found on every engine.

With their extensive motor industry, research and testing resources (compared to Bradshaw's shoestring operations), a prototype NSU-Wankel engine was being road tested in an NSU Prinz Sports chassis during late 1961. Dr Wankel announced in late 1962 that following extensive trials, the first production car would be formally announced in early 1963; but 1963 came and went and the first production NSU Wankel Spyder did not appear until 1964. NSU launched their new NSU Ro80 saloon in late 1966, but this used a co-axial, twin rotor engine which developed 50bhp from an equivalent 500cc piston engine. Some 37,204 Ro.80s had been produced by 1977, but so problematical were the engines that many owners replaced them with the compact Ford Taunus V-6 engine.

Although Daimler-Benz had partially cured the tip-seal problems, they soon abandoned their work on a Wankel replacement for their famous 2.2 ltr 6-cylinder in-line engine and it was left to the Japanese Mazda car company to fully develop the Wankel from 1967 experimenting with their 1108 sports car; but they too had a long way to go before resolving tip-seal wear; this was eventually solved by Norton motorcycle engineers. Some two million Mazda engines later, the now extremely reliable rotary engine remained in production until recently.

Despite NSU's intentions for the Wankel engine, the first motorcycle to be so powered was the German Hercules W2000 of 1975. This was later joined by the Japanese Suzuki RE5, but neither were a commercial success. Meanwhile in Britain, the Wankel twin-rotor project under development by BSA/Triumph was inherited in the early 1970s by Norton-Villiers-Triumph who successfully developed a Fichtel ünd Sachs design in 1978 for the Norton 'Interceptor 2' which was sold almost exclusively to the Police and Ministry of Defence as the 'Interpol 2'. Professional police riders appreciated the 'vivid', if very uneconomical, acceleration up to a top speed of 130+mph; the turbine like power surge came in around 8-9,000rpm. The engine's low weight of barely 90-lbs contributed greatly to the Interpol's incredible handling. Despite the benefits of long service intervals (many examples having clocked up over 400,000miles of trouble-free operation), the engine still required much more development before it was felt suited for the civilian market - a market which, sadly, was never fully exploited. Unfortunately its effeminate 'meowing' simply did not sound like a 'proper' British motorcycle and only about 100 civilian models were made.

While Norton exhibited examples of the engine for controlling Thorn-EMI mobile radar units, RPV (remote piloted vehicles), light aircraft engines and marine engines, their greatest reward came by licensing it in 1983 to Teledyne-Continental Inc in America who produced many thousands for micro-light aircraft and military target-drones in the 1980s.

Even the Ford Motor company produced a Wankel engined concept car, but the cost of retooling and development was deemed unacceptably high and they left its development to Mazda, in whom Ford had become a major shareholder.

In the aero-engine field, Curtiss-Wright gave up their Wankel rotary engine development in 1980; that project was taken over by John Deere Lycoming, but that too ended in 1989. Today, only Wankel and the Russian VAZ plant are still developing rotary aero-engines.

21
Twin-Toroidal Gas Generator

With a proven application in jet fighters, such as the Gloster Meteor, engineers around the world began looking at the gas-turbine's potential in motor-vehicles, trains, boats and power generators. Britain's Rover car company led the field with a gas turbine engine in 1950. This, the early success of Wankel's engine and Bradshaw's ability to demonstrate a working single chamber toroidal engine, proved to sceptics that there was indeed an alternative to the conventional overhead or side valve motor car engine.

Bradshaw was now determined to incorporate the considerable advantages in power, efficiency and compactness inherent in free-piston and gas turbine engines into his toroidal engine and in early 1956, submitted a patent application for a twin toroidal gas-turbine design. However, as the output was by a gas turbine rather than direct crank action, it could not be called an 'engine' under patent rules, so he christened it a 'gas generator' but to add to the confusion, he also at times referred to it as a 'gas cycle' engine - which was more correctly his Pulsation Motor! He was now however more convinced than ever that his improved, compact and lightweight twin toroidal 'gas-generator' engine would be superior to GM's 'Hyprex' free-piston engine and Rover's gas turbine, making the Bradshaw 'engine' a world leader.

Free-piston engines

Gas turbines use the exhaust gases from an exploding mixture (usually kerosene) to turn a turbine blade whose output shaft provides motive power through rotation of a motor-car axle or a propeller, as in a turbo-prop aero-engine. A gas turbine engine has no valves, gears or crankshafts - it is a 'free' engine.

Turbines are in fact centuries old - windmills and water-mills stand testimony to this - but the first practical use of exhaust gases to drive a turbine came in 1908 with the Swiss engineer, Dr Bulch, directing exhaust gases from a diesel engine to a turbine which then transmitted power through a shaft. In the 1920s, Swedish engineers used diesel exhaust gases to power a turbine in a boat which turned the propeller, saving considerably on boilers, chimney stacks and propeller shafts.

By strict definition, a 'free piston engine' is one where a pair of loose pistons move freely within a cylinder, be it straight or toroidal; ie. they are not attached to connecting rods as in a conventional motor car engine. As these pistons are propelled towards each other, they compress a volatile charge which on being ignited, forces the pistons apart and in so doing, expel the cold, ambient air behind them which purges the combustion chamber and drives an exhaust driven turbine.

The 'free-piston engine' was fully developed by Raoul Pateras Pescara de Castelluccio in the 1920s. Like his contemporary, Granville Bradshaw (1888-1969), Pascara (1890-1966) also had an interesting, pioneering life.

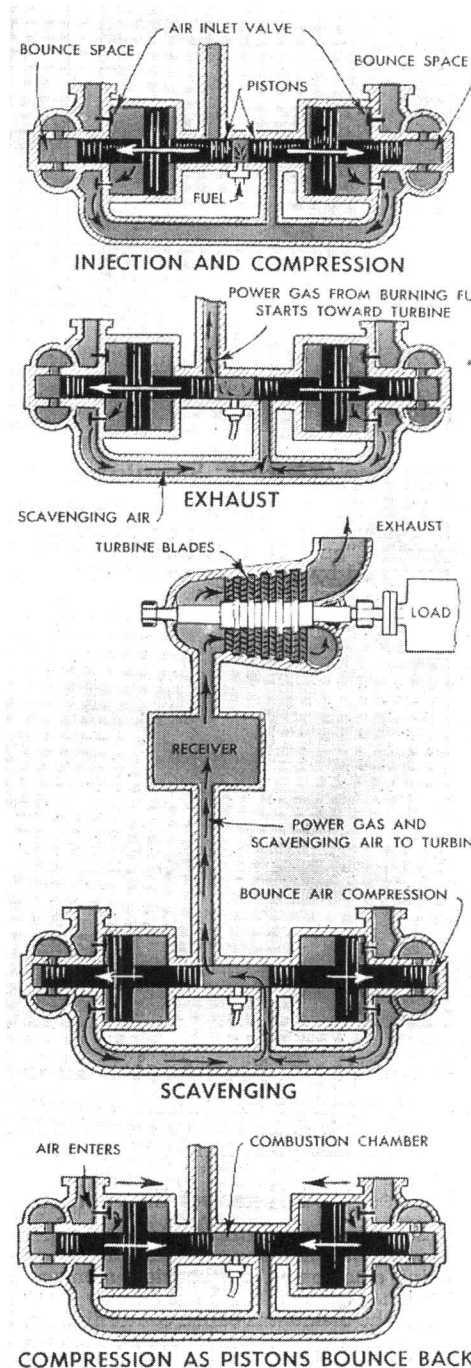

General principle of a free piston diesel engine showing scavenged air purging and assisting in driving an exhaust turbine output shaft

Born in Buenos Aires, Argentina, he moved to France to study engineering and joined the laboratory of Gustave Eiffel (of Eiffel Tower fame), where he worked on compressed-air propelled torpedoes. Moving to Italy, he trained as a lawyer but also designed a sea-plane for the Italian Navy. As early as 1919 he was working on the theory of helicopters and developed a co-axial, contra-rotating, variable cyclic pitch, 4-bladed twin bi-plane rotor design; it took to the air in April 1924 and set a new altitude record of 2,400ft in 4 mins 11 secs, but he abandoned further development in 1925 and in 1929 co-sponsored his brother Enrique and the Spanish Government in the short lived 'Fabrica Nacional de Automoviles' motor car business which developed an American inspired straight-8 powered luxury car, much like Macklin's 'Railton', but the Spanish Revolution of 1931 brought that project to an early end! By then, Pescara had already begun work on his free-piston compressor and in 1933 established his Pescara compressor business in Luxembourg.

Pescara's concept was developed at the Alstrom works in Belfort, close to the Swiss border, but with France falling to the Nazis in 1940, production ceased. Some of the technology passed to the German Navy who already had diesel powered free-piston air compressors in their U-boats for charging cylinders to fire torpedoes by compressed air. Immediately after the war, Alstrom resumed their development work while in Switzerland, the Sulzer Brothers began work on a free-piston gas turbine engine. In Britain, Henry Meadows Ltd, the heavy car and lorry engine makers, gained an exclusive licence to develop Pescara-type free piston engines (with pistons up to 6" in diameter) to power a turbine final drive.

While Meadows undertook all prototyping and production work, they formed a joint venture with the Associated British Engineering Group, called the Free Piston Engine Company, to undertake its design and marketing. ABE included several famous engine and transmission companies: British Polar Engines Ltd, Controllable Pitch Propeller Co Ltd and Parsons Engineering Co., (the British pioneers of steam turbines). The Free Piston Engine Company was based at Meadows' Park Lane Works, Wolverhampton, next door to Guy Motors. Meadow's first engine was to be developed in conjunction with Alstrom in France and Alan Muntz & Co Ltd in Britain, (who were already producing similar 'gasifiers') and was scheduled for completion by the end of 1957 but their development came to an end when Henry Meadows Ltd was taken over by Jaguar Cars Ltd in 1964, joining their Guy Motors subsidiary.

When in October 1956 Col Reg Gray heard from Granville of Meadows' work, he offered to promote his 200hp twin toroidal 'gas-generator' design (then under a provisional patent) to his friend, Sir Henry Spurrier at Leyland Motors as well as to Perkins diesels. Gray cautioned Bradshaw of their potential reaction: 'You come up against the old trouble - one designer trying to condemn the work of another'.

Other engineers around the world were also persevering with free-piston designs. In 1956, a Scotsman, Harry Anderson, prompted by the announcement of Bradshaw's 'Omega' engine in 1955, approached Granville offering him his provisional patents on a design for a static toroidal, free-piston diesel engine/compressor which turned a turbine output shaft. Anderson claimed his design overcame the reciprocating connecting rod sealing problems by using a sun and planet gear to expose a sleeve valve on the inlet and exhaust ports, which allowed gases in and out of the chamber as necessary. His provisional patent was not granted. It is doubtful that Bradshaw took any interest in Anderson's design as his own engine was already well advanced.

Gas turbine cars

Both the Rover Car Company and General Motors, in America, began development of a gas turbine car engine in 1948. Austin Motors began work on their gas turbine engine in 1949, testing it in an Austin Sheerline saloon. Sir John Black's Standard car company also developed an industrial gas turbine engine in 1957. Even the French and Italians developed gas turbine car engines in the mid 1950s, but the dominant companies in the field were American: Chrysler, Ford and General Motors all of whom prototyped futuristic cars of which Dan Dare of *The Eagle* comic would have been proud.

It should be remembered that Frank Whittle's gas turbine jet aero-engine design of 1936 was co-developed under government control by Power Jets Ltd, the Rover Car Company and British Thompson-Houston in the early war years and that the technology was given, *gratis*, to General Electric in America in the interest of winning the war. Rover later exchanged their jet engine technology for Rolls-Royce's development of a truncated V-12 Merlin Spitfire aero-engine, which became the Rover 'Meteorite' V-8 powering tanks and the mammoth Thornycroft 'Mighty Antar' oil field pipeline and tank transporter.

Rover's first gas turbine car of March 1950 (reg no. JET 1) used a modified Rover 75 chassis, powered by their 1S/60 60hp engine. It easily reached 85mph with the compressor rotating at 35,000rpm, yet the engine's cruising speed was 50,000rpm! In 1952, a modified engine was tested at Jabbeke in Belgium where it reached 152mph. Three further cars were built: the T2 and T3 of 1956 and T4 of 1961. The T3 had an 2S/60 engine which developed 120hp and consumed 14mpg at 60mph. These were followed by the new Rover-BRM racing car of 1963 which sucessfully entered the 1965 Le Mans.

Gas-generator

Bradshaw's 'gas generator' engine comprised a small, forward mounted toroidal diesel (compression ignition) engine, supercharged by a large toroidal free-piston compressor driven by a common shaft. The diesel engine's exhaust gases then drove a turbine providing motive power while its concentric crankshaft from the oscillating pistons, drove, for instance, a dynamo or pump.

This idea was not entirely new, for in 1950 Napier had produced their 'Nomad' compound flat 12cyl two-stroke diesel aero-engine whose exhausts powered a 3-stage turbine which drove a 12-stage axial flow supercharging compressor. But Bradshaw's concept of a toroidal compound engine was secured under a joint Bradshaw and Bromega patent, (No. 856619, laid in May 1956 granted on 21st December 1960). Patent applications were also lodged in France, Sweden, Italy, Belgium, Germany, Switzerland, Canada and the USA; it was suggested he also applied in Australia and Holland. The German patent office initially rejected the application as they considered there was little difference from earlier patents, but Bradshaw pointed out that earlier designs used gears for transmitting the power from the oscillating pistons to the output shaft; these gears constantly chattered and broke teeth, whereas his new patents used two concentric shafts directly connected to the pistons which eliminated any chatter which now made the engine commercially viable.

By March 1960 Bradshaw was convinced that such a modified Pescara type free piston 'gas generator' turbine engine would see off the new Wankel engine. A drawing dated 26th April 1960, produced when convalescing through 10 days of illness on the Isle of Wight, shows a 16" diameter, 250bhp diesel gas generator engine weighing only 200lbs; it is coupled to a triple blade 'Lucas' gas turbine, driving an epicyclic reduction gear with optional reverse and transmission brake for a clutchless, 'two-pedal' (accelerator and brake) automatic transmission car. A smaller 150bhp diesel version was also proposed with the annotation, 'the worst driver cannot stall the engine'. These he sent in confidence to twenty motor-car companies, including Daimler, Bristol and his old friend, Mr Everden, Chief Designer at Rolls-Royce.

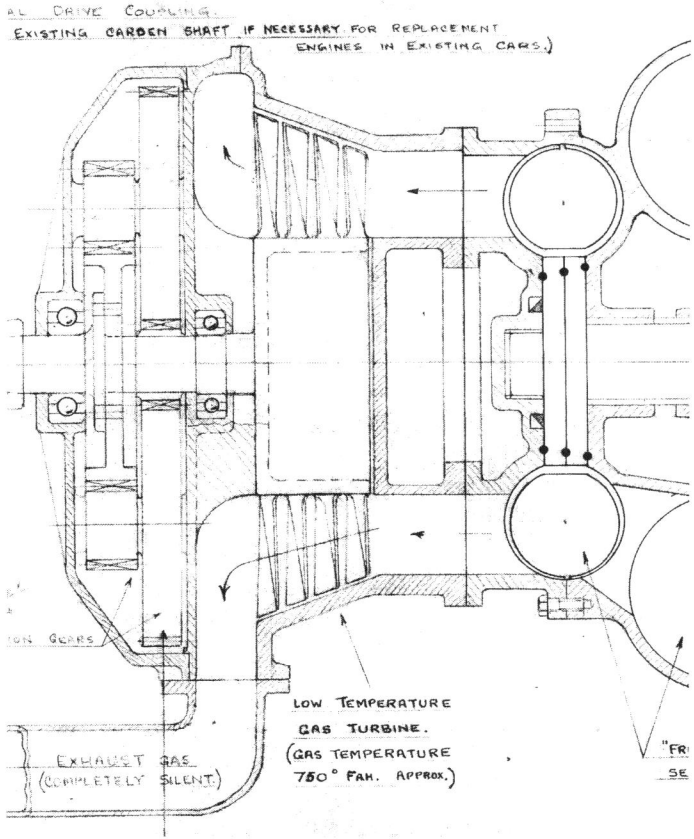

2-pedal diesel twin toroidal gas turbine car engine of 1960 driving a reduction gearbox and output shaft (GB)

It is clear from his letters that he was finally seeing the light at the end of the tunnel over the toroidal engine, but his relentless endeavours had once again brought him perilously close to bankruptcy just when he was in urgent need of a motorcar to road test the engine, then being built for him in Preston. By late autumn 1960, Reg Gray had secured for him an old Ford Prefect. Bradshaw removed the engine but retained the flywheel and transmission. On first firing, the newly build twin toroidal engine soon seized through too fine a tolerance in the plain bearings, for he had not allowed sufficient expansion of the different metals. This was soon resolved by hand scraping of the bearing surfaces, unfortunately, these archives end in November 1960, so there is no further record of the engine's progress.

Gas generator aircraft engine

In September 1958, Bradshaw proposed a 220bhp, 10.8 litre twin toroidal chamber 'gas generator' engine for motor vehicles, trains or helicopters. Indeed, later designs incorporated trunions within its casing, so that it could be rotated from the horizontal (for forward propulsion), to the vertical for lift and STOL (short take-off and landing) aeroplanes

As earlier mentioned, in November 1959, Granville heard from an old technical journalist friend of his, David Scott, who had seen an American design similar to Bradshaw's. He was keen to help promote Bradshaw's gas generator engine through his colleagues in America where interest in gas turbine and free piston engines was strong. Bradshaw realised that 'it was my old P&M toroidal' so, he immediately applied to extend his original patent to America under a UK Provisional Patent, No. 030957/58. However, the US Patents Office rejected it in March, probably on the same grounds as had the UK Patent Office.

Nevertheless, David Scott was still actively promoting Bradshaw's designs to major US companies and by April 1960 he had approached the Eaton Manufacturing Corporation, a huge motor vehicle transmission group. Encouraged by Scott's enthusiasm, in January 1960 Bradshaw offered his 200hp twin toroidal gas-generator design to various North American companies including AVRO-Canada and the Curtiss-Wright Corporation in New Jersey to whom he also offered a 500hp, twin planet gear action engine weighing only 200lbs; this was followed by a proposed 16 litre, 600bhp engine. He duly offered Curtiss Wright the sole US rights for £10,000, but with no further interest forthcoming (probably as they already had a licensing agreement with Wankel) and, frustrated by their apparent contempt at having to pay a quite sensible royalty, Bradshaw withdrew his exclusive offer on 14th July, 1960. He also tried to get his old friend Tom Sopwith, now running Hawker-Siddeley, interested in this more powerful engine, but Bradshaw's request for Tom to find an aero-engineer to refute his technical claims to the engine was never taken up!

David Scott also knew Peter Masefield of Hawker Siddeley and was more than happy to promote Bradshaw's engines to him. By October 1960, things were looking up after Peter Masefield started to take a keen interest. In fact as early as February 1960, Bradshaw had approached the Brush power generator business and the Petters Diesels company at Staines; both companies were part of the Hawker-Siddeley group. Brush were interested as they saw their future lying in free-piston gas turbine engines, particularly as they had the ability to run on any hydrocarbon fuel, as Bradshaw well knew, but they would not take up Bradshaw's design as he had not yet secured UK rights to the Pescara free-piston engine! Bradshaw now wisely decided to promote the gas-turbine output benefits, rather than draw attention to Pescara's accepted principles.

Petters were in fact quite keen on the Wankel rotary engine but remained cautious of its now well publicised piston seal problems. Seeing his opportunity, Bradshaw confidently offered J C Dacombe at Petters' Hamble works revised drawings of the proposed Bromega 2.5 ltr diesel gas generator toroidal

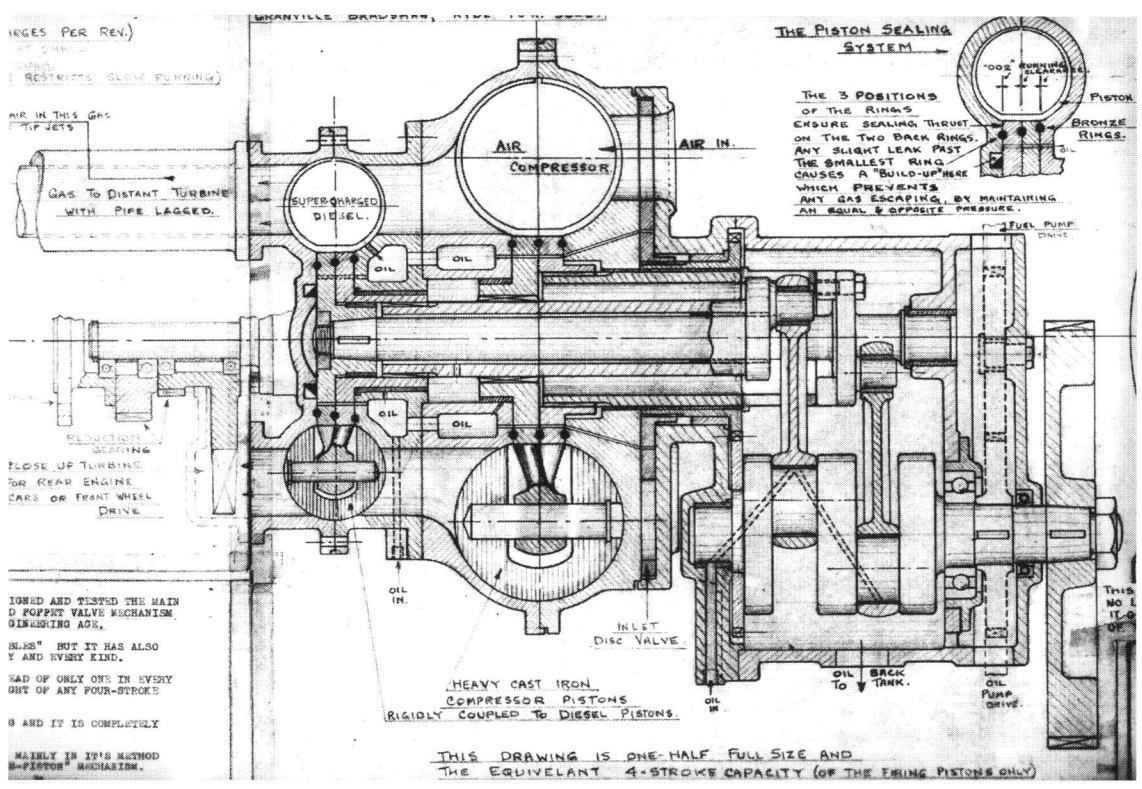

The larger free-piston toroidal compressor supercharges the smaller diesel toroidal engine and uses an exhaust gas turbine output (GB)

engine, a capacity well suited to their diesel powered generator sets. Smaller diesels would be better suited to their 'Thermo King' lorry mounted refrigeration units, built at Hamble. Petters' interest was aroused and in April 1960, they offered him £100 for limited rights to the engine, which Bradshaw accepted.

After a thorough examination of the toroidal rotary engine principle using an example made at G E Motors of Bembridge, Petters summed up in a letter to Bradshaw in May 1960 that 'we have come to the conclusion that the problems which would be involved in bringing the engine to a commercial unit would be long and expensive in resolving. It is of our carefully considered opinion that other novel forms of engine are likely to be more attractive'. They were almost certainly now pinning their hopes on the Wankel and their interest in Bradshaw's engine ended. Petters returned the demonstration engine with a note that the pistons were broken. It was immediately evident from damage in the combustion chamber, that the engine had been tested to destruction at around 6-7,000rpm, and as neither the pistons nor piston rings were returned, Bradshaw suspected foul play! The engine was duly restored by H A Wills & Co. Coventry-Climax managed to destroy a further prototype in October 1961!

There is a report that a Bradshaw toroidal engine was tested to destruction at Harry Weslake's works at Rye, but whether in an 'approved manner' or in association with Coventry Climax or Petters is not known, for Bradshaw makes no mention of Weslake in his diaries.

The final years

Despite his growing wariness with the project, its continual sapping of his limited resources and the rapid onset of old age and ill health, Bradshaw persevered with the project but took partial relief in January 1961 when Harry Louis persuaded him to restart his series of 'Jottings' for *MotorCycling*. Later that month, Bradshaw received the good news that his American patent for his twin toroidal gas-generator had been granted on 17th January 1961 as US Pat No: 2,968,290. This greatly boosted his confidence and in May 1961, the BBC contacted him for information following a question on their *Any Questions* radio programme, which was answered on the 4th May *Any Answers* edition. A demonstration film then appeared on the BBC TV *Tonight* programme with Cliff Michelmore - unfortunately these no longer exist in the BBC's archives.

By late 1961, once again in ill-health, he had persuaded Lucas to allow him to use their new electronic ignition system for the toroidal engine, but this made little difference to the engine's running and it increasingly became clear the answer lay in the simpler compression ignition diesel cycle.

In December 1961 came news from Col Gray that contact had been made with both Standard-Triumph and Sir Henry Spurrier of Leyland, who were both interested in the gas-generator engine, probably to counter Daimler-Benz's involvement in the Wankel engine. Leyland did indeed develop a short lived 350/400hp gas turbine developed from the Rover 2S/350/R engine in 1968; this 38ton GTW tractor unit was the only British, commercially successful, gas turbine driven vehicle, albeit built in very limited numbers.

Also in late 1961, with Bond cars sales in the doldrums through the adverse Purchase Tax, Col Gray secured the European rights to the glass-fibre, twin hulled, American Power-Ski - a self propelled 'ski', powered by an outboard motor. This required a suitable engine and full manufacture in Britain. Bonds optimistically geared up for 1,000 Power-Skis per annum, but sales proved dismal and barely 100 were actually made. They also revived an earlier Bond glass-fibre boat of 1950, rebadged 'Sea Ranger' for 1962, but even fewer of these were built. Granville then proposed that Harry Wills should buy the Power-Ski rights for the Isle of Wight and Channel Islands, fit a Bradshaw toroidal engine to provide water jet propulsion and abandon go-carts. Needless to say, it fell on deaf ears! Ironically, in the autumn of 1962, a small Wankel engined high speed 'tug' was launched in continental Europe as a self-propelled training device for the small niche market of professional water skiers.

Decline

While Bradshaw's last twin toroidal design appears to have been drawn up in May 1962 for a 500cc motor car engine, now at the age of 75, he was still talking enthusiastically to clubs and institutes, responding willingly to all questions and eagerly demonstrating his engines. His optimism never wavered in spite of those many setbacks and the increasing criticisms now being levelled at his many designs, but regardless of his undiminishing enthusiasm, his age and ill health were now taking their toll and his son, Geoffrey, urged him to leave it to the professionals in the industry to iron out the remaining teething troubles.

Following the death of Francesa's son-in-law in March 1964, and with the short-term tenancy at Holmbury now up, she and Granville rented Sunnyside at South Harting, Petersfield so she could be closer to her daughter, but this move merely increased her own frustration at Granville's 'failure' to conclude the engine's development and provide a return on her own quite considerable and unselfish investment in him and the engine. This, understandably, soured their long, loving and caring relationship and in May 1964, they parted for good. His son, Peter, paid for him to lodge temporarily at Southdowns Hotel, Trotton, Rogate near Petersfield.

Granville's overwhelming desire to finish his career 'with an engine that will make all 4, 6 and 8-cylinders obsolete' was not to be. He became ever more frustrated with his worsening health and reliance on others for his mobility and funds. In a letter of August 1964, Granville reported to Peter that 'the (unidentified) manufacturer has a larger factory in Portsmouth' which would be easier for him to reach from Petersfield. Who that manufacturer was however remains a mystery.

He had already submitted a manuscript for *Jottings from an Inventor's notebook* which he had written to help potential inventors and submitted to Newnes. He now hoped to complete his autobiography for Temple Press and he earnestly hoped significant advance royalties for both his books would allow him the opportunity to return to the Isle of Wight to see his toroidal gas generator engine finally completed. But they were never published, though many excerpts from this now lost manuscript were later used by Geoffrey for his mini-biography which appeared in the Brooklands Society's *Gazette* in the 1980s.

In 1965 Granville moved into Peter's home in Hitchin where he, for once, seemed at ease and content with life, still enjoying a quiet lunch-time drink at the local pub. However within a couple of years, he began to find it quite a struggle to walk to the pub and then, suddenly, he began to feel truly tried and, as if giving up on his life, retired to his bed, barely eating. Two weeks later, on 13th April 1969 he died peacefully in his sleep at the age of 82. Sadly, his death brought forth few obituaries.

Of his father, Geoffrey observed that, "It has often been said that he did not receive the commercial rewards his brain warranted, but at least some of his designs are preserved for posterity in museums the world over, and amongst such household names as Stephenson, Watt, Mitchell, Barnes Wallis, Rolls and Cody, to name but a few."

The end?

Looking at Granville's simple demonstration model, one can't help but be mesmerised at its workings and potential of simple rotary engines. Even the only successful rotary design, the Wankel, has had a very limited market success probably due to the public's misunderstanding of its working, preferring the known reliability of conventional internal combustion engines.

But the toroidal engine and compressor design is far from dead for, like the alchemist's dream of instant gold, engineers around the world still strive to make the toroidal engine yield its elusive power which had defeated Beck, Esselbé and Aumeteyer. Included among later generations of engineers working on experimental free or gear driven piston, or vane, toroidal engines and compressors are those from Kauertz in 1967, Tschudi (1967), Rocha-Cano (1971), Taurozzi (1972), Farrokhzad (1991) and Kaloustian (2003).

To his credit, Bradshaw's determination over many years, working almost single handedly on a shoe string budget, had produced a working, if unrefined, engine whose commercial success was killed by the Wankel which had received huge financial backing from the German, American and Japanese industries.

Toroidal Engines Ltd was wound up and dissolved in April 1975.

Granville Eastwood Bradshaw in 1956 (GB)

22
Conclusion

Patentee Extraordinaire - A Personal View

Funding is the major bane of any inventor's life and often they have to rely on their inventive mind to create money spinning ventures. Sir Hiram Maxim, of machine gun fame, was unable to the raise funds to build his imaginative steam powered, multi-winged aeroplane, so he built a fair ground roundabout which brought in a good return. Dr Townend, of 'Townend ring' fame, designed a gyroscopic toy trick-bicycle which sold by the hundreds; Alfred de Quervain Colley built Britain's first tape recorder, but production was thwarted by problems with the magnetic oxide emulsion. While this was being resolved, he made his fortune with the famous solar powered, bobbing 'drinking duck' toy.

Bradshaw's fertile and imaginative mind provided novel solutions, often years ahead of their time to a wide range of challenges; as a result he was able to build up a bank of unexploited ideas from which he could draw when money was tight. In fact such was his extraordinary gift for inventing, that it was said by the Patents Office that he had lodged more patent applications in any one year than any other person; but though he was one of Britain's more prolific inventors, he had 59 full patents to his name, far fewer than Frederick Lanchester, a true pioneer of motoring, with over 100.

In an *Autocar* article of 20th May, 1938 Granville wrote that it is 'more difficult to find the thing to invent then it is to invent it'. An idea would be triggered by something, even momentarily, and he would be off to his drawing board to formulate his idea, calculating in his mind the many mathematical permutations and, with slide-rule always to hand, working out ratios at lightning speed; he once wrote an unpublished book on the slide-rule.

He could become quite obsessed with a problem and, knowing that it could be solved, became so absorbed in his invention that he often missed meals. His daughter, Pamela, recalled how when living near Brooklands, she and her mother tried to drop him his lunch from the footbridge as he raced around the concrete track and how he often forgot all about time, arriving home for dinner many hours late, much to the annoyance of his house staff. As a young bride during the war, Gwen the wife of his son Geoffrey, recalled how at her first dinner with her new father-in-law, Granville suddenly left the table, still chewing his meal. When he did not reappear, she feared he was ill, but an hour later he returned from his drawing board and resumed his cold meal, oblivious of the time. But he was at times difficult to handle as he became ever more frustrated with those involved who failed to understand the problem, but worse, it drove him to poverty on more than one occasion, sacrificing his potential income on its success in order to try out a further improvement. This, though, is true of most inventors!

Patents

A provisional patent cost a hefty £1 in Edwardian times and stayed at that price until World War II. It granted protection for one year whilst awaiting submission of the full patent which, if granted, secured 16 years protection of the patentee's invention, no matter how weird it was! It is much more difficult to secure a patent these day, through much tighter controls and tests of practicality than it was in the *laissez-faire* days of yesteryear when a mere idea alone could be patented. Of course, not all provisional patents succeed: despite much research, many were deemed unoriginal, wholly unworkable or were withdrawn through the inventor having discovered a flaw or a far better idea, abandoning the patent in the process. Unfortunately, abandoned provisional patents are rarely recorded in any depth.

It has also to be borne in mind that a patentee faces a considerable dilemma. If he perseveres with his invention to near perfection without lodging a provisional patent, he risks the concept of his invention coming into the public domain, thereby making it unpatentable - his endeavours would have been in vain. Yet, if he lodges a provisional patent and appends his product with 'Patent Applied For' (or more mischievously with the prefix, 'World'), it affords some stalling of competitors as its full specification is not open to scrutiny. By this means the inventor has a chance of demonstrating and selling his idea without going to full patent, which would then be open to legal challenge.

Bradshaw freely admitted that, 'the inventor needs a convincing argument and persuasive tongue - but if it is technical he must be capable of answering the highest technicians'.

Many well known and respected engineers and companies, such as Phelon & Moore, Ford Motors, Curtiss-Wright of America, had no hesitation in regarding Bradshaw as a respected consulting engineer, accepting his design rights or offering him all the assistance he needed, often openly backing him more than once; they can't all have been gullible! He once challenged his old friend Tom Sopwith, then at Hawker Siddeley, to send down his best engineers to debunk the 'Omega' toroidal engine, but no one came or attempted to disprove his claims.

He also freely admitted in a letter in the early 1960s, that in the internal combustion engine field, patents were a waste of time for there were some 500,000 patents on file which, by clever combination of several, legal Counsel could claim the patent was a mere rearrangement of several others - in other words, 'there was nothing new under the sun'. The new patent must then relate only to a previously unintended application, which is why the majority of patents granted to patentees are 'Improvements upon' an original idea in their attempt to exploit a loophole in an earlier patent's highly complex specification. For that reason Bradshaw, like most engineers, always wrote his own full patent specification and drew his own illustrations, for he better understood its complexities and potential applications than did a non-technical, but legally conversant, Patent Agent.

He continued in that letter saying:

'I can state with conviction that there is not a Bromega patent for the toroidal engine that would win in a High Court of Justice if any firm went out to fight it - but they do not want to, and this is the reason: By the time any new engine has been built and tested and perfected, it has cost so much money that any firm wishing to employ it would rather pay a reasonable royalty in order to save the time and money. They want to buy what is much more important - ie: the know how'.

This is certainly true in the engineering industries, where companies are bought purely for their intellectual know-how and patents, and not for their physical assets.

But, as Bradshaw observed in another letter,

'...the lone inventor is trapped between the devil and the deep blue sea for, if an inventor approaches a technical man who appreciates its merits, that man will invent something better, while a commercial man will claim there is a 90% chance it will not work or sell, simply because he lacks imagination and fears the unorthodox and unknown'.

Yet, paradoxically, no one would back an invention without legal protection afforded by a provisional patent to protect his investment even though a patent is no protection against infringement, except when tested in a High Court! Most investors therefore protected their interest by acquiring a provisional patent subject to the invention being proven as workable, at which point they would elect to follow it through to a full patent as 'the applicant'; the inventor was of course still credited in those letters patent. For that reason, in his later years, Bradshaw rarely took out a full patent himself and relied heavily upon a confidentiality clause, and his copyright, against infringement. Few inventors however could carry the cost of litigation to seek damages from unscrupulous investors, especially from those abroad, and though he succeeded in obtaining several foreign patent rights, these often required clever rewriting of the specifications.

History has shown that in the majority of cases, by the time the invention reached the market place as a commercially viable product, the patent's protected life had half-expired. As a consequence the revenue received was often much less than expected as others had by then stepped in with improved ideas, opening up new markets and leaving the hapless inventor to rely on the kudos of 'the original and best'. A further insight into this problem came in a letter to his nephew, Lt Col Reg Gray at Bond Cars in August 1960, when Bradshaw justifiably blamed the patenting system for unacceptable delays which unwittingly encouraged patent agents and patentees to make extravagant claims such as 'world patents applied for' in order to secure at least some income as they attempted to sell their invention.

He also compared his desperate plight for royalties with that of Rudolph Diesel who hawked his compression ignition 'diesel' engine (patented in 1892) around the world progressively getting into greater debt with very few rights secured. Then in 1913 and on the point of financial success with the Royal Navy planning to use them in their submarines, Diesel mysteriously disappeared overboard from *SS Dresden* during its passage to England. Did he commit suicide under the stress of financial pressure, or was he murdered in a sinister plot by business and political interests? Today, the diesel engine is the engine of choice for business and the armed forces.

In L T C Rolt's absorbing autobiography, *Landscape with machines*, Rolt captures well the resentment and animosity between theorists and practical men when he related his father's own maxims:

'A pinch of practice is worth a pound of theory: the curse of all works is the designer without workshop practice: any fool can tighten a nut with a pencil!'

Bradshaw cannot be accused of these faults, yet a major hurdle which all inventors faced was production engineering. Bradshaw freely acknowledged that, though he was a skilful 'hands-on' engineer, he was not a production engineer, so he preferred to rely on their skills and inventiveness in interpreting his 'General Arrangement' drawings and recommending the necessary modifications to enable production at an economic cost. Of this, Bradshaw wrote: 'When one designs for a manufacturer, it is the manufacturer who likes to have the final say' and it has to be said that on more than one occasion, respected aero-engineers observed that Bradshaw's hand-built ABC Gnat, Wasp and Dragonfly aero-engines performed much better than those built by the mass-production engineering specialists - but Granville Bradshaw still got the blame for the engine's failings!

More often than not, it was the crucial fine tuning stage which proved most costly in time and money as 'the light at the end of the tunnel' never seemed to get any closer; at which point the accountants or shareholders pulled the plug on a project - just as the inventor believed it was about to bear fruit. This was well demonstrated with the toroidal aero-engine. But, hopefully, the wise inventor would have secured an income, at the manufacturer's cost, as their consultant engineer during its development. In fact Granville made a very good income from this and the outright sale, or rights to his ideas, together with reasonable royalties. In 1953, he calculated that these had brought him over £600,000 (around £60m today!) before the war - though this figure is felt by family members to be exaggerated. Yet, following his bankruptcy in 1936, his post war years were an endless battle of having to borrow from family members to pay off debts fuelled but his relentless drive to complete the toroidal engine within his lifetime. Unfortunately for Bradshaw, this revolutionary engine had just reached its final fine tuning stages when the funding was withdrawn, coincidental to the launch of the well funded, but equally troublesome, Wankel rotary engine.

A flawed genius?

His alert and fertile mind led him to become easily bored as he pondered yet another idea which probably accounts for some of his more bizarre designs failing to live up to their expectations, leading him into ridicule.

His old friend 'Ixion' observed in *The Motor Cycle* in 1932 that:

'He is a brilliant genius of the type which produces hosts of original ideas. Mated with a canny Scots engineer as works manager, Bradshaw would make a fine team; but his nature was always too ardent for him to take serious in the drudgery of test and production.' He went on to say, 'You cannot estimate men of genius by ordinary canons... and that, ... his habit was to build everything far too light for commercial purposes'.

Bradshaw was not alone in having his prototypes fail - every inventor experiences these! The supercharged BRM racing car, for example, had received huge financial backing from Joseph Lucas Ltd, Rubery Owen Ltd and others as well as technical help from Rolls-Royce; but despite this, in preparation for their first race at Silverstone in 1950, it suffered cracked cylinder liners and shattered half shafts through incorrect hardening of the steel, yet BRM still went on to lead the field in racing-car development in Britain!

Although his business letters constantly reveal his frustration at sceptical businessmen (and lawyers!) who were not engineers, there is no evidence that he set out to deceive them, yet Bradshaw was often blamed for a company's downfall despite there being no direct evidence. It is, after all, the company directors who must exercise due diligence in their company's affairs and it is equally true that companies associated with Bradshaw, such as Sopwith Aviation Ltd, Gilbert Campling Ltd and ABC Motors Ltd, all went into voluntary liquidation within months of each other in late 1920 when they were far from bankrupt. That is an option which still remains open to companies today who wish to change direction and capitalise on their assets.

In contrast, under Harper-Bean, ABC Motors (1920) Ltd was forced into receivership by its creditors, as was the Belsize company who built the Belsize-Bradshaw, yet Bradshaw had no managerial involvement in either company. In fact he rarely held, or had any interest in holding, directorships other than to set up, promote and then sell the business as a going concern. He established companies to exploit his patents and such activities led him to become a millionaire in the 1930s, but as Bradshaw freely admitted, he was no businessman and it was this failing alone which allowed him to become unwittingly caught up in the Clarence Hatry swindle.

In 1960, Erwin Tragatsch, of motorcycle history fame, wrote that Granville was 'a superb technician', 'a genial and clever designer (who) makes the mistake of entering into, then open creations, which are in need of more development'.

In the enlightened 1980s, Brian Woolley, a former two-stroke motorcycle engineer who had worked for Scotts and Villiers but was by now a motorcycle journalist, was 'enraged that anyone could - in his wartime talks and articles - be so self-deceiving as was Bradshaw'! He wrote his personal corrective in a series of 'Persuasive Mr Bradshaw' articles which appeared in *Classic Motor Cycle* magazine. He aimed to denounce the 'slavish Bradshaw-boosting by Ixion, Cyclops, Torrens and above all by Bradshaw himself'. The magazine's editors appended the articles with headers asking if Bradshaw was 'a fraud', 'a charlatan', 'a plausible rogue'. While Woolley did not write those headers, he maintained in letters to me that Bradshaw's exaggerations were 'near pathological... many of his 'inventions' and patents were rubbish and many of his ideas on engineering were down right wrong' - like, for example the use of machined steel cylinder barrels which were in fact widely used in British and French aero-engines before Bradshaw adopted them in his! Woolley was not immune from aborted projects, as was demonstrated by failed new engines he developed at Scotts!

After his diatribe on Bradshaw, Brian Woolley conceded in letters to me, his praise for Bradshaw's ABC motorcycle as 'unquestionably one of the most brilliant designs of all time'. He also praised his Panther ohv engine and admitted he had come to admire the man and his work. But as Woolley's articles had set out to see through Bradshaw's public *persona* and show that he was 'a fraud', 'a charlatan', 'a plausible rogue', they would all be too readily accepted by those who bore him a grudge and who had the benefit of hindsight, happily dismissing the unrevealed history in his personal archives and now recorded in this book.

Bradshaw was certainly 'persuasive' when it came to charming his way into the affection of his many lady friends, which caused problems with his married life! He was after all a very handsome, rich and confident young man.

Certainly in the early 1960s, now well into old age, in ill health, poverty and deeply frustrated at trying to get the toroidal engine finished before the grim reaper struck, his letters do show signs of his memory playing tricks, with exaggerated claims in an attempt to win an argument. In one letter of September 1964, for example, in issuing a 'misfeance' *(sic)* claim, he said that he had been a Solicitor's Chief Clerk for a number of years. In fact he had only been a boy clerk in a solicitor's office, though this fib is little different from bluffs pulled by others in the same situation.

It is certainly true that Granville embellished some of his theories and claims in letters to prospective business clients, the more so in magazine articles in which their editors enjoyed playing their part by encouraging him to be controversial and thereby boost the magazine's sales and readership! 'Carbon', writing in *MotorCycling* of February 1944, likened Bradshaw and his 'Bradshavian' character to George Bernard Shaw, playing the same role in motorcycling as GBS did in theatre:

'One must always remember that Granville Bradshaw has the knack, in which I have no doubt that he takes great delight, of saying things that always sting his critics into incendiary, if not explosive replies. GBS can hardly make a public statement without shocking a very considerable number of people and Granville Bradshaw seems to work on the same plan.'

The Autocar, of 3rd May 1940, put it more gently;

'Our friend Granville Bradshaw, the designer who so often puts the cat among the pigeons and, when chided, manages to convince us that the cat wasn't real and the pigeons were dead anyway...!'

On reflection, most of Bradshaw's exaggerations are no more serious than an advertiser's tantalising 'invitation to treat' bluster - and to take that legal jargon two steps further, we all have a duty to ourselves and others of due diligence; but at the end of the day, it is always a case of *caveat emptor* (let the buyer beware) usually being ignored in favour of a commercial gamble to gain a lead over a competitor for a quick profit - for which the inventor cannot be blamed!

Men from the Ministry

Perhaps the most famous example of this was the prototype ABC Dragonfly radial aero-engine of early 1918 which, having passed its trials with flying colours, was ordered into mass production at the government's insistence despite both Bradshaw's and the Royal Aircraft Establishment's pleas to allow it to be fully developed first. The Dragonfly debacle lay entirely at the feet of government ministers (documented in the Walton Motors appendix) who had still not learned their lesson from earlier, but less well publicised, aero-engine disasters such as the Sunbeam Arab, Siddeley Puma and then the Siddeley Tiger V-12 prototype of 1918. In fact Siddeley and their predecessors, Beardmore and BHP, had a very poor record of aero-engine development, encountering endless problems in their designs and use of porous aluminium castings. Pre-production Dragonfly engines revealed serious faults when subjected to extreme conditions, especially the destructive 'synchronous torsional vibration' which had also seriously affected the earlier Sunbeam Arab with fatal consequences. But while synchronous torsional vibration was not understood by engineers at the time, Bradshaw still got the blame for it on his Dragonfly, though thankfully there were no fatal consequences.

It was only in the retrospective post-war calm that the ABC Dragonfly was heavily slated. In the 1930s, Major George Bulman (later director of engine production in World War II), expressed that, in his view, the Dragonfly would have lost the war if it had continued into 1919. W O Bentley states in his biography,

'That the war ended in November, before this radial (Dragonfly) reached the squadrons in any numbers was merciful... It still succeeded in killing several good men, among them the brilliant test pilot, H G Hawker. His engine caught fire and he went straight in.'

In fact, as this book shows, Bentley's claim is unfounded, for while Peter Leigh and Harry Hawker were the only pilots to die at the controls of a Dragonfly powered aeroplane, no blame was ever attributed to the engine in either Hawker's *post mortem* or the Air Accident Investigation report! Bentley also falsely claimed in his biography that: 'The final order for BR.2s was 30,000', a figure far in excess of reality. Bill Gunston states in his brief *World encyclopaedia of aero-engines* that his 'long talk with Cpt Norman Macmillan and Major Oliver Stewart MC left him in no doubt that the ABC Dragonfly would have necessitated a frantic re-engining programme for thousands of aircraft had World War I lasted into 1919'.

If the Dragonfly engine was truly such a disaster, one would reasonably expect both Macmillan and Stewart, as World War I fighter pilots and post-war test pilots, to report this calamity in their several books on the development of aviation - but they make no mention of it. Without doubt the Dragonfly's lack of proper development would have made it a disaster, but then who knows what damage the fast, Dragonfly powered fighters could have inflicted upon the enemy had the Ministers allowed it a proper development programme (like that which the Bentley rotary enjoyed)? All officials concerned admitted that when running properly - and they certainly were by 1920 - the ABC Dragonfly was a truly impressive engine. It is significant that an example of Bradshaw's Star 40hp in-line is at the RAF Museum, Hendon, while an ABC Gnat, Wasp and Dragonfly exist in the Science Museum, South Kensington, London - quite an achievement for such a widely debunked aero-engineer.

Unfortunately for Bradshaw, the Dragonfly represented the memorably glamorous side of dare devil fighter-pilots and had been tagged as 'the engine which would win the war', whereas the barely remembered failures of the Sunbeam Arab and Siddeley's Puma and Tiger aero-engines, were for the lowly, unglamorous bombers.

One thing for which Bradshaw's 'failed' Dragonfly ought to be credited was the successful standardisation by the RAF in their post-war programme of large radial aero-engines and also the new, statutory, 50 hour endurance acceptance trials in both Britain and America. Exhaustive post-war research into the Dragonfly's 'failings' also forced a complete redesign of Roy Fedden's Cosmos radial aero-engines, which had been sidelined by the Government in favour of the Dragonfly. These greatly improved Cosmos engines, now built by Bristol, remained in service with the RAF for over 30 years.

Bradshaw's critics should also consider the plight of Barnes Wallis, an apprentice engineer at John Samuel White, ship-wrights and aeroplane makers. He then joined Vickers and designed their dirigible airships which later employed a revolutionary geodetic lattice framework for the ill fated R100 and later Wellington bomber - an idea ridiculed at the time by engineering experts! Wallis then tried to convince the Men from the Ministry that he could drop a 5 ton bomb onto a lake, make it skip across the water and detonate it precisely 30 feet underwater, hard against the dam's wall to breach the Mohne Dam! How ridiculous can one get?! Unfortunately for both Wallis and Bradshaw, neither suffered fools gladly and both made enemies among those in high places who knew little about engineering or science.

As Granville once wrote, 'all the great inventions were unorthodox'.

Appendix 1
Walton Motors

Bradshaw's claim to being awarded 'around £43,000' for the ABC Dragonfly is a fair claim, but needs qualification.

Government documents at the Public Records Office, released in 1972 under the 50 year rule, reveal that the Ministry of Munitions of War intended to select only Rolls-Royce in-lines and ABC Dragonfly radials to satisfy their 1919 air-offensive to bring the war to an early end. There were potential orders for some 100,000 engines.

The documents also expose how, having initially offered very favourable royalty terms, the Government systematically tried to reduce their liability after the war by denying the agreed royalty terms, despite them having been minuted. The documents also show how they failed to expedite an honourable compensation to ABC and Walton Motors for their prototype development work on the Mosquito and in getting the Wasp and, more particularly, the Dragonfly, into mass production.

While Granville Bradshaw is never mentioned in these documents, it was clear to him that his ABC Hersham works could not possibly meet the projected 'tens of thousands' of orders without a major factory expansion. As it was, those few ABC Gnat and Wasp engines that had been produced appear to have been assembled at Selsdon Engineering. In addition his entire machine shop production capacity, currently committed to the auxiliary engines for the army, would have to be turned over to aero-engine prototype development.

It was on this basis, and following unofficial advice from the Ministry of Munitions and the Air Board (sanctioned by the Treasury), that in late 1917, with funding from Samuel Waring, Walton Motors Ltd was formed as a subsidiary of ABC, expressly to take over the rights, development and pre-production of the ABC Wasp and the Dragonfly aero-engines and prepare for the construction of a factory and office extension to meet the Ministry's proposals, revised in May 1918, for '20,000 engines, 7,000 as soon as possible and the balance in batches'. But, it later became clear that ABC itself would only be required to produce up to 12 Wasp and 75 Dragonflies per annum, with the bulk manufactured elsewhere, for which ABC/Walton Motors would receive a small royalty.

ABC Auxiliary engines

As a consequence, production of the ABC auxiliary engines was brought to a halt in January 1918 and the Hersham works was transformed into an experimental and pre-production machine shop. Thereafter, while ABC would still make certain components, the assembly and production of auxiliary engines for Lyon & Wrench Ltd passed to Fiat Motors Ltd in London, to whom ABC supplied some £2,286 worth of parts and spares. Fiat had developed successful in-line and V-12 airship engines, and motor car engines at their Italian factory, but had only imported cars into Britain at the time.

By this time ABC had supplied 784 auxiliary engines, to a value of £47,984, which included a very reasonable profit margin of 31% (£19/2/0d per set). In 1919, ABC submitted a compensation claim of £16,054 for the enforced abandonment of the auxiliary engine contract and its associated loss of forward profits, but they got back just £10,000 as the Government would not award costs for loss of future profits from cancelled production, cost of spares, or costs incurred in a necessary expansion to the Hersham factory for, being war-time, any new factory construction required government approval - and that planning permission was refused to ABC. (This possibly relates to the purchase of the Hersham Lodge estate for a new factory.)

Unbeknown to Walton Motors, this was the start of a protracted dispute over royalty payments and hardship claims, complicated by the unique ABC/Walton Motors tie-up and, along with post-war Dragonfly development work, it all made this a somewhat unusual and complex affair. Such was the delay in receiving compensation, that it led to a desperate presentation in March 1921 to Stanley Baldwin, PPS to the Chancellor of The Exchequer, to expedite payment of royalties and hardship awards. This intense, bureaucratic nightmare, was not finally resolved until October 1921. This presentation records the background to this sorry saga.

Walton Motors Ltd

Walton Motors Ltd was formed in November 1917 with £200,000 capital, in £1 shares, to purchase or otherwise acquire and manufacture aero, motor and other engines under agreement with ABC Motors Ltd and Granville Bradshaw. But other than his role as design engineer, Granville was not directly involved in the daily affairs of Walton Motors Ltd, whose financial backing and driving force had come from Samuel James Waring, of Waring & Gillow Ltd., the famous furniture manufacturer, who had made his factories fully available on government contract work for the War Office and Admiralty. He was later honoured for his war work with a Baronetcy.

Walton Motor's directors were Ernest Noel, Cpt Ronald Charteris RFC (of ABC, he had joined the Society of British Aircraft Constructors in September 1917 and by July 1918 had been promoted to the Technical Branch RAF), Sqd Commander *(sic)* Reginald L G Marix DSO, RN (who as a Flight Lieutenant bomber pilot in 1915, had patented a pilot's safety harness while at RNAS Eastchurch) and E Metts (formerly Metz, a Dutchman who had gained British citizenship). T A Dennis of ABC was their company secretary.

Waring had earlier established the British Aerial Transport Company (BAT) taking on Frederick Koolhoven as the designer of their new, all plywood, biplane fighter (described earlier), powered by Bradshaw's experimental 6-cylinder 'Mosquito' radial. Exactly who approached whom over the Mosquito's development is not now known, but Waring was sufficiently impressed with its potential that he earnestly supported Bradshaw in the development of the ABC Wasp to meet the new Air Board specification; it was this which led directly to the Government ordering development of the 9-cylinder ABC Dragonfly.

Royalty payments

The Ministry of Munitions of War had accepted that the Dragonfly was: 'the best and most economical (engine) that had yet been evolved and that enormous numbers of them would be ordered (for the planned 1919 air offensive). In fact for 1919 nothing but Rolls-Royce and this firm's design would be used. The Ministry had not allowed them (ABC) to develop the production side but confined them to developing the engine. Their remuneration would be royalty on the engines ordered from other firms.'

Further documents record the Treasury's concern at this 'huge gamble' on an untried engine, but Sir Philip Henriques, Assistant Financial Secretary at the Air Board, recognised that 'war conditions render it necessary'. But at least the treasury recognised that, as a major bonus and one which hugely influenced the decision to order the Dragonfly in volume, was the ABC's simplicity in design and manufacture allowing it to be built in shadow factories at a unit cost of around £600 compared to comparable designs from other companies of around £1,000 each. As Waring later pointed out, this represented a huge saving to the treasury of £10-£15m for their proposed 1919 air offensive.

With the Treasury recommending a one off payment of £100,000, plus a small royalty of £5 per engine actually delivered and all except £40,000 (corrected from £400) of this to be subject to the Excess Profits Tax of almost 80%, the government entered into negotiations with Walton Motors in May 1918 over royalty payment terms.

Yet, both the Government's Aircraft Production Department and Sir Arthur Duckham opined that this was nowhere near sufficient recompense given the terms agreed to foreign engine makers. Even Sir Philip Henriques recognised that the Government 'had practically commandeered' ABC's design before it had had its proper prototype development and that 'it was taken from them really before it was completed and altered under (government) supervision, which Walton Motors contend had reduced its efficiency'. Accordingly, ABC had not had an opportunity to protect and register their design rights or patents, before the Government freely made them available to (unidentified) aero-engine manufacturers elsewhere in France (Gnôme or le Rhône?), Italy (Fiat?) and America (Wright-Martin?), for which, as friendly Governments, Walton Motors would only receive a single, flat fee.

Waring was justifiably far from happy with this as they stood to lose significant potential royalties, but there was another catch; the War Break Clause meant that on cessation of war, ABC/Walton Motors would have to bear the full cost of all cancelled orders and work without recompense!

Waring very properly refused their miserly offer and duly made out a very strong case for fair recompense, demanding a guaranteed royalty of £35 for the first 5,000 engines ordered, £15 for the next 5,000 and £10 thereafter - all of which should be free of Excess Profits Tax, and that Walton Motors suffer no losses incurred on cancelled orders. These proposed terms were for British rights alone; those for America were to be separately discussed - or, as ABC had already spent around £40,000 on development work on both the Wasp and Dragonfly, he would accept a lump sum of £600,000, with a token royalty per engine.

The Ministry refused but then offered him £50,000 for the work so far done, plus £50,000 for all future work with a £5 royalty per engine. To this, Waring quickly pointed out that, after the Excess Profits Tax, it equated to a 14 shilling royalty per engine despatched! He demanded parity with the other aero-engine makers producing for the War Office and, to strengthen their case, Ronald Charteris advised the Government that ABC had already received a substantial offer for the sole rights to the Dragonfly from (an unidentified) 'Mr. Bagshawe at The Leeds Foundry'. (Could this have been Isaacson's or possibly the Redrup/Hart/Vickers concern, or even perhaps The Leeds Forge locomotive works?).

Furthermore, as ABC/Walton Motors would now only be prototype developing the engine, and would be barred from its manufacture, they would lose all potential profits for their work when the engine entered full scale production. Implying the Treasury was being incredibly miserly in its royalty offer, Waring pointed out that his proposed terms would prove a major cost saving to Government as the cost of the Dragonfly was half that of other engines and would therefore make a saving of £10-£15m on the latest proposal for 20,000 engines. In this he was supported by Sir Henry Fowler, Assistant Director General and Controller of Aircraft Contracts, and in fact the most eminent engineer on the Government's aero-engine selection committee. Fowler pointed out that ABC had made themselves and their workshops fully available to the government and that 'had other engines been pushed forward as much as the ABC it is probable that they might have been as good' inferring, one suspects, that other aero-engine makers were unable or unprepared to meet the Government's immediate specifications.

Royal Commission of Awards to Inventors

By now it was 1919 and while the Treasury continued to balk at settling payment terms, the Government's Lubbock Committee was already assessing royalty claims from other manufactures. The Government had also set up a separate Royal Commission of Awards to Inventors to judge awards on usage of patents, designs and for 'hardship' claims incurred due to the war effort.

In the meantime, ABC and Walton Motors had been bearing the full development costs for the Wasp and Dragonfly prototypes. Compensation for the Wasp alone was estimated by Waring at £100,000 before tax and, given that they had also undertaken considerable development after the war on the Dragonfly in accordance with Government instructions, both ABC and Walton Motors were now on the brink of bankruptcy, through no fault of their own. They now lodged a hardship claim at the Royal Commission for compensation, with Waring pointing out that some 300 staff were on the point of being made redundant at a time when Britain's unemployment levels were an embarrassment to the Government. The '300' was misquoted as '3,000' in the Ministry's presentation to the Chancellor of the Exchequer, which now made Government departments suspicious of the validity and accuracy of Walton Motors' claim - which all added to the delay!

Hansard reported that in the House of Commons on 9th December 1920, Cpt Colin Coote MP raised the matter of the non-payment of royalties to Walton Motors by the Royal Commission of Awards to Inventors, pointing out that the Disposal & Liquidation Board was now selling off working engines (which embodied ABC design and patents), cheaper than ABC would be able to make them for the post war civilian market. He then asked when would ABC get its money, for by now Walton Motors' claim for compensation was 'in the order of £2.5 million'? Winston Churchill, for the Ministry of Munitions, replied that it was up to the Commission, and that many of ABC's patents had not been granted.

There then followed 'a major dispute' with both the newly formed Air Ministry and the Admiralty Board firmly pushing all responsibility for the matter onto the Ministry of Munitions of War, who in turn, devised a defence against paying ABC/Walton Motors monies rightly owed! For their part, The Royal Commission could only consider 'hardship' claims, not royalties, cancelled potential orders or the necessary factory costs towards meeting the potential 'tens of thousands' of anticipated aero-engine orders.

The matter remained unresolved and by early 1921, both the near bankrupt Treasury and the Royal Commission were still doing their utmost to minimise awards. They now dismissed compensation to Walton Motors for 'an alleged communication' by Sir Henry Fowler of ABC's unprotected designs to Italian and French aero-engine makers. Then, the Air Ministry joined in claiming that in fact only 800 of the potential 20-30,000 engine orders had in fact been delivered (records indicate that of the 11,050 Dragonflies built, 1,147 were delivered) and on that basis, the Government should therefore only have to pay one quarter of the aero-engine development costs! To this, Walton Motors firmly pointed out that had the initial order been for so few engines, the unit cost including its development, would have been far higher, making the engine an unattractive proposition to both the government and Walton Motors.

By February 1921, the Royal Commission was now made acutely aware that if Walton Motors' hardship claim was not soon resolved, they faced bankruptcy and the Government would have to bear the brunt of increased unemployment costs. As the Lubbock Committee had now finished sitting, the Government was finally prepared to sanction a hardship award through the Royal Commission, of £20,000 *ex-gratia* for development work by ABC, bringing that total to £30,000 simply to avoid them laying off 300 staff - the royalty claims would have to be settled later. Once again Samuel Waring refused this paltry sum, for an award of £40,000 had earlier been agreed as the minutes proved! The Ministry of Munitions then raised its offer to £35,000 in full settlement.

At this point, in March 1921, Waring pressed Sir Philip Henriques to take up the matter up directly with Stanley Baldwin at the Chancellor of the Exchequer's office. Baldwin was very concerned that Walton Motors/ABC would be forced to close that month due to the Government's failure to make a proper award and, now that the Lubbock Committee had ceased dealing with claims, he urged the Royal Commission to settle the matter urgently. In June 1921, the Treasury duly advised the Royal Commission that the Lubbock Committee would not have entertained the following: loss of goodwill; losses arising from not being allowed to build a factory or losses arising from not being allowed to manufacture engines; they recommended that the hardship claim should be dependent on the royalty claim - which of course had still not been heard!

In September 1921, the Disposal & Liquidation Board now contrived a defence to further reduce the award. In essence, and despite minuted meetings, they now denied everything that Walton Motors had ever claimed over the engine; they denied that ABC would have been awarded a contact to build even 300 engines; and claimed that: 'The Dragonfly engine never was, and was incapable of being produced so as to be a satisfactory aeroplane engine. The said engine was and is unsafe in use. The said engine further, even if of use for war purpose would have been useless for commercial aviation'. The facts however proved otherwise.

Legal opinion, sought by the Government, now encouraged the Treasury to try and wriggle out of a fair payment! Tellingly, the opinion recognised that the Government was on a sticky wicket unless it could offer a sound basis for attacking Walton Motors' claim at such a late stage, but added that 'it is difficult to know what sort of evidence can be called for the Crown'. To further delay the hardship claim, the legal opinion advised that urgent legal counsel should be sought, but with the lack of further 'certain information' coming forward, Counsel finally recommended that the Royal Commission's hearing must proceed without delay; the date was duly set for 5th October 1921.

On 8th October 1921, The Royal Commission on Awards to Inventors recommended a royalty award of £28,000 in addition to £20,000 already paid (to ABC), for the use by the Crown, past present and future, for the Wasp and Dragonfly engines and for ABC's designs and the following patents applications:

Patent	Year	Subject	Published
3338	1917	Hardened roller bearing	Incorporated into Pat. No 128383 & 129248
12420	1919	Hardened roller bearing	Published as Pat. No 129248
12421	1919	Hardened roller bearing	Published as Pat. No 132472
11062	1917	Roller bearing cage	Abandoned
2629	1918	Bolt	Published as Pat. No 133093
6286	1918	Wrist pin for connecting rod	Published as Pat. No 142147
11315	1917	Cowling	Published as Pat. No 139827
18994	1917	Cowling	Published as Pat. No 139827
8185	1918	Valves	Published as Pat. No 132827
7864	1917	Connecting rod	Abandoned

Having settled the royalties, the Royal Commission held that no hardship award should be made. At this point, the Army Council stepped in, for they were incensed that Walton Motors and ABC were to be recompensed and demanded that an outstanding sum for £4,755 be withheld from Walton Motors' award for work the Army had done on ABC's behalf for (unidentified) 'Steering units'. Remarkably, while this was a claim against ABC, and not Walton Motors, the money was nevertheless withheld by the War Office until protests from Walton Motors' solicitors led to the War Office being instructed in November 1921 to pay it back...

... and then the tax-man gleefully started to rub his hands at the prospect of recovering much of the award through 'direct assessment'!

Having now paid for the 'future use' of ABC's patented roller bearing cages, the Air Ministry released some to The Hoffman Manufacturing Company of Colchester, the famous bearing makers, in the spring of 1922. Whether Hoffmans then freely profited from Bradshaw's designs in not known.

Neither is it known to what extent this whole awards debacle was instrumental in ABC selling out to Harper-Bean in 1920; nor is it clear what happened to Walton Motors for by October 1921, neither Ernest Noel nor E Metts were listed as directors but, having served its purpose, and doubtless all of their ABC aero-engine rights being passed to T A Dennis' new ABC Motors Ltd in 1923, Walton Motors Ltd was wound up in January 1925.

Appendix 2
The Family

William Septimus Bradshaw

William and Alice Bradshaw's sixth surviving child was William Septimus (Granville's father), born in Chorley on 9th April 1847; he became a commercial traveller and moved to Woodhead Road, Ecclesall, Sheffield where he met and married on 21st November 1872, the 22 year old Annie Virginia Mathews of Washington Road, Sharrow, Sheffield. Her father, John, was a veterinary surgeon of Lord Street, Sheffield. They returned to Lancashire, where he set himself up a clockmaker/repairer and jeweller at 2 Victoria Buildings, Fishergate, Preston. They had no children. Sadly, Annie Virginia died of acute peritonitis on 29th June 1878, aged 27 years.

William Septimus married again in 1879 to Annie Eastwood, who bore him five children in Preston: Ruby (b.1870), Violet (b.1881), Ewart Gladstone (b. 16/6/1884), Granville Eastwood (b. 8/12/1886) and Henrietta, (Hettie or Ethel b.1888 d. 12/10/59). Interestingly while Ewart was tall and slim, Granville was quite short, yet his own son, Geoffrey, was tall and thin. The family home was at 148, Friargate in Preston's town centre but by 1889, they had moved up the road to 59, Tulketh Crescent, Ashton-on-Ribble and by 1891 to 138, Friargate. William then became an 'Optologist' (with a London Diploma) making and fitting spectacles and, with Ewart's help, operated as Bradshaw & Son, opening further shops in St.Annes and Fleetwood. It appears he later made motoring goggles.

Annie died in 1904 of double pneumonia, aged 55 years. William remarried and had a son, John who, it appears, lived mainly in London. By 1910 the family had moved to 7, Park View, Tulketh Road. William Septimus died on 30th September, 1915 and was buried at the Methodist chapel, Wellington Road, Ashton.

Ruby, a dress maker, became a fairground clairvoyant, although Ewart did his very best to dissuade her by offering her an attractive allowance! Hettie eventually married a Mr Granfell of Preston who had a large motor factors.

Ewart Gladstone Bradshaw

Ewart married three times; his first wife Annie bore him Anne, John, Leonard, Ewart Junior and Peter. After she died, he married Lillian Walmsley in 1944; she was related to the Walmsley ironmonger and engineering family. Lillian bore him Adrian. Ewart then married again to Dorothy who bore him Vivian, Robert, David and Guy. Ewart died on 5th June 1959 and was buried in his father's grave on 8th June 1959. Dorothy survived him.

Anne married Reg Gray; they had two sons, George Peter and Andrew Charles.

Granville Eastwood Bradshaw

Granville married Violet Partridge (born Walsall 2/11/1888) in Wolverhampton in Spring 1911 - this was said to have been 'a marriage of convenience to secure Granville his job' - most probably as a form of security for the ABC partnership. Violet was a regular visitor to Brooklands where it is believed her brother raced; she was also an early aviatrix, though she did not take her RAeC certificate and little is known of her flying experiences. She bore Granville two children; Pamela (also known as 'Joan' b 21/6/12 d 13/5/95) and Geoffrey Lionel Guy, (b 12/4/14 d 26/4/98), but the marriage came to an end after Violet petitioned for divorce in 1916. Violet moved to Eastbourne, where she lived to just over 100 years of age; she died at Easter 1989. Joan attended the Eastbourne College of Domestic Economy and later became Mrs Macajone, living in both Eastbourne and London. Geoffrey attended Brighton College school as a border and it appears his holidays were spent at school, staying with his mother at her hotel lodgings for only a couple of days at Christmas. He trained as a draughtsman at a large electrical engineering company.

The marriage with Violet faltered when Granville fell in love with Muriel Mathieson (b. 5/4/1896 or 1897), a driver in the FANYs (First Aid Nursing Yeomanry); they married in 1927. Her wealthy family had connections in the City of London. Though a mismatch of characters, she proudly bore him twin girls born on 24th January 1918: Vyvien Hope Granville ('Fluff') and Pamela Eleanor Granville ('Pammy') and a son, Peter Frewin Granville (b. 26/3/20). The children were privately educated by a governess until Granville's bankruptcy. Although he and Muriel were still living together, they eventually separated and with a small inheritance from her aunt, she and the children moved to Downs Cottage, Crawley Down, West Sussex in early 1938.

Meanwhile, while working for the Admiralty, Granville Bradshaw had met a lady driver from the Admiralty's driving pool, Francesca Beavan King (b. 20/3/1903, d. 11/1992), who drove regularly for Lord Louis Mountbatten. As their close friendship developed she would soon became his housekeeper and constant companion.

During the war, Muriel let Downs Cottage and moved to Bromley to care for her widowed uncle; his daughter was an Almoner at St. Thomas's. Muriel organised the WVS (Women's Voluntary Service) canteen service for doctors at St Thomas's Hospital, where she and Granville were Governors. Muriel later emigrated to South Africa where she died, around about the same time as Granville.

Both Vyvien and Pamela joined the London Fire Service on outbreak of war but an influential relative at the War Office managed to get them into the WRNS (Women's Royal Naval Service) as Petty Officers to train at Bletchley Park in the operation of the captured Engima cipher machines; they were sent to operate from Ceylon during the later war years. Pamela contracted TB in Ceylon and was returned to Britain by hospital ship arriving in Britain in January 1946. By sheer chance, she convalesced briefly in a hospital in Preston and, through Ewart's considerable generosity, enabled Muriel, Peter and the family to stay close to hand until she was transferred to Battle, East Sussex where she died a few days later, on the 22nd March, 1946. She is buried at Salehurst, East Sussex.

After the war, Vyvien was the Personal Assistant to an exhibitions designer before she emigrated to Nairobi, Kenya working for the British Government. She married David Lunan, a game hunter/farmer. After the Mau-Mau uprising, David's farming brother, Tiny, sold up and emigrated to Australia. Meanwhile, Vyvien and David had set up a dried flower business, exporting to Europe, but after a few years they sold the business to a Dutch company and moved to Australia to help Tiny who was by now in very poor health as a result of a serious war wound. Vyvien was quite unsettled in Australia and returned to England, settling in Hitchin where she died on 24th November, 1998.

With Granville's worsening financial plight, Peter was withdrawn from prep-school in 1932 and worked, underage, for his father servicing the amusement machines for which Granville had secured a concession. However Muriel's great aunt, Ethel, intervened and ensured Peter went back to school. He eventually gained a place at Churcher's College, Petersfield but in an attempt to improve his chances at school, an operation to correct a squint became septic, almost making him completely blind. After months of recovery, he left college and became an articled clerk in a City based Chartered Accountant's office, lodging in Earl's Court, London until being called up into the Army's Pay Corps where he gained the rank of Corporal; he repeatedly turned down the chance of a commission. Later based in Leicester, he met his wife-to-be who was working in a reserved occupation. He was transferred to Chelsea Barracks, from where he was demobbed. Now living in London, he qualified in 1947 as a Chartered Accountant and having married Millicent Hunter, left in 1950 for Newmarket, Suffolk with their children Stephen (b. 1947) and Charlotte (b. 1949) returning in 1957 to Hitchin now with their youngest children Simon (b. 1952) and Alison (b. 1955), where he founded Bradshaw, Johnson & Co in 1957 at Frewin House, Highbury Road, Hitchin. Although Granville often relied heavily on his children for financial support in his later years, Peter wisely kept out of Granville's business affairs. He died on 13th May, 2006 and is survived by Milly and their children.

Geoffrey Lionel Guy Bradshaw

In the early 1930s, Geoffrey was living at 5 Palm Court, Angmering-on-Sea, West Sussex, moving to 34, Hillingdon Road, Uxbridge in 1934 and then in 1936 to 27 Streatly Road, London NW6 later that year. By 1939 he was at 35, Curzon Street, London W1 and later at 50, Marylebone High Street, London W1.

He served as an Lieutenant in the Royal Artillery in Italy during the last war and often recalled an incident at one briefing when it was suggested, tongue in cheek by a superior officer, that with the name 'Bradshaw', he would know the train times to the new posting. To the astonishment of all, Geoffrey rattled off times and connections, for unbeknown to the officer, Geoffrey had taken the precaution himself of earlier checking the times!

Married to Gwen Amy Beecher-Goater (b. 16/12/19) in July 1943, he was invalided out of the Army in December 1944 and was off sick for quite some time. Unable to work, he stayed mainly with his in-laws at Farnham, Surrey. His interest in radio and electronics allowed him to repair the odd radio for friends. Like his father, he also developed several inventions, although none were patented.

By early 1949, Geoffrey, his wife Gwen, daughter Angela Dawn (b. 16/4/44) and son Antony Jon (b. 11/11/47), had moved to 27, Avenue Road, London NW8, joining the Avamore Engineering Co of 104 High Holborn, London WC1 as Sales Manager. By 1951, now with their youngest daughter Josanne Margaret (b. 29/1/51), they had moved to No.7 (then in 1952 to No.8), Stirling House, Barrow Hill, Marylebone, London NW8, by which time he was working as a Sales Engineer for Purdy Machinery Ltd of Gower Street, London NW1.

In 1967, when visiting the octogenarian Granville Bradshaw, Courtney Edwards, now of *The Sunday Telegraph*, was amazed to be told of some 65 modifications Geoffrey had made to his 100,000 mile, 1963 Ford Zodiac Mk.3. These included a radiator over-flow reservoir, a carburettor accelerator pump cut-out switch to prevent excess fuel being pumped when 'kicking down' his self-fitted Borg-Warner automatic transmission, improving fuel consumption. For safety, he modified all switch-gear to be operated by the same hand, while an auxiliary headlight thumb operated dip-switch was fitted to the steering wheel. He also fitted 4-way hazard warning lights and intermittent windscreen wiper action, long before they became the norm.

In 1973, Geoffrey developed a combined ball type gear-lever knob and switch which could be fitted to any gear lever. It was operated by twisting slightly off centre to engage or disengage, electrically, the

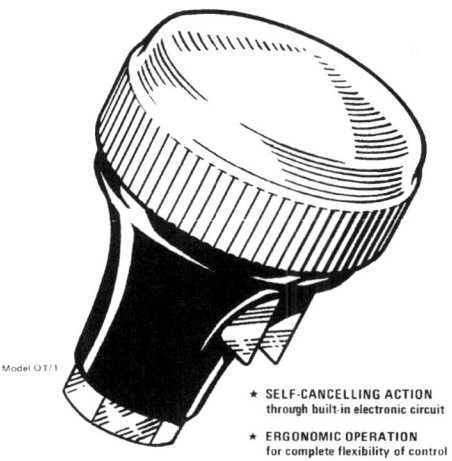

ADVANCED AUTO AIDS
(Geoffrey Bradshaw)

solenoid engaged Laycock type overdrive unit allowing the driver full control at all times compared to the more usual dashboard mounted stalk switch. His 'Gripswitch' was also available with built in headlight flasher to forewarn the driver ahead when overtaking; a common practice on the Continent. He set up a new company, Advanced Auto Aids Ltd at his home in Leatherhead, but although tested by one major motor manufacturer, nothing came of it. Soon after, Triumph adopted a gear-knob mounted slide switch for their overdrive equipped cars.

Geoffrey died on 26th April, 1998; Gwen died in 9th July, 2001; they are survived by their children.

Appendix 3
Addresses

Granville moved frequently to be near his work. Known addresses and approximate dates are:

Years	Address
1914-22	Darby House, Sunbury on Thames
Late 1922	30, Hill Street, Kensington, W.1
Apr 1923	90, Jermyn Street, London SW.1 (Offices)
1925 - 1927	6A, Kensington Crescent, London W.14
May 1927	5, Beauchamp Place, Brompton Road, London SW.3
1927	14, Horton St, W.3 (shared the same telephone number as Kensington Crescent WEStern 3598
Oct 1928 - 1934	16, Sloane Avenue, Fulham Road, London SW.3)
Apr 1929	Trafalgar House, 11-13 Charing Cross, London SW.1) (Offices)
Nov 1929	11D, Canton House, Regent Street, London SW.1)
1930	Hyde Croft (later known as Lowfield Park), Crawley
Jul 1931	59, Park Street, Camden Town, London
1930s	Sunbury Court, London
1936	16, Sloane Avenue, Kensington, SW.3*
1937	Downs Cottage, Crawley Down, Sussex (Muriel's home)
1937	12, Wigmore Street, London W.1
1938	Hanover House, 14 Hanover Square, London W.1
1938	16, Sloane Avenue, Kensington, SW.3*
Jan 1942	Lovelace Lodge, Woodland Drive, East Horsley, Surrey
1948	Lake Cottage, North Chapel, Petworth
Feb 1952	The Elms, Audlem, Cheshire
Feb 1953	Noxa, Audlem, Crewe
Mar 1953	12, Victoria Parade, Broadstairs (Francesca's home)
May 1953	11, Nevern Square, London SW.5
Jun 1953	1, Glazebury Road, London W.14
Nov 1953	Little Heckfield House, Heckfield, Basingstoke
1954	Hill Cottage, Eversley, Hants
1956	Westerton Cottage, Westerton, Chichester
Jun 1958	Sea Garden, Fishbourne Green, Ryde, IoW
Oct 1958	Wooton Bridge, Ryde, IoW
Feb 1959	The Dolphins, Fishbourne, Ryde, IoW
May 1959	The Grey Cottage, Pelham Fields, Ryde
Aug 1956	York Cottage, Church Road, Leatherhead (Geoffrey's home)
1963	2, Holmbury Hill, Holmbury St. Mary, Dorking
Apr 1964	Sunnyside, South Harting, Petersfield
Jun 1964	Southdowns Hotel, Trotton, Rogate, Petersfield
1965-69	Peter's home in Hitchin

* As recorded in bankruptcy hearings.

Appendix 4
Patents

Unfortunately we are now unable to determine what his 'abandoned' and 'voided' patent applications relate to.

Aviation

Patent No	Laid	Subject
21903	25/09/09	Flying machine - Void
2136	27/01/11	Flying machine - Abandoned
2229	28/01/11	Piston rings - Abandoned
7861	28/03/11	Connecting rods - Abandoned
10891	08/05/12	Flying machine - Abandoned (joint ABC)
21572	23/09/12	Rotary engine - Abandoned
18194	11/08/13	Internal combustion engine - Abandoned
19144759	21/02/14	Crankshafts - Abandoned
191424190	17/12/14	Bearings (joint ABC) **
1114	23/01/15	Bomb sets - Abandoned
11056	30/01/15	Propulsion - Void
102186	23/11/16	Internal combustion engine (I.C.E) cylinders (joint ABC)
104483	08/04/17	I.C.E bearings (joint ABC)
105216	05/04/17	I.C.E bearings (joint ABC)
106944	27/11/16	Aircraft propeller (joint ABC)
107742	12/07/17	Carburettor (joint ABC)
111681	13/12/17	Carburettor (joint ABC)
128232	26/06/19	Bearings (joint ABC)
129248	10/07/19	Bearings (joint ABC)
132472	18/09/19	Bearings (joint ABC)
133093	09/10/19	Securing studs and bolts (joint ABC)
137874	29/10/20	Bearings
139827	18/03/20	Cooling radial engines
142147	06/05/20	Connecting rods
142516	13/05/20	I.C.E engines
152383	07/10/20	I.C.E engine cylinders

** Not all of his patents are fully recorded

1916102186	Laid 15/02/16 Granted 23/11/16. Refers to machined steel, finned cylinders and copper plating of same for improved heat dissipation. Refers to the practice by others of brazed-on cooling fins.
111681	Laid 30/08/16 Granted 13/12/17. Is for a conventional carburettor with an inclined variable jet in an inner venturi to prevent excess fuel flow at very high feed rates.
24190	Laid 17/12/14 Granted 05/04/17; 105216 and 104483 both Laid 27/11/16 Granted 08/03/17 are respectively variations on retaining roller bearings on a crankpin by sleeve and maintaining separation of roller bearing connecting rods, which share a common crankpin, by spacer rings.
106944	Laid 27/11/16. Refers to a cupped housing and resilient mounting of propeller blades to engine boss
107742	Laid 18/12/16 Granted 12/07/17 is a variation of 111681 above.
128232	Laid 07/03/17 Granted 26/06/19. Refers to end of shaft roller bearing flange plates for attachment to fixed bodies
129248	Laid 07/03/17 Granted 10/07/19. Refers to oil galleries drilled in crankshaft web to supply big end roller bearings
132472	Laid 07/03/17 Granted 18/09/19. Relates to carbon case hardening of mild steel parts into and onto which roller bearings are fitted such as gear wheels and chain sprockets.
133093	Laid 14/02/18 Granted 09/10/19. This is a captive flanged set screw, bolt or stud in which either the flanged head or casting cavity (which ever is harder) is milled, broached or splined so that in tightening the bolt head, it is bedded properly and does not turn such studs are often used on modern brake drums. A joint GB/ABC patent
137874	Laid 24/12/17 Granted 29/01/20. Relates to captive spaced rollers in a connecting rod roller-bearing assembly. A joint GB/ABC patent
139827	Laid 07/08/17 Granted 18/03/20. Refers to a ducted aero-engine nacelle which directs air to the stationary radial cylinders and heads
142147	Laid 13/04/18 Granted 06/05/20. This design shows how piston connecting rods are joined to the crankshaft on short throw engines by a knuckle joint the patent particularly relates to lubrication channels and oil ways. A joint GB/ABC patent
142516	Laid 21/01/18 Granted 13/05/20. Relates to improved, compact radial engines whereby the pistons and cylinders are elliptical which increases the surface area exposed to air flow, it takes into account closeness of bearings and the need for skirted pistons etc.
152383	Laid 07/06/19 Granted 07/10/20. Relates to the insertion of a highly heat conductive, finned spacer ring of copper, aluminium etc., between the cylinder head and barrel in order to dissipate heat away from over head exhaust valves. A joint GB/ABC patent

Motorcycles and motorcars

Patent No	Laid	Subject
19144503	01/02/14	Cycle frames - Abandoned
19144504	21/02/14	Cycle frames - Abandoned
19145400	03/03/14	Sidecars - Abandoned
191411310	07/05/14	Cushion drive - Abandoned
191424189	17/12/14	Motor cycle springing (joint ABC)
191424191	17/12/14	Cycle forks (joint ABC)
191500090	05/07/15	Cycle handlebars
113345	21/02/18	Springs (joint ABC)
119480	13/12/17	Motorcycles (joint ABC)
132827	18/03/20	I.C.E cylinders (joint ABC)
132952	1919	Motorcycle rear springs (joint ABC)
139570	11/03/20	Motorcycles etc
139834	18/03/20	Motorcycles etc
140154	25/03/20	Motorcycles
151032	13/09/20	Motorcycle frames
164458	06/06/21	Vehicular lamps
169255	21/09/21	Oil cooled engine
188087	09/11/22	Change speed gears
194907	22/03/23	I.C.E
223324	23/10/24	Fluid flow passages I.C.E
225275	27/11/24	Two stroke cycle I.C.E
230145	02/03/25	Power generating system
13472	12/05/37	Prov pat; Torque equaliser
512514	19/09/39	Transmission reverse brake
614040	12/08/48	Cylinders and valve gear (Bumblebee)
856619	21/12/60	Gas generators

191424189 Laid 17/12/14 Granted 2/12/15. This is a simple spring rear frame using a leaf spring off the saddle bracket.

191424191 Laid 17/12/14 Granted 16/9/15. On this a front leaf spring is fitted to the steering head and a forward vertical stay connected by a balance beams through which the axle passed, to the girder fork.

191500090 Laid 05/07/15 Granted 20/01/16. This is for a ducted, hollow head stock allowing control cables to be passed through the headstock for brakes etc and having an adjustable cup and cone bearing.

113345 Laid 07/03/17 Granted 21/02/18. Refers to laminated leaf springs clamped at the their ends and separated by interleaves where necessary to improve damping. The important point is that the centre is waisted and thinned to improve axial flexing.

119480	Laid 13/12/17 Granted 10/10/18. Refers to a rear spring frame using a spring arrangement as in 24189 above for shaft driven motorcycles (in this case a fore-and-aft twin). The engine is mounted on the swing arm pivot and thus moves with the rear wheel.
132827	Laid 7/8/17 Granted 18/3/20. Relates to an exhaust valve guide extended above the cylinder head to assist in cooling
132952.	This is a double laminated leaf spring arrangement of 24189 above
139570	Laid 1/2/19 Granted 3/11/20. A variation of 24191 and as fitted to the ABC motorcycle
139834	Laid 12/06/18 Granted 18/03/20. This refers to a bent steel plate which forms the protective front leg shield and underplate for a horizontally opposed motorcycle engine with longitudinal crankshaft. A joint GB/ABC patent
139570	Laid 01/02/19 Granted 11/03/20, Refers to a laminated leaf spring motorcycle front girder fork suspension. A joint GB/ABC patent
140154	Laid 22/01/19 Granted 25/03/20. This is a variation of the mud shield underplate which has openings (louvres or shutters) for improved air cooling on a transverse twin engine.
151032	Laid 13/03/19 Granted 13/09/20. This is the symmetrically splayed tubular motorcycle cradle frame for a horizontally opposed twin.
164458	Laid 04/03/20 Granted 06/06/21. A vehicle lamp casing with protected glass, in which lettering may be displayed and which projects a coloured light
169255	Laid 21/06/20 Granted 21/0-9/21. Relates to an oil cooled engine, primarily automotive, in which almost the entire cylinder barrel is encased is within an enclosed crankcase and cooled by oil.
188087	Laid 10/08/21 Granted 09/11/22. This is form of clutch brake to retard the gearbox shaft speed to improve the engaging of the gears. It used a conical thrust member and predates synchromesh.
194907	Laid 22/02/22 Granted 22/03/23. Refers to a pumped oil supply to the cylinder to lubricate and cool the piston at its upper stroke and to direct an oil jet at the exhaust valve chest and guide; it was used in the Panther motorcycle engine.
223324	Laid 25/07/23 Granted 23/10/24. Relates to a spiral inlet manifold lying tangentially to the valve stem/port which causes in-flowing fuel:air mixture to spiral around the inlet valve and by that turbulence, improve atomisation and combustion. This too was used in the Panther motorcycle engine.
225275	Laid 27/08/23 Granted 27/11/24. Refers to two stroke cycle engines with stepped piston in which the larger lower half acts on a sleeve valve which pumps oil to the upper cylinder for lubrication.
230145	Laid 28/11/23 Granted 02/03/25. This is a three cylinder two stroke engine combined with a three cylinder steam pump; the crankshafts were not coupled. The entire cylinder barrels and heads are within a large water jacket which acted as a boiler, the generated steam then acts on the steam-pump and regulates the 2-stroke engine's carburettor. The patent points out that in descending hills the expansion engine augments pressure in the steam chamber! A regenerative engine or perpetual motion engine; but was a working example ever made?

13472	Prov pat laid 12/05/37. A provisional patent for a bicycle crank type torque equaliser to assist the cyclist's efforts to take the pedal over top dead centre when at full load, for example hill climing, requiring greatest effort.
512514	Laid 01/03/38 Granted 19/09/39. This is a gearbox/transmission band brake which prevents a vehicle rolling back on hill starting
614040	Laid 02/07/46 Granted 12/08/48. Cylinders and valve gear plate for Bumbleebee engine
856619	Laid 21/05/57 Granted 21/12/60 was a joint patent with Bromega Ltd for a twin toroidal supercharged diesel 'Omega' engine

Fasteners

Patent No	Laid	Subject
23043	13/06/07	Adjustable spanner
501548	01/03/39	Screws
502348	13/03/39	Screwdrivers
511033	08/08/39	Improvement to screws
513686	19/10/39	Improvement to screws

190623043	Laid 18/10/06 Granted 13/06/07. Refers to an adjustable sliding jaw, cam operated wrench. A joint GB/John King patent
501548	Laid 03/09/37 Granted 01/03/39 relates to thief-proof woodscrews driven in by a special adaptor engaging with a tapped centre
502348	Laid 13/09/37 Granted 13/03/39 relates to a 'Yankee' type screwdriver with adaptors for screwing in and extracting thief proof screws as per 501548 above
511033	Laid 07/12/37 Granted 08/08/39 is a variation of 501548 using friction clutch drive
513686	Laid 17/03/38 Granted 19/10/39 is for screw with raised head of irregular shape requiring a matching socket screwdriver

Fairmile Marine

Patent No	Laid	Subject
547279	20/08/42	Stanchions
549585	27/11/42	Adjustable pitch propellers
578515	02/07/46	Power shaft connector
581760	24/10/46	Reverse gear for propellers

547279	Laid 09/06/41 Granted 20/8/42. Relates to telescopic ships' stanchions; a joint patent with Fairmile Marine

549585   Laid 27/06/41 Granted 27/11/42. Relates to water cooled and lubricated propeller shafts with simple hand controlled variable pitch propeller blades; a joint patent with Fairmile Marine

578515   Laid 07/08/42 Granted 2/746. Relates to coupling joint for two shafts using a wedge lock device; a joint patent with Fairmile Marine

581760   Laid 23/06/41 Granted 24/10/46 is for an epicyclic reverse gear box with clutch, linked to the throttle to permit rapid reverse of propeller shafts on launches and ships; a joint patent with Fairmile Marine.

Amusements

Patent No	Laid	Subject
167923	25/08/21	Ash tray
171810	01/12/21	Wardrobe
260641	08/11/26	Advertising device
288744	19/04/28	Coin operated amusement
306858	21/02/29	Coin operated amusement
317136	15/08/29	Amusement game
325353	20/03/30	Pinball machine
328749	08/05/30	Noughts and crosses machine
345960	27/03/31	Form of shove ha'penny
380896	29/09/32	Bowling game
329789	29/05/30	Passenger carrying railway

167923   Laid 18/06/20 Granted 25/08/21 relates to a smoker's ash tray with divided compartments to trap and hide the ash

171810   Laid 06/09/20 Granted 01/12/21. This was a portable wardrobe rack and curtain jointly patented with Robert Rood

260641   Laid 06/05/25 Granted 08/11/26 a game of skittle which flagged up advertisements

288744   Laid 24/01/27 Granted 19/04/28. This was the 'Little Stockbroker' gambling machine

306858   Laid 21/11/27 Granted 21/02/29. A variation of 288744

317136   Laid 15/05/28 Granted 15/08/29. A form of shooting range with scoring device

325353   Laid 14/01/29 Granted 20/03/30. A pinball machine

328749   Laid 23/03/29 Granted 08/05/30. A noughts and crosses game machine

345960   Laid 27/11/29 Granted 27/03/31. A form of shove ha'penny using special pucks riding on air

380896   Laid 29/07/31 Granted 29/09/32. A bowling game with a bobbing target

329789   Laid 22/04/29 Granted 29/05/30. A passenger carrying railway on a helter-skelter principle but with missing rails

Science

Patent No	Laid	Subject
295611	13/08/28	Photographic booth
322503	09/12/29	Heart beat indicator
437517	28/10/35	Stereoscopic apparatus
437518	28/10/35	Stereoscopic apparatus
437519	28/10/35	Stereoscopic apparatus
460655	01/02/37	Stereoscopic apparatus
12827/53	1953	Prov pat system of lighting
4360/53	14/2/53	Prov pat Stereo television
5478/53	27/2/53	Prov pat Stereo television
6091/53	05/3/53	Prov pat Stereo television
7143/53	16/3/53	Prov pat Stereo television
7472/53	18/3/53	Prov pat Stereo television
8670/53	30/03/53	Prov pat Stereo television
9546/53	08/4/53	Prov pat Stereo television
10903/53	1953	Prov pat Stereo television
12827/53	1953	Prov pat Stereo television
657/55	08/01/55	Prov pat Stereo television

295611   Laid 11/05/27 Granted 13/08/28. A discrete photographic booth with the photographer working behind the facade.

322503   Laid 09/07/28 Granted 09/12/29. A mechanical spring balance heartbeat indicator

437517, 437518, 437519   Laid 26/04/34 Granted 28/10/35 all relate to stereoscopic photography using prisms to form multiple 12-15 line images on special sheets; their frames, and moving device for the camera. (This involved six patent applications in 1934: provisional patents: 12556, 12557, 12558, 15708, 16821, 25903 and 24774 no details survive).

460655   Laid 30/05/35 Granted 01/02/37. A variation on 437519

4360/53   Laid 14/2/53 'Improvements in three-dimensional moving pictures' ie. the use of screens

5478/53   Laid 27/2/53 'Improvements in devices for making and showing stereoscopic moving pictures' ie. the mounting and movement of cameras in the scanning box

6091/53   Laid 05/3/53 'Improved means of taking and showing stereoscopic moving pictures' relates to electronically controlled illumination of the screened object to mimic the 405 lines of TV screen

7143/53   Laid 16/3/53 'Improved means of transmitting stereoscopic moving pictures' a modification of 6091/53 above

7472/53   Laid 18/3/53 'Improvements in stereoscopic moving pictures' relates to electronic camera and encoded mosaic type images unscrambled by electronic means to present 3D images

8670/53	Laid 30/03/53 'Improvements in stereoscopic moving pictures' which adds to the electronic imagery but using colour filters in each camera to give full colour spectrum for colour 3D viewing by using a very fine series of coloured screens on the television receiver
9546/53	Laid 08/4/53 'Improvements in 3D moving pictures' covers a back-projection screening box for electronic scanning
10903/53	'Improvements in colour television' is a simpler version of the full colour spectrum television receiver screen (8670/53); this application is for three prismatic lined screens using the three primary colours
12827/53	A system of prisms and reflective surfaces to direct light into remote areas of a room etc.
657/55	Laid 08/01/55 improved scanning box for 4 camera set up via a vertical screen with fine v-shaped lenticular lens screen on the receiver to unscramble the encoded image.

Appendix 5

Walter Lawson Adams

The enigmatic Walter Lawson Adams was born in spring 1874 to Amos Lawson Adams (1846-1895), a miller, and Emma (neé Nobbs) a dressmaker. The 1881 census records Walter as a scholar, living with his mother at her parents in Loddon, Norfolk, while the 1891 census shows the family together in the village of Great Ryburgh, near Fakenham, where Amos was a 'flour miller' which, *prima facie*, suggests him working at the nearby post-windmill at Colkirk, rather than F & G Smith's Great Ryburn water-powered cake and feed-stuff rolling mill. This had been converted to steam power in 1875 and re-equipped with a modern rolling system in 1893 by Whitmore & Binyon, a long established millwrights who, having erected many windmills, were now concentrating on steam beam-engines made at their Wickham Market iron works for driving mills, rather than relying on the unpredictable wind power.

Walter is recorded in the 1891 census as an 'engine fitter'. Indeed, he was articled to Whitmore & Binyon at Mark Lane, London where it appears he studied; their engineering side closed 1901.

Cycle Maker

For reasons unknown, Walter is next found in Bedford where, with a capital of £50, he established what appears to be bicycle business. Then, in autumn 1894, he wed the previously married Susan 'Susie' White, 8-years his elder (she gets younger at each census!), but through his ill-health, they moved from Bedford to Suffolk taking up residence at 319, London Road, Kirkley, south Lowestoft and now, with a capital of only £5, set himself up as a cycle agent and repairer at 51 High Street (also recorded at No. 52). Interestingly, a bankruptcy order is recorded against him in November 1896; this he put down to his customers "being a bad lot" and "damaging his machines" for which he had been unable to gain compensation, made worse by his solicitor's fees. Despite pawning his bicycles at well below cost price, he was unable to meet his £302 liabilities.

Drawn steel-tube bicycle frames relied heavily on cast lug-joints, traditionally pegged or brazed, but around 1894 George Sanders of Small Heath, Birmingham, had pioneered the use of bamboo for bicycle frames. By using modified lugs, the bamboo was inserted under pressure and retained by various methods; these he patented in June 1895, including a tapered mandrel to splay and trap the end (this technique is still used today in hydraulic hose connectors). His Bamboo Cycle Company was duly formed in Small Heath with Pickering Phipps JP (a 'gentleman', doubtless his backer who also appears to have been in partnership with Thomas Phipps Dorman - possibly a relation - in Northampton; see later).

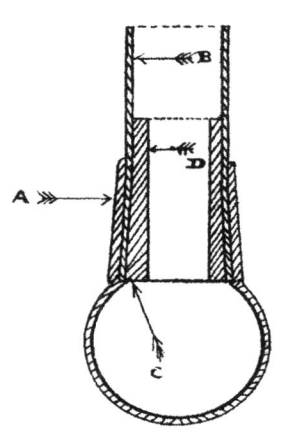

Adams' patented cycle frame joint

While their bamboo framed bicycles soon secured a small niche market, they also patented that year a process of making imitation bamboo tubes from steel; several such bicycle frames were duly made.

Possibly inspired by this, Walter Adams lodged his first patent in 1897 at Lowestoft as a 'cycle engineer and maker' for an expanding tubular steel frame joint system for 'cycles, motor cars and machinery', granted Pat No. 29230 in 1898. By slitting the tube at its end and pressing it home over a tapered mandrel in a frame lug, it mimicked the expanding fox-wedge system for blind housings in joinery and that partly covered in George Sanders' patent. Later that year, Adams submitted a joint patent with Dr Arthur Wellington of Lowestoft for an improved football bladder inflation valve (Pat No. 12443) which also had applications in bicycle inner tubes.

'Adams' car

In an interview with the *New Haven Evening Register* of 21st September, 1924, Walter reports that through "extensive experience with various well known firms, I started to build and market the Adams cars and engines, many of which were sent all over the world". This is a mystery, both in identifying the "various well known firms" and his Adams' car (undoubtedly a quadracycle or cycle-car). But in 1900, Walter and an Edmund Minott Davey submitted a patent application (Prov. Pat. 23402/1900 "internal combustion engines"); which however was then abandoned, so no details survive. Little is known of Edmund or a Thomas Thurlow Smith (a later co-patentee) but they both appear to be London based so perhaps Adams knew them from his apprenticeship days? Then in June 1900, Susie lodged a patent (granted as Pat No. 10599) for an eccentric (cam) driving both the exhaust valve (timed to maximise exhaust gas expulsion) and also circulate, by pulses, coolant through a water jacket. She is recorded in this patent under S Adams Ltd, 319 London Road, Lowestoft, 'Motor Manufacturers', possibly to get round Walter's undischarged bankruptcy.

There are no records of his 'Adams' car, so it is likely he and Susie sold their designs to others. As it was later reported, "he had some experience with boat engines at Lowestoft, and this knowledge had led to several water-cooled Dorman motorcycle engines", which suggests he may have offered himself to J W Brooke & Co of Adrian Ironworks, Lowestoft, boat builders, who made their first two-stroke engine in 1899 and, under John W Mawdsley Brooke, their first car, a 3-cylinder, in 1901. But did they take on Adams' designs? We simply don't know, but it should be noted there is no connection between Walter Lawson Adams and the American Adams-Farwell car company, nor with Arthur Henry Adams (another American) and his Adams Manufacturing Company of London (electrical engineers), their Antoinette aeroplane concessionaire of 1908 and their 'Adams' motor cars built at Bedford between 1905-1914.

Northampton

The *New Haven Evening Register* goes on to report that after selling his "Adams car and engine concern... I then designed the 'Precision Engines and Motorcycles' (many thousands of Precision engines and motorcycles with a few modifications, are being sold even today in England)". His letter-heading of the late 1920s indeed states he was: "Designer, inventor and patentee - Precision Motor Cycle engines", but modern motorcycle encyclopaedia dismiss the Precision motorcycle as being 'primitive and powered by a Belgian 211cc (62 x 70mm) Minerva engine' yet his claim of "thousands of..." has led to some enlightening detective work!

The 1901 census shows him and Susie in lodgings at Mrs Lamberts' apartments, 24 Wood Street (now under the Grosvenor Centre), Northampton. He is recorded as 'Motor engineer' (later appended 'cycle') and a 'Worker' (ie. employed, rather than self-employed or an employer). Most towns (and many villages) now had a cycle maker or two, by why the move to Northampton, the centre then as now of Britain's boot and shoe industries?

In fact he had been engaged by the Dorman Engineering Company of Mayorhold, Northampton. Originally the Patent Manufacturing Company (stampers and piercers of metal sheet) they had made a wide range of small components for the trade in the 1850s but by the early 1880s, their Francis Wileman had designed and built a sewing from stamped and pierced steel, devoid of any castings. The business soon became the Dorman Sewing Machine & Engineering Company, owned by Thomas P Dorman (this appears to be Thomas Phipps Dorman, possibly an accountant) and managed by Alfred Henry Dorman. They produced 'Dorman' sewing machines (winning four Gold Medals between 1887-1890) from their fully equipped 4-storey, court-yard factory and machine shops at Mayorhold, sited between Broad Street and Bulls Head Lane (Wileman's home and office was in Bulls Head Lane). Their showroom was at nearby 4, Sheep Street.

In 1890 their Dorman Engineering Company division began manufacture of safety-bicycles, under the 'Whirlwind' name, in four frame styles including a diamond frame (*sans* saddle down-tube) with a sprung front fork, mainly in 24" or 26" frame sizes but others were available to order. It appears that in the late 1890s, the Dormans and their business partner Horace Walter Dover passed control to Spencer Downing. While Alfred appears to have retired, Thomas appears to have continued in practice with Pickering Phipps at P. Phipps & Co Ltd, Northampton. Horace stayed on at Dormans for a while, but also established a separate cycle component company, Dover Ltd (given as 2-16, Victoria Park in 1901 and later Park Road, St James). Although known locally as 'gearcase' makers, they actually made bicycle chain guards.

Dorman's 'Whirlwind' bicycles proved very popular (many went to the British army) and by 1897 they had acquired the major interest in Sanders and Phipps' Bamboo Cycle Company; Pickering Phipps duly joined Dorman Engineering at Northampton but in 1898 moved with Horace Dover to Dover Ltd to develop use of celluloid and cellulose coated steel products (for which process they secured a patent); Dover soon became famous for their celluloid enclosed chain guards and later became part of Bluemel Brothers, a familiar name in cycle parts and their famous celluloid bicycle tyre pump.

In late 1900, Dormans decided to offer a clip-on motor-cycle engine 'of their own manufacture'. The design of this $1^3/_4$hp ($2^5/_8$" x $2^5/_8$" : 67 x 67mm) aioe, with one piece cast head and barrel and a crankcase cast with a frame bracket, is credited to a Mr Roebuck (believed to be John William Roebuck).

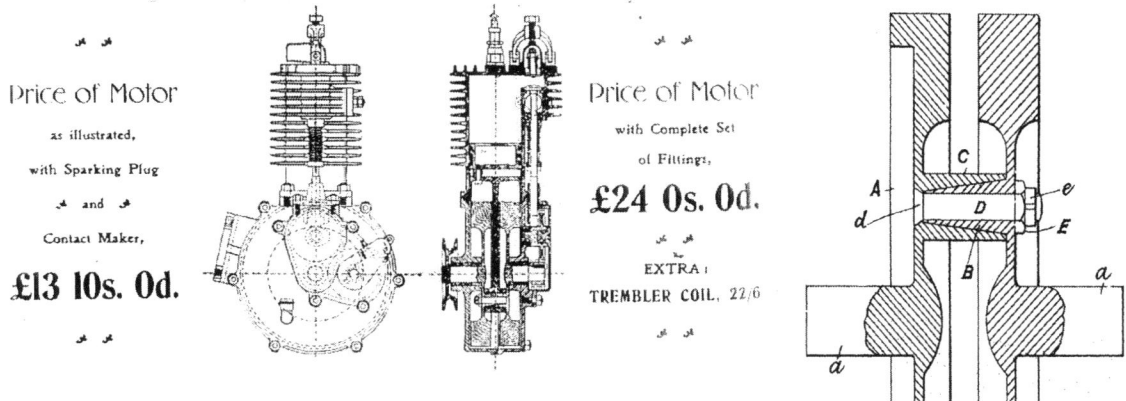

DEC $2^1/_4$hp aioe engine of 1901 featuring the taper fit, cast crank-pin halves as in Downing's patent

Marketed as the 'DEC Whirlwind' engine for self-fitting by cyclists for light pedal assistance, the basic engine cost £13 or, fully equipped for direct fitting, £24. The bare castings were also available by which others were able to develop their own engines, just as were many early engines deDion or Minerva based. The 'Whirlwind' engine sold very well and several were bought by Triumph for testing.

Its success prompted a move in 1901 to complete motorcycles (from £40) with a range of single cylinder clip-on 'DEC' engines; the $1^3/_4$hp was duly followed by a 2hp aioe (believed to be $2^1/_2$" x $2^1/_2$"), then a $2^1/_4$hp sv, all with underslung exhaust. These used spray carburettors. Of interest is that the flywheel and crankpin were cast as one with tapered male and female inter-fitting crank-pin halves drawn together and locked by bolt, for which in January 1902, Spencer Downing lodged a patent, "Improvement in cranks for motors" (granted Pat No. 302 that December).

Dormans had meanwhile acquired an Enfield quadracycle frame to develop into a cycle-car for which they engaged the highly experienced Walter Adams whom Alan Burman advises in his brief history of the 'Whirlwind': "had some experience with boat engines *(at Lowestoft)*. This knowledge led to several water-cooled DEC engines... The Whirlwind Launch Motor was a 2-cyl ($4^1/_2$" x $5^1/_2$") with reverse gear drive", to which Alan added that one of the first was fitted to a Thames launch for the millionaire Vanderbilt family. Their new DEC 3hp sv water jacketed engine of 1902 was undoubtedly designed by Walter Adams and likely incorporated Susie's water pump design, but the 'Whirlwind' 3hp water cooled quadracycle project floundered, possibly because Walter had left by summer 1901 to form his own Precision Motor Company. The quadracycle was sold off in 1902, fitted with a deDion engine, but the 3hp water jacketed engine survived for their new 1903 season motorcycle, with its specially designed down tube and looped bottom bracket frame. The prototype was much used by Spencer Downing for pacing racing cycles at the Northampton Country Ground.

Dorman however was by now being brought to its knees by a slump in demand for their bamboo-framed cycles, forcing a creditors' meeting in January 1903. Spencer Downing then put his Dorman

Dorman water cooled engine

Dorman motorcycle fitted with the water cooled engine

Engineering Company into voluntary liquidation and ceased trading in March, followed by a major factory auction on 6th and 7th April. Sufficient parts existed to allow a handful of motorcycles to be built locally. Spencer then moved to Moseley, but remained in the motor industry. The Mayorhold factory was demolished in August 1970. (Please note there is no connection between this Dorman and W H Dorman & Co of Stafford, as claimed in modern motorcycle encyclopedia).

'Precision' engines

In 1901 Walter submitted a patent on "explosion engines" (Prov Pat. 13072/1901), but this too was abandoned. However his Precision Motor Company is first recorded in summer 1901 at 79, Derngate, the residence of a Miss Smith, close to what is now Spring Gardens in the town centre; it became his address by 1902. It is quite possible that he rented a workshop or stables at the back and we could speculate that any engineering work was done locally.

Announced that summer, his 'Precision' engine had a one piece cast cylinder, head and valve chest (which differed from the DEC) with a frame-clamp cast with the crankcase for mounting to the front down-tube of the bicycle, just like the DEC, driving the rear wheel by belt. Developing 2hp, it offered 35mph. An exhaust heated air intake improved atomisation in the jetted carburettor, fitted with a rotary throttle.

A 'Precision' motor-cycle appeared in late 1902; this had a second, parallel, down-tube from the headstock to a crankcase lug below the cylinder; the base of the modified crankcase then bolted to a bottom bracket lug, allowing rapid removal. Whether a twin or a water cooled engine

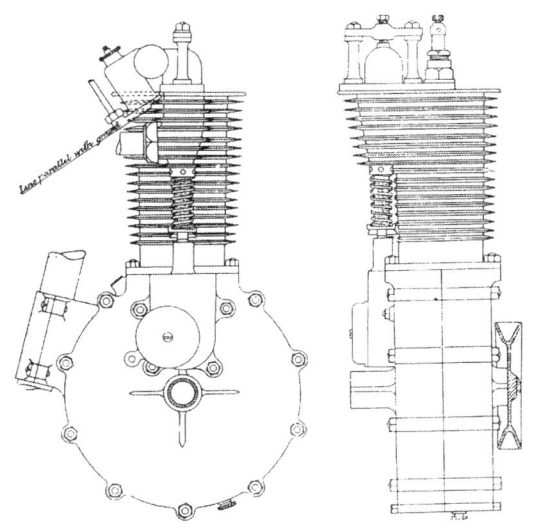

Adams' Precision engine of late 1901

was developed is not known. Of interest in Adams' engine were his cast flywheel and crankshaft 'connected by a single pin' to which his third abandoned patent application "combination crankshafts and flywheels for multi-cylinder explosion engines for single track vehicles" (Prov Pat. 19916/1902) undoubtedly related. This was also abandoned; perhaps it was a failed attempt to avoid Spencer Downing's ostensibly similar patent of January 1902 for his combined cast flywheel and crank-pin but, as an abandoned patent, no details survive other than it was submitted jointly with Thomas Thurlow Smith. Was he related to Miss Smith of 79, Derngate?

A conundrum

Contemporary local 'classified ads' record 'Precision' engines fitted to various makes of bicycle frames. How many 'Precision' engines and motorcycles were made before they reportedly closed in 1906, is not known. However Adams declared in the American newspaper that he had "disposed of my controlling interest in this concern" and that "many thousands of Precision engines and motorcycles with a few modifications, are being sold even today in England".

To whom he sold his interest and what happened next is a mystery as there is little evidence of the business in local papers or motorcycle journals, yet a 'Precision' brand reemerged in 1912. It is entirely possible - but by no means certain - that Dorman's assets and engine designs may have gone to two Northampton cycle makers and agents, Douglas Gainsford and Frederick Smart who had, from 1902, been attaching engines to bicycles and selling motorcycles. From 1903, they produced a similar range of 'their own engines': 2hp aiov, $2^{1}/_{4}$hp sv V-twin and 3hp water cooled engines with castings believed to have been made by Dover Ltd. In May 1905, they registered themselves as the Advance Motor Manufacturing Company at their Louise Road works, and evolved into an important engine maker for land, sea and air.

It could also be that in selling his controlling interest in Precision, Walter received royalties. Had the rights and assets been sold to Advance, accounting for Precision's closure by 1906? Could it also be that when in 1912 Advance abandoned engine production and relocated to their new factory at Kingsthorpe Road as component makers, those rights to Precision passed to Frederick Baker who began production in 1912 of his own 'Precision' brand of engines and motorcycles, mostly for export? Baker's 'Precision' two-stroke engine was made for him under contract, quite possibly by Beardmore, leading to the post-war 'Beardmore-Precision'. Also in 1912, Ernest Smith & Woodhouse offered their 'Regal' motorcycle powered by 'Precision' engines for which Charles Green then offered water-jacketed, re-built 'Precision' engines, as his 'Green-Precision'. Could such a convoluted (and quite probably seriously flawed!) theory thus explain Adams' claim of "many thousands...", or had he wrongly associated these sales with his own, earlier, 'Precision'? This author would certainly like to know!

Redbridge

While living in Northampton, Walter became well known to the local Police and Magistrates, but for entirely the wrong reason: Drunk and disorderly in Mayorhold, July 1901 - fined 10/- inc costs; 'furiously driving a light locomotive' on the Billing Road on 13th June 1902 - in this case the Magistrate acted 'with leniency', fining him 20/- costs; driving a motorcycle without lights in Stoke Goldington, 12.10am 17th August, 1902 - fined 30/- with 15/- costs. Walter and Susie did not stay long in Northampton!

He declared that he then "spent some time on the Continent studying French, Italian and German automobiles. When I returned I sold my developed patents to Messers David Brown & Sons of Huddersfield and took a position with them as Chief Engineer and Assistant Works Manager".

There is a newspaper record from Lichfield in 1903 of a Walter Lawson Adams as chauffeur to a Cpt Haig. Had he toured Europe in 1903 as Haig's chauffeur - or is this a chance namesake, of whom their are several? To what the 'developed patents' he sold to David Brown relate is not known, but while David Brown were best known as gear-makers, they had produced parts for Ralph Lucas' 'Valveless' twin-cylinder two-stroke engine and car of 1901, which they then assembled 'with modifications' between 1908 to 1914. There is a further abandoned patent by Adams (Prov Pat. 18236/1904) for 'explosion engines'; here again, no details are known. It is quite likely that, as did Granville Bradshaw, Adams sold his 'Patent Pending' ideas without formal disclosure to maximise his financial return.

In 1904, he and Susie moved to 'Yewtrees' (also known as 'The Yews'), Redbridge, Southampton from where he operated as W L Adams Ltd, Motor-car Manufacturer. Redbridge had been a significant boat-building centre, still with a large sawmill and chemical works, but there is no trace yet of the works from where: "In 1905 I commenced to study aviation and started in business... making my eight cylinder dirigible balloon engines for the British Government (exhibited in 1905) at the same time as making hydroplane engines". That he designed and built a handful of small engines (possibly 'Precision' based) is quite likely; a 10-12hp engine (most likely a twin, possibly based on the 'Whirlwind' launch motor) was reported as being fitted in 1904 to the *Bluebottle*, securing it a Class win in an endurance trial on Southampton Waters in 1905.

The development of his 'Redbridge' marine V-8, the 'Aeroplane' V-8 engine and his association with Granville Bradshaw and ABC Ltd is told earlier in this book - to which the reader is now referred.

W H Dorman & Co

While Granville Bradshaw was in his element at ABC, Walter appears to have been turning away from aero-engines. He had already been approached by W H Dorman & Co of Stafford to buy his aero-engine design and patents and was even offered the position of "Manager, Chief Engineer and designer" in their new Aero-engine Section of the Dorman Internal Combustion Engine Department (Dormans merely describe his position as 'Engineering and Sales Expert'). It is quite possible that the War Office's dismissal of the value of aeroplanes in warfare prompted Adams to finally sell his W L Adams Ltd Redbridge Motor Works to Dormans in May 1912. Then on 24th July 1912, the Aeroplane Engine Company went into voluntary liquidation. Adams' 90°, 60hp 4" x 4$^{3}/_{4}$" V-8, parts and production machinery went to W H Dorman & Co and duly became their short-lived '1912 Pattern Dorman Engine', while his other aero-engine designs appear to have gone to ABC.

But he did not stay long with W H Dorman for he is next reported as being a 'Consulting Engineer in London' developing a small 12hp (25bhp) 4-cyl water-cooled lightweight cycle-car engine for use in a new 1913 Mendip and the fledgling Day-Leeds cars. His own account says he left Dorman "to keep in touch with the aviation side of the industry and designed some improved aviation engines and started what is known today as the A.B.C - Adams, Bradshaw and Charteris - with a factory at Bournemouth and Brooklands track. I spent two years at Brooklands track, and gained very valuable experience in and developing my own engines and aeroplanes and kept in close touch with the work of neighbouring firms such as Vickers, Bristol, Sopwith, A V Roe etc. While there I acted as consulting engineer to Job Day of Leeds and managing director to Mendip Motors Ltd designing these two very successful light cars and engines. Both these cars were well into production in August 1914 when the war broke out and I received a commission and was attached to the Royal Naval Air Service".

Here his chronology appears awry, as it suggests he returned to ABC. There appears to be no further record of him or Susie at Bournemouth after 1911, nor in London. Brooklands could, at a stretch, be called 'south west London' but his whereabouts remains unknown. Neither is there a clear trace of

Susie after her petition for divorce in 1915, undoubtedly for his adultery, for in 1912 Walter Adams had met Ada Plater (b. 1883 in Thame, daughter to John, a baker, and Elizabeth Ellen, née Belson). Ada had become a domestic servant to the famous Beerbohm theatrical family in Marylebone, London, in 1911, but how Walter and Ada met is not known.

Mendip motor-car engine

Around 1912/early 1913, Adams designed a lightweight cycle-car engine which he clearly intended offering to automobile (and possibly aeroplane) engineers and possibly to European and American companies.

It was announced in September 1913 by C W Harris & Company's 'Mendip Motor & Engineering Company' at Chewton Mendip, near Wells, Somerset followed that November by Job Day & Sons of Leeds for their new light-car to replace their earlier 8hp V-twin cycle-car. The Day-Leeds used a long stroke 10hp version of Adams' engine, originally of 59 x 100mm 1096cc, but production models were 64 x 100mm 1287cc - just to confuse matters in 1919 both *Autocar* and *Motor* quote this as 1266cc, however it is believed a 1266cc Turner engine had been fitted in 1914.

Mendip had already built some steam waggons and, from 1911, 30cwt vans and light lorries prompting their move to manufacture their own engines, but what the working relationship was between Adams and Mendip is unclear for in the American newspaper report, Adams described himself as Managing Director of Mendip, and Consulting Engineer to Job Day, as well as "designing these two very successful light cars and engines". That there was a close association with Mendip is reinforced by Ada giving birth to their daughter, Joan Winifred, at Chewton Mendip on 16th October, 1913; their son, also Walter Lawson (usually known as 'Lawson'), was born at nearby Wells in late 1914. Meanwhile, Charles Harris had sold his interest in the Mendip car business in late 1913, and left in 1914, so it is quite possible that Adams did later became Managing Director, overseeing the new Mendip factory at Henham, Bristol in 1914 wherein was designed and built the new Mendip car.

The first 'Mendip' 12hp, 4-cyl water-jacketed 1,092cc (64mm x 85mm) engine developed 25bhp, however production engines were of 1,255cc (67mm x 89mm) rated at $11^{3}/_{4}$ hp and developing 20bhp, which kept it outside the lower, under 1,000cc, cycle car taxation class. A more powerful 1,312cc (67mm

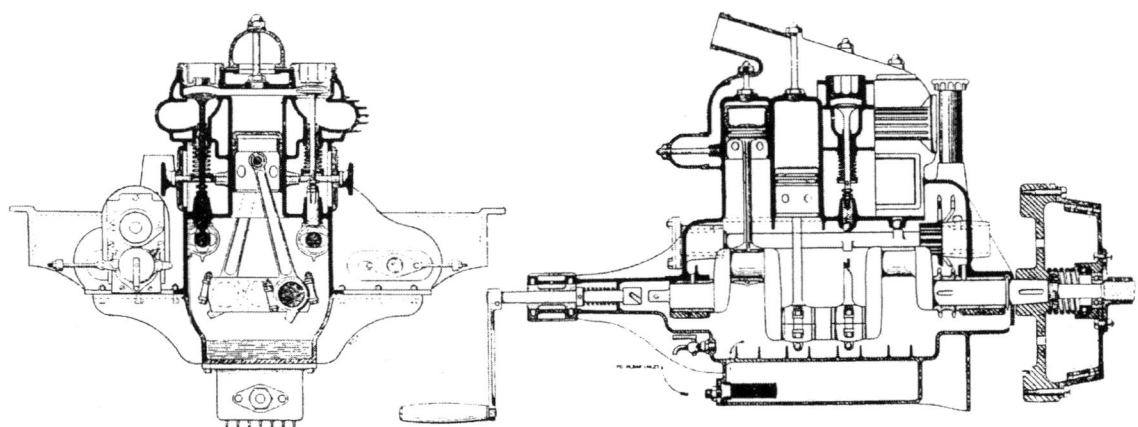

Adams' twin cam side-valve engine for the Mendip light car. Note the 3-point mounting, extended crankshaft housing and the two magnetos making it ideal for car or aeroplane use

The Mendip light car with Adams' engine

x 93mm) engine was proposed, but that used in the Day-Leeds car was of an even longer stroke (64mm x 100mm) and variously described as 1,266 or 1,287cc. His twin-cam, pushrod side valve, T-head format and horizontally-split cast alloy monobloc design, with integral timing chest and splash lubrication, and separate magneto (for ignition) and dynamo (lighting) displaced either side of the crankcase (V-8 aero-engine fashion), was remarkably advanced for its time, yet it raised little comment in *The Light Car and Cyclecar* report.

Casting the monobloc, and forging the crankshaft, was entrusted to a respected Belgium foundry - this, it is said, at the behest of Mendip's manager, George Thatcher, although Adams may have recommended it following his earlier Continental tour. The 'Mendip designed' gearbox and worm-drive axle castings came from a Manchester firm, but were assembled by Mendip; these may also have been influenced by Adams. Their first car, looking much like a small Bullnose Morris, was rolled out in March 1914 using light C-section 5% nickel-steel chassis members, designed by Adams, but pressed in Belgium. However, supplies ceased soon after Germany invaded the Low Countries in August 1914 and systematically took over Belgium, destroying much in their path.

Who this 'respected Belgium foundry' was, is open to conjecture but given the Antwerp based Minerva (supposed) connection with Adams' Precision engine; the David Brown connection with Adams and D-B being concessionaires of the Belgian made SAVA of Antwerp sporting-car between 1912-1914, it could quite possibly be either of these two Antwerp companies (SAVA was taken over by Minerva in 1923) - or another independent foundry. Both Mendip and Day-Leeds production ended in early 1915 and resumed post-war, but when surviving stocks dried up, Mendip offered Alpha or Dorman (believed to be the KNO) engines, while Day-Leeds resumed production of the original engine using castings and forged components now 'from a Manchester firm'.

If, as Adams implies, he had returned to ABC after Dormans, it is open to debate whether the Mendip twin-cam engine reflected Adams' and Bradshaw's aero-engine experience and whether it was also intended for light-aeroplanes. This is hinted in *The Light Car and Cyclecar* report of 1913 which states that Mendip "also produced an entire power unit, consisting of engine, clutch, three-speed-

and-reverse gearbox, gate change, clutch, brake and accelerator pedal". Indeed, valve-gear trains vary considerably among engine makers; many adopted push-rods from a low mounted cam (the Benz 4-cyl in-line and V-12 aero-engine from 1913 used twin cams, like the Mendip) while some V-engines had a low mounted single cam, yet others adopted chain or bevel gear and shaft drive to an overhead cam on each block or, on narrow angle V-engines, a central single overhead cam. The provision of twin magnetos (one in reserve due to their unreliability) mounted either side of the crankcase was not uncommon in aero-engines, but such design features in Adams' Mendip design were not usually found on a motor-car engine.

America

Walter Lawson Adams, Bradshaw and Charteris dutifully came to defend King and Country. As a manufacturer, Granville Bradshaw produced engines during the war, while Ronald Charteris served with the RFC. Adams was commissioned in July 1915 as Lieutenant, Royal Naval Volunteer Reserve and served in Section E of the Admiralty Air Department, Victor House, London before being seconded to the Royal Naval Air Service. Ada and the children moved to Thame for safety.

His much respected knowledge of aero-engines saw Adams despatched to America as a representative of the British Government's interests and to learn American mass-production methods. The *New Haven Evening Register*, referring to him as Lieutenant Commander, put it more colourfully: "Therefore the first Lord of the Admiralty called upon Mr. Adams, designer of a number of successful automobile motors and a leading figure in European aeronautics, to represent Great Britain in America. Answering to the call of his country, Mr. Adams reported to the Admiralty office and was asked to organise two national factories for airplane engines in England, the Walthamstow works for Gnôme and LaRhône motors and the Clerget factory at Chiswick. Applying his wide knowledge of engineering and shrewd executive ability to the problem, Mr Adams had both plants under production in a short time and was profusely complimented by the air board. He was then told to prepare for service in America. This was in early 1915, when this country was still neutral and submarine warfare was in its infancy. Air raids were infrequent, too, since that method of fighting was almost unknown".

He undoubtedly did act as Admiralty Office in-charge in the Glen Curtiss Aeroplane & Motor Corporation contracts in America, but his 'organising' of engine production of Gnôme rotaries at George Holt-Thomas' Peter Hooker Ltd factory at Walthamstow, and of Clerget rotaries at Gwynnes in Chiswick, later under the auspices of W O Bentley, needs clarification. Holt-Thomas established the Gnôme Engine Company as the British agency in 1913; his first British-built 80hp Gnôme was completed by April 1914 for the military trials and was then put into production by Daimler, coming on stream that November. Large scale production of the 100hp Gnôme at Hooker's began in 1916, followed by the Clerget at Gwynnes.

Under the American newspaper's headline, "Whole German Army Couldn't Stop Adams", they continued in true *Boy's Own* tempo stating that the German spy system had flashed news of Adams' impending departure to Berlin and that "the train on which the engineer was leaving London for the port of embarkation was supposedly known to none save Mr. Adams and the First Lord of the Admiralty", and that the train was bombed by "a fleet of German aeroplanes" a few hours out of London. Several passengers were killed. Walter Adams miraculously escaped, although all of his luggage and credentials were lost. (There appears to be quite some journalistic licence here, but it undoubtedly relates to the raid on August 17th by three Navy Zeppelins, of which only Zeppelin L10 reached Leyton (having mistaken it for London), killing nine, wounding 48 and destroying a railway station. The first bombing raid by aeroplane was not until that November). Perhaps he had been visiting the nearby Walthamstow Gnôme factory?

On returning to the Admiralty, he was offered a leave of absence for three months to recover, but he refused and "caught the next train for Liverpool where he boarded the SS *Arabic* for America. But the trouble did not end there; the Germans were still determined to hold up the British air program at any cost and to them that meant keeping Adams away from the production centres in America". The SS *Arabic* was a White Star Line trans-Atlantic liner built in 1903. Barely a day at sea, she was torpedoed on 19th August, 1915 by the German submarine U-24 in an unannounced attack, 50 miles SW of W off the Old Head of Kinsale, Cork (U24 survived the war). The SS *Arabic* sank in 15 minutes with 44 lives lost, but 380 were rescued by a British destroyer the next day. Neutral America protested strongly and Germany agreed thereafter not to attack without warning... but in January 1917, Germany declared war on any neutral ship trading with Britain, bringing America into the war on 6th April, 1917.

Back at the Admiralty, they insisted he take leave, but once again he refused and three days later boarded a ship bound for America. US Immigration papers record that he had embarked the *Rochambeau* at Bordeaux and arrived in New York on 6th September. (To avoid torpedo attacks, the regular *Rochambeau* Le Havre - New York service had been transferred in 1915 to Bordeaux). Ada and the two children joined him on 21st December, having sailed safely from Liverpool on the SS *California*.

Lt Walter Lawson Adams, RNVR

Safely ashore, Adams began supervising the construction of all aeroplanes and parts destined for the British forces and "immediately got the Gnôme Motors factory into production at Long Island City and at the same time increased the production of Curtiss motors at Hammondsport *(New York)* and their various plants about Buffalo". Here again the newspaper gives a misleading impression. Small scale production of the Gnôme had already begun at the General Electric Company's General Vehicle Company factory in Long Island City, but they were unable to mass produce the 150hp version to meet British demand, so production of that was destined to pass to William C Durant's General Motors.

Curtiss however were well equipped to meet British demands. Developed in 1907, the Curtiss 'L' water-jacketed V-8 aero-engine evolved into the 'O' with the 100hp OX-5 V-8 (4" x 5") being licence-built by Austin Motors in March 1914 for the Curtiss flying-boat entered in the Circuit of Britain race. The Curtiss engine was then selected by the War Office, tested in the Avro 545 and fitted to some production DH.6 elementary trainers from 1917, as an alternative to the OX-5 powered Curtiss JN4 trainer. The DH.6 was described by A J Jackson in his definitive *DeHavilland aircraft since 1915* as "utterly viceless" but the OX-5 engine was not particularly reliable, suffering poor quality control and valve operation to which many crashes on trainers were attributed. Inspired by the Curtiss V-8, Austin then launched their own 250hp V-12 design at the 1920 Olympia show, using an ABC 6hp Auxiliary engine, coupled to the crankshaft, as a starter motor.

Knox V-12

Adams spent his spare time in America closely examining and reporting to the Admiralty on American engine developments and manufacturing practices. He also inspected the production of some 75,000 Canadian designed 'Splitdorf' magnetos to meet the huge allied demand, now that the superior German Bosch magneto was no longer available.

It would have been out of character had he not also tried personally to persuade American engine makers to develop his preferred 'V' aero-engine designs for the British war effort. In fact, the government did grant him permission to design the Knox V-12, built by the Knox Motor Co of Springfield, Massachusetts (founded by Harry Austin Knox who had left the business by 1905). This narrow, part-aluminium, 60° V-12 weighed 1,430-lbs and had a 7" stroke, compared to $4^{3}/_{4}$" of Adams' last ABC V-8 design.

In the newspaper report, he claimed the Knox was "the first 12-cylinder engine to develop 475hp and pass all allied government tests". One of these push-rod V-12s was indeed submitted to US Navy trials but, as reported in *Flight* of June 14th, 1917, it only achieved 353hp, not the 475bhp claimed. Nothing more was then heard of the Knox V-12, or the General Ordnance Co of Connecticut (a major naval arms manufacturer who made depth-charge throwers for the Royal Navy) who had also begun construction in spring 1917 of a 200hp V-8 ($4^{3}/_{4}$" x $6^{1}/_{2}$") water-cooled cast aluminium-alloy, wet-liner, aero-engine. Could this have also been influenced by Adams?

Meanwhile, Lt Walter Adams had returned to England with his family aboard the American Lines SS *Kroonland*, docking at Liverpool on 5th November 1916. In January 1917, he lodged a US patent, in his capacity with the Admiralty, for an interesting wet-liner cylinder retainer plate for radials or V-engines, with the liner secured by huge ring-nuts to ensure a water-tight fit, bringing to an end external water jackets; it is not known if this was used in the Knox. (He finally submitted a UK patent for this wet-liner cylinder design in January 1918, granted Pat No. 112462 in May 1918).

Whether he returned to America during the war in his capacity with the newly formed Air Board is not known, but with war declared by America on 6th April 1917, a joint British, French and Italian

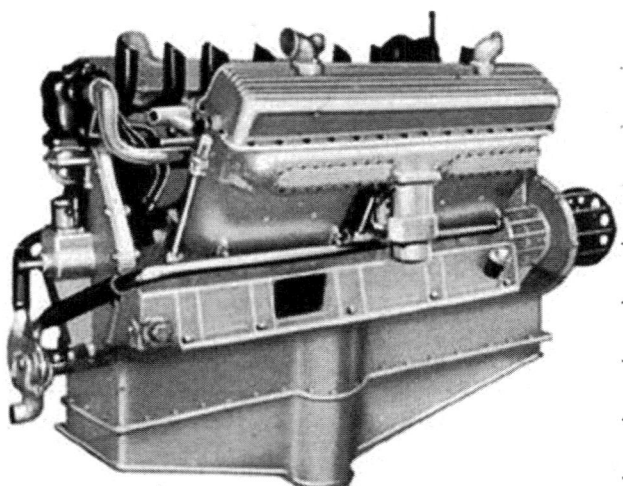

The aborted Knox 60° V-12 of 1917

Commission to the US Aircraft Production Board sat between 28th May and 1st June, 1917 to determine the allied's aero-engine needs. American mass production methods had already delayed progress of 'foreign' engines, as each component part's production drawings and metallurgy had to be assessed and necessary modifications made, so the decision was taken in late July to concentrate entirely on a single, all-purpose, 'All American' Packard/Hall-Scott designed 45° 400bhp Liberty V-12, incorporating the best of European and American design and technology.

The prototype was claimed to have been designed and built in 28 days, passing its trials on 25th August 1917 and with that, it brought further development of the Charles B Kirkham designed Curtiss V-12 (5" x 7") of 1916, and many other companies' efforts, including the Knox, to an early end. Kirkham was the former Chief Engineer of Aeromarine Corporation. As with Granville Bradshaw's ABC Dragonfly, the Liberty (and its component parts) was designed to be built by motor-car makers 'from Connecticut to California' and, just like the Dragonfly, its reliability varied considerably between makers. In his *Aircraft Piston Engines*, Hershal Smith records that, "A good Liberty was extremely reliable, a bad one pitiful"; it was of some comfort to Bradshaw that the Liberty also suffered severe torsional vibration problems, only later resolved!

Laxtonia Engineering

On their return in November 1916, and after a brief stay at the Langham Hotel, London, the family took up residence at 'Keswick', 22 Athenaeum Road, Finchley.

With his experience of American manufacturing practices, in 1917 he bought into a partnership with Herbert Robert Wade at the Laxtonia Engineering company, Burton Street, Peterborough. Perhaps they had worked together at Redbridge, for Wade is a common Hampshire surname? He later bought a half-interest in the Precision Brass foundry, Wolverhampton, of which little is known, and the Bury Iron Foundry, Ramsey, Huntingdon (this appears to be the former Hughes & Kimber 'Britannia Iron works' type foundry) to where he moved the family.

Now working from St John's Lodge, Thorpe Lea Road, Peterborough, Laxtonia's work concentrated on motor engineering and components. Adams patented under his own name in September 1917 an improved universal joint (Pat No. 112402) in which six steel balls transmit power, much like today's constant velocity joints. An improved version to drive magnetos appeared in 1919; this now had rotational adjustment to allow accurate ignition timing. Also announced was an inlet-manifold water valve which allowed steam from the car's radiator to enter the induction system, improving fuel atomisation and fuel consumption by up to 25%. Atomised fuel carried on water vapour or steam injection has long been known to improve an engine's performance (especially noticeable with diesel engines in fog), yet has never been adopted in production engines.

When Herbert Wade retired on 25th March 1919, Adams became sole proprietor, concentrating on his new, compact, Laxtonia V-12 'double six touring car' engine which clearly incorporated many his war-time ideas for a lightweight V-type aero-engine - in a sense 'his' V-12 Knox given a new lease of life. However while the push-rod Knox was of 24.4-ltr (4.75" x 7"), the ohc Laxtonia V-12 was only 6.9-ltr (3" x 5").

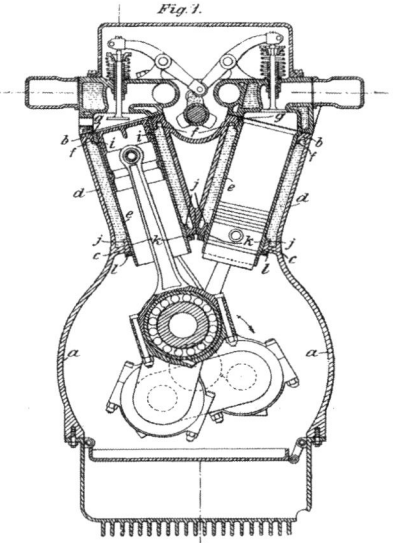

Adams' war-time patent for a wet-liner system on a 30° V-aero-engine, as used on his post-war Laxtonia V-12

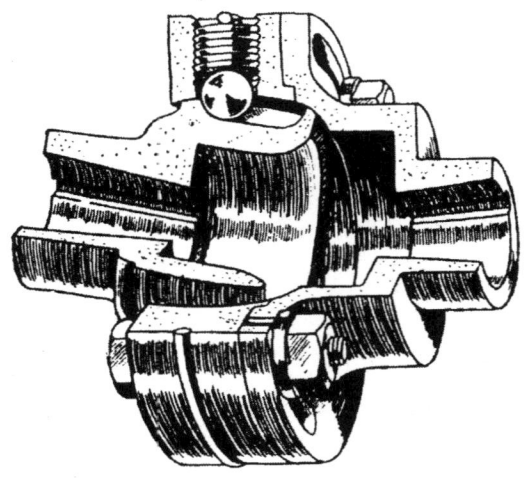
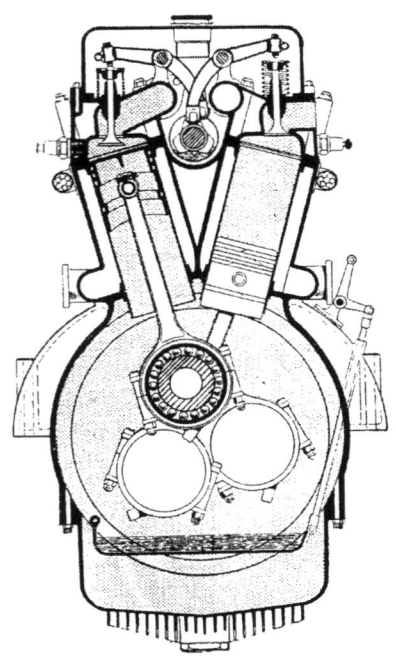

Above, Adams' patented flexible joint (Pat. No 112402) was improved with rotational adjustment. Right, Laxtonia V-12 cross-section. Note similarity with the war-time patent

Of narrow 30° V-pattern, the wet-liner cylinder block was only 12" wide making it no larger than a normal Ford 3-ltr car engine and, being of cast-alloy, as light in weight. The alloy monobloc and cast-iron head were exceptionally smooth and elegant with a full width cover over the head-cum-oil-bath rocker box, hiding the spur gear driven single overhead cam, a departure from the Knox, but commensurate with his war-time patent drawings (Pat No. 112462) for his wet-liner mounting system. However the Laxtonia's liners had a conical beading to ensure perfect fit at the cylinder head; the lower section again being threaded externally for a massive ring nut-which firmly clamped the liner in situ, making a watertight joint. Ignition was by coil and two distributors.

Of note were the caged roller bearing big ends and unsupported, 3-throw crankshaft held only by 4" wide roller bearings at each end, with a ball-race thrust bearing. As with his earlier V-8s, the crankshaft was hollow but now only to reduce weight and whip; this raised concern by *Autocar* over whip and imbalance but were discounted by Adams by virtue of its relatively short unsupported length, compared to many other engines, the large diameter and a very stiff crankshaft. Adams did agree that the 30° V-configuration was not ideal (60° being considered optimum) for smoothness but the firing pattern provided in effect two impulses per 120° of crankshaft rotation, the same as a conventional 6-cylinder engine, but offered greater smoothness and power through a more uniform torque. Problems with imbalance would have undoubtedly been further addressed in his proposed double-V 24-cylinder engine which, alas, remained on the drawing board.

Details of the engines were reported by *The Motor* and *Autocar* in February 1920 followed by *The Engineer*, but while a V-12 prototype had clearly been built, there was no indication that it had been road tested and no performance data was given. The reports suggest its intended use was in motorcars as a 60° V-12 "would necessitate an extremely wide engine, which would not only call for a bonnet of greater width than is desirable, but would also render the engine as a whole less simple (eg: two overhead camshafts would be needed) and less accessible". It is also clear from the reports that Adams

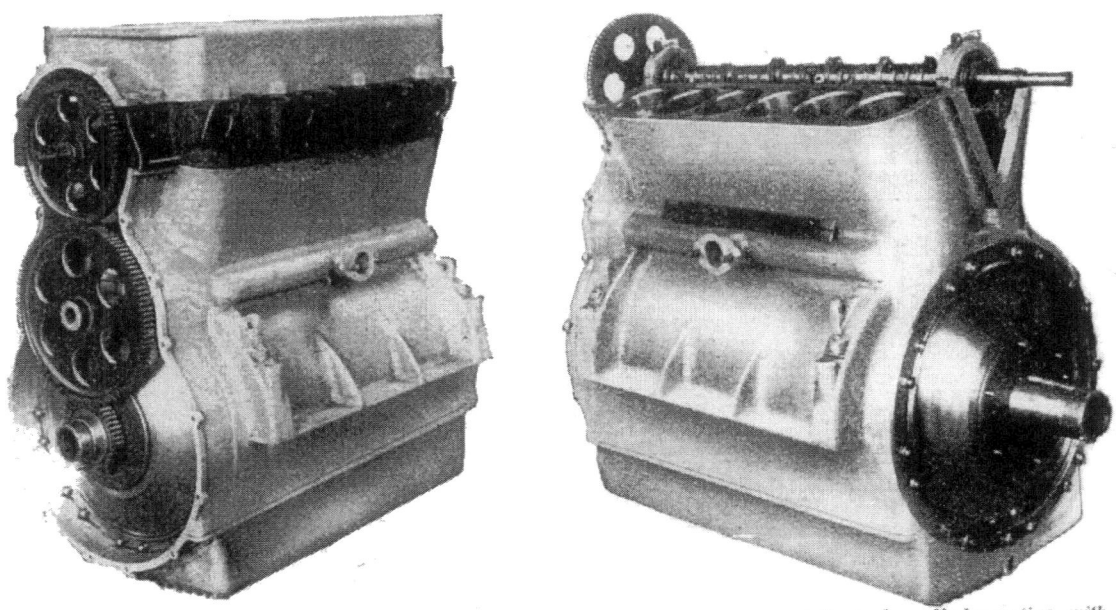

Adams advanced Laxtonia V-12 6 ltr prototype embodied many of his war-time ideas for a modern V-engine. the monobloc casting was particularly smooth. It is not known if the engine was fully tested

was uncertain of the V-12's fate; should he put it into production, or sell the rights? As there is no further record of it, nor of Laxtonia, it is presumed he sold them.

Returning to America to seek his fortunes, he and the family sailed from Southampton aboard the White Star Line RMS *Homeric*, arriving in New York on 8th June, 1922.

New Haven, Connecticut

Back in America, he briefly "served on the engineering staff of the Wright Aeroplane Company, the Aeromarine Corporation of America and the Durant Motor Company *(part of General Motors)*" all of whom had produced aero-engines for the war effort. He then moved to New Haven, Connecticut, taking up residence in 1923 at 388 Orange Street, New Haven, to become Production Manager of the new Driggs Ordnance Company's car factory at 50 Whitney Avenue.

The Driggs Ordnance & Manufacturing Co of New York was a major arms manufacturer, but their pre-war Driggs-Seabury Ordnance Corp of Sharon, Pennsylvania group subsidiary had offered the 'Vulcan' (27hp 4-cyl; 1913-1914) and 'Twombly' (4-cyl L-head; 1913-1915) light-cars, briefly followed by a 12hp 4-cyl light-car which then became the short-lived 'Sharon'. They then acquired the 'Ritz' cyclecar (4-cyl L-head, V-twin), exporting several to Britain in 1915.

Driggs resumed car production at their new New Haven plant in 1921, and offered, under the slogan "Built with the precision of ordnance", a three-model 104" wheelbase line-up (coupé, saloon or touring) but, being relatively expensive, only around 150 had been built when production ended in 1923 whereupon Driggs turned to a 108" w/b taxicab, powered by a 12hp 1,600cc (66 x 114mm) L-head engine. Adams now left Driggs.

Adams Motor Corporation

The New York Times records on 23rd November 1923 that Adams had been made President of the Advanced Motors Corporation on August 21st 1923 following the enforced resignation of Salvatore Barbarino over his misstatements to fellow directors over the issue of shares in Advanced Motors to a company promoter, not connected with the company, leading to an injunction on their issue on 2nd November; this was now lifted. It also stated that Advanced Motors "had a well equipped plant at New Haven and had brought the Richelieu Motor Co".

How and when he became involved with Advance is not known; he almost certainly knew many of those involved from his war work and recent return to America. The Richelieu Motor Car Company was formed in Ashbury Park, New Jersey in October 1921 by former Duesenberg employees; Deusenberg were most famous for their racing cars and engines but had produced marine and aero-engines in the war, mainly the A44 4-cyl ($4^{3}/_{4}$" x 7") and briefly in 1918, a V16 as well as a Bugatti-King H-16 paired straight-8. Their war surplus A44 engines went to the Rochester Motor Company whose standard Rochester-Duesenberg 4-cyl engines went, among others, to the newly formed ReVere Motor Corp for their luxury cars, built in Logansport, Indiana between 1917-22. ReVere's designer, Newton van Zandt then joined Richelieu but their ambitious luxury car was overpriced and they became bankrupt in early 1923. Their assets were acquired by a consortium of investors led by Salvatore Barbarino (Louis Chevrolet's former racing mechanic and driver) and clearly included Walter Lawson Adams; their new company, Advanced Motors Corporation, was based in Stamford, Connecticut. Plans were now laid for a high quality leRoi 4-cyl powered luxury car but following the legal dispute over shares issue, and his forced resignation as President, Barbarino left to form his own Barbarino Motor Corporation in Port Jefferson, New York in early 1924; it went into receivership in 1925.

Meanwhile, with Adams at the helm and having left Driggs, he formed the Adams Motor Corporation (of which he was President and Treasurer); it is quite possible this was simply the renamed Advanced Motors Corp.

Based at the 128, Hamilton Street works, New Haven (his home and office was now at 100 Clifford Street), he now had ambitious plans for his own 'Adams Air Cooled Six'. The newspaper report continued: "The new car will include a number of innovations entirely unknown to the automobile world, Mr. Adams declares, and he confidently states that when the New Haven built car reaches the market it will make this city a centre of interest in the automobile industry". While nothing is known of his air-cooled engine, innovative the chassis certainly was for he had developed for it a unique springless rubber suspension system.

Rubber shock absorbers

Connecticut was the industrial heartland of America's vulcanised rubber industry. New Haven born Charles Goodyear had developed the technology which, through the United States Rubber Co (better known today as Uniroyal), centered on Nuagatuck, north of New Haven. His technology was brought to Britain by Goodyear's British agent, Stephen Moulton (his grandson, Alex, developed the Mini's Hydrolastic suspension and his Moulton rubber suspension bicycle). After his death, Stephen's business merged with George Spencer's, forming George Spencer Moulton Ltd, specialising in compression type shock absorbing suspension aids to leaf-sprung railway carriages.

Adams doubtless saw the potential in combining the contemporary adjustable hydraulic elbow shock absorber joint (later with friction discs in the Andre type) with the resilience of rubber and in June 1924, applied for a US patent (US 1515716), lodged as a British patent in June 1925 (granted Pat

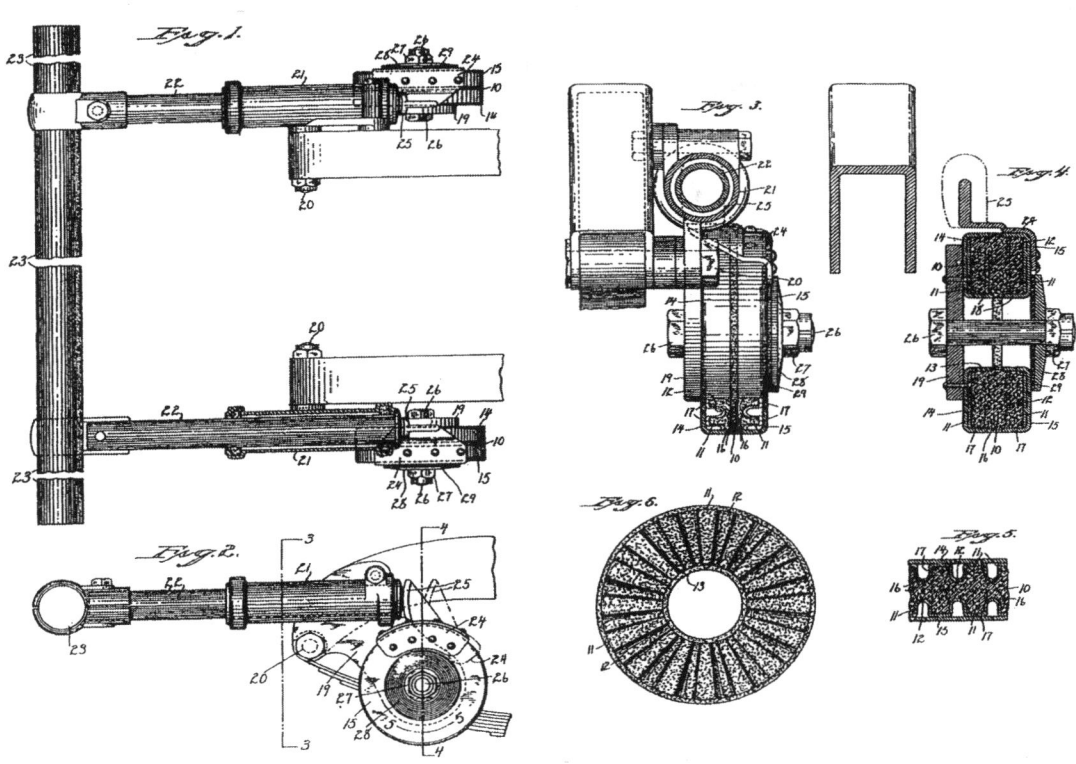

The bumper mount incorporated Adams' torsional rubber suspension system

No: 236243) for a maintenance free, cantilevered, bonded India-rubber torsional suspension arm, dispensing with conventional leaf springs. Although Leyland had developed a steel rod torsion bar springless suspension in 1921, followed by Andre Citroen, Adams' design appears to be the first use of bonded rubber anti-vibration mounts (such as in 'Silentbloc' bushes) as a torsional 'shear resistance' springless and lubrication-free suspension medium. The arm featured a circular, ribbed rubber core sandwiched between two similarly ribbed steel plates, one mounted to the chassis, the other to the suspension arm. He duly formed the Adams Springless Suspension Corporation in 1925 and modified a car chassis with the new system; it was photographed supporting 17 men, including a confident Adams. A second, modified and bonnetted, chassis followed which appears to be a Richelieu rather than the very similar Driggs.

This patent was followed by another for a rubber shock-absorber mounting system for car-bumpers, incorporating the torsional arm device. Applied for in America in September 1925, it likewise was granted a UK patent, Pat No. 258292 in 1926.

He and the family returned to Southampton aboard the RMS *Homeric*, arriving on 12th September, 1925; Ada and the children moved to her sister Alice's home at 22, East Street, Thame, Oxford. Curiously, the ship's manifest shows Walter Adams recorded as of The Bettee Play Co. of 27, Phillipot *(sic)* Lane, London (possibly to produce a child's perambulator?), but there are no records of this company, nor the address, or its variations. Never-the-less, he did establish his Adams Patent Suspension Company Ltd at 5, Fenchurch Street, London EC.1. Among the eight directors were those with interests in the far eastern rubber plantations, such as Sir Edward Rosling.

Adams' suspension system on what appears to be a Richelieu (or early Driggs) chassis with (inset) 17 men standing on a later chassis. Walter Adams is 5th from left (Popular Science)

A Ford car was duly acquired and fitted with Adams' patent suspension system and bumpers. Tested by *The Motor* in late 1925, they gleefully reported that over a very rough pot-holed road, limiting most cars to 10mph, the Adams system allowed 20-30mph in smooth comfort. Being lubrication free and lacking shock absorbers, they confidently predicted early adoption by car makers - the bumper could be fitted retrospectively - but they didn't.

Adams finally returned to London aboard the American Lines SS *American Banker*, docking on 17th October, 1926; his address being shown as Alice's in Thame. He and Ada are then found in the 1927/28 telephone directories at 14, Alexandra Grove, Finchley, but little more was heard of his system or company. It appears he and Ada then separated; Ada returned to Thame while Walter moved to the village of Gamlingay, near Sandy, where he lodged his last patent in December 1931 (Pat No. 396377, granted July 1933) for a self-returning compressed rubber hinged joint, "especially applicable to steering connections of vehicles, boats and aeroplanes", such as a track-rod and steering ball joint which benefited from its built-in damping of road shocks and added self-centering safety in the event of a puncture. He duly formed the Adamite Safe Steering Joint Company at Gamlingay, but like his suspension system, little more was heard.

He continued to offer his services as 'W Lawson Adams, M.S.A.E, Consulting Engineer' but faded quietly into obscurity and is believed to have died in Hammersmith in 1945. Ada died in Oxford in 1968.

After the war, George Spencer Moulton Ltd redeveloped the idea of bonded "rubber torsional spring hinges", for which various patents were sought, in automotive applications. Their Moulton 'Flexitor' units were first extensively tested around 1950 by the army for trailer suspension; the results so impressed Austin Motors that they adopted steel torsion bar suspension on their new Austin Champ military 4x4 field car and gun tractor while the 'Flexitor' rubber suspension (with hydraulic dampers - a vast improvement over the Land Rover's spine jarring car springs, saved only by its coil-sprung seats!) was adopted for their new Austin Gypsy civilian 4x4 field car of 1958. The 'Flexitor' system also appeared on the Bond Minicar, and is today widely found on light trailers.

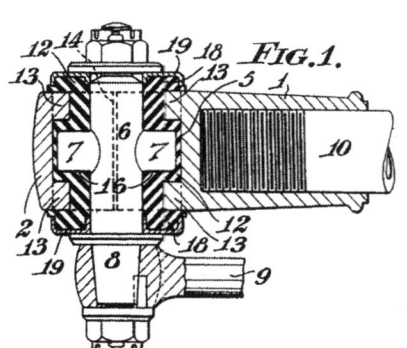

Adams' last patent, his Adamite Safety self-returning bonded rubber ball-joint

Bibliography

The following sources have been used in researching the material for this book.

An account of partnership - Industry, government and the aero engine - M C Neale
Action Stations Nos. 1-10 - various authors - PSL
Aero-engines - G A Burls
AVRO aircraft since 1908 - Roger Jackson
Aircraft of the 1914-1918 War - Owen Thetford & E J Riding
Aircraft piston engines - Herschel Smith
Armstrong Whitworth aircraft since 1913 - Oliver Tapper
The Autocar
The Automobile
Aviation - the creative ideas - Oliver Stewart
The Bean - Jonathan Wood
Bert Houlding: TT Pioneer - Paul Ingham
Blackburn aircraft since 1909 - A J Jackson
Boulton & Paul aircraft - Alec Brew
Bristol Aircraft since 1910 - C H Barnes
British aircraft 1809-1914 - Peter Lewis
British Car Factories from 1896 - Paul Collins & Michael Stratton
British piston aero-engines and their aircraft - Alec Lumsden
British Civil Aircraft since 1919 - A J Jackson
British Flight Testing - Tim Mason
British Military Aircraft serials - Bruce Robertson
British Research and Development aircraft - Ray Sturtivant
Brooklands Gazette
By Jupiter - Bill Gunston
Colonel Cody and the Flying Cathedral - Garry Jenkins
The Classic Motor Cycle
deHaviland Aircraft since 1915 - A J Jackson
Development of piston aero-engines - Bill Gunston
Devoid of Trouble - DOT 1908-1978 - Ted Hardy
English Electric Aircraft and their predecessors - Stephen Ransom and Robert Fairclough
Farnborough - 100 years of British aviation - Peter Cooper
Flying Start - M H Goodall
Flying Units of the RAF - Alan Lake
Fraudsters - Michael Gilbert
Gloster Aircraft since 1917 - Derek James
The Great No.1 Factory - Don Williams
Handley-Page aircraft since 1907 - C H Barnes
Harry Ferguson - Colin Fraser
Hawker - L K Blackmore
Hawker aircraft since 1920 - Francis Mason

History of British Aviation - R Dallas-Brett
A History of Brooklands motor course - William Boddy
Illustrated Encyclopaedia of Motorcycles - Erwin Tragatsch
Janes all the worlds aircraft
Janes historical aircraft 1902-1916
The knife and fork man - Bill Fairney
Landscape by machines - L T C Rolt
Laurie Bond - Nick Wotherspoon
Martlesham Heath - Gordon Kinsey
Motor Cycle
Motorcycling
My Life and my Cars - W O Bentley
Names with Wings - Gordon Wansborough White
New encyclopaedia of Motor-cars - G N Georgano
Noel Macklin - David Thurley
Powered vehicles in the Black Country - Jim Boulton
Pure Luck - the biography of Tom Sopwith - Alan Bramson
The Panther Story - Barry M Jones
The Rotary aero-engine - Andrew Nahum
Saunders and Saro Aircraft since 1917 - Peter London
Shorts aircraft since 1900 - C H Barnes
Sopwith - The man and his aircraft - Bruce Robertson
Story of the British Light Aeroplane - Terence Boughton
Supermarine aircraft since 1914 - C F Andrews and E B Morgan
Trojan - Can you afford to walk? - Eric Rance and Don Williams
Warplanes of the first world war - J M Bruce
Wheels of misfortune - Jonathan Wood
Wilbur and Orville - Fred Howard
Wings over Brooklands - Howard Johnson
World encyclopaedia of aero-engines - Bill Gunston
The National Motor Museum Beaulieu
The Patents Office
The Public Records Office

INDEX

A

ABC: 291
  All British Company Ltd: 15, 17-19, 291, 293
  All British (Engine) Company: 18-19, 25, 52, 56, 101, 174, 268-270
  ABC (Motors) 1920 Ltd: 82, 86, 130, 155, 168-175, 262, 267
  ABC Road Motors: 103-105
  ABC vs contract-built engines: 41, 52, 55, 79, 81, 261
  Société Francaise des Moteurs ABC: 119, 123, 168-169
  4-cyl in-line: 18, 20, 26-27, 35, 85, 108
  5-cyl radial: 59, 166
  5-cyl rotary: 35, 43
  14-cyl radial: 59, 82
  auxiliary engines: 29-32, 35, 49, 78, 84, 103, 105, 123, 266-267, 295
    engine starter: 30
    Firefly engine: 105, 111, 125, 170
  Brooklands works: 25, 28, 31
  carburettors: 19, 42, 51, 53, 167
  cycle-car: 86, 105, 165-176, 180
    cycle-car engine: 167
  Dragonfly: 14, 15, 29, 36, 39, 46-48, 51, 52-59, 61-64, 67-72, 74-83, 107, 117, 149, 170, 177, 232, 264, 266-270, 297
    Dragonfly Mk.II: 62, 73-75, 84-85, 117
  Gadfly: 59-60, 166
  Gnat: 20, 35-41, 48-50, 55, 59, 61, 68, 264
    Gnat Mk.II: 41, 42, 84-85, 122-123, 173
  Hersham Lodge: 29-30, 168, 170-171, 174, 267
  Hersham works: 29-30, 32, 51, 96, 103, 125, 167-168, 171, 174-175, 266-267
    cycle-car factory: 170-171, 175
  Hornet: 39, 91-94, 130, 131
  Mosquito: 39, 47, 49-50, 61, 65-66, 266-267
  motor-cycles: 35, 98-106
    flat twin motor-cycle engine: 20, 38-40, 86-87, 94, 98, 105, 127, 130, 134, 151, 166
  'Revs' telegraphic address: 101
  Robin monoplane: 89
  Scorpion: 39, 85, 87-91
    Scorpion Mk.II: 88-91, 131-132, 151, 176, 244
  Skootamota *see* Skootamota:
  Sopwith-ABC motor-cycle *see* Sopwith-ABC:

  V aero-engines: 18-20, 23, 26-27, 35, 51, 96, 150, 291
    lightweight: 19, 20
  Wasp: 15, 49-57, 59, 65-68, 73-74, 77-78, 80, 82, 264, 267-269
    Wasp Mk.II: 51, 52, 57, 62, 64-65, 67, 77-78, 84-85, 122-123, 266, 270
ABC Motors Ltd: 84-96, 130, 175, 271
  auxiliary engines: 93, 94-96
    Bee V-4: 95-96
  contract engineering work: 95-96
  Super Sports cycle-car: 172, 175-176
AC cars: 117, 140, 125
Adams-Farwell radial engine car: 149, 286
Adams, Walter Lawson: 14-19, 23-25, 28, 36, 62, 285-302
  Ada Plater: 292, 296, 299, 301-302
  Adamite Safe Steering Joint Co: 302
  Adams, W L Ltd: 15, 18, 291
  Adams Motor Corporation: 300
  Adams Patent Suspension Co: 301
  Adams Springless Suspension Co: 301-302
  4-cyl in-line aero-engine: 17
  air-cooled six engine: 300
  auxiliary exhaust ports: 15, 17
  bankruptcy: 285
  Bournemouth: 16, 291
  carburettor: 15
  consulting engineer: 291, 302
  crankshaft air cooling: 15, 17, 24
  cycle-car engine: 291, 292
  New Haven, Connecticut: 299
  patents: 286, 290-291, 296, 298, 300-301
  Redbridge Motor Works: 15, 18, 20, 25, 290-291
  rubber suspension/shock absorber: 300-302
  steering joint: 302
  Susie Adams (née White): 285, 288, 291
  S Adams Ltd: 286
  trans-Atlantic crossings: 295-296, 299, 301
  V-8 aero-engine: 16-17, 19, 24, 62, 130, 291, 296
  V-8 marine engines: 15-17, 24, 291
  voluntary liquidation: 19
  Yewtrees, Redbridge: 15
AD Navyplane: 34, 58
ADC Cirrus aero-engine: 93-94
Admiralty: 36, 58, 83, 95, 100, 146, 151, 193, 199-205, 207, 267, 269, 296
  Air Department: 19, 33-34, 36, 47-48, 83, 294-295
  Admiralty/P&M toroidal engine: 204, 231, 240-241, 254

  Admiralty Rotary AR.1 aero-engine: 34, 48, 83
  3-cyl radial engine: 202, 204
Advance Motor Manufacturing Co: 290
Advanced Motors Corporation: 300
Aerial barrage/mines: 199
Aerial Derby: 67, 71, 75, 117, 118
Aerial Target/Torpedo aircraft: 36-40, 145
  Hewitt-Sperry AT: 38
Aero Engines Ltd: 94
  Sprite: 94
Aeromarine Corp: 297, 299
Aeronautical Inspection Department: 53, 55, 80
The Aeroplane: 87
Aeroplane Engine Company: 14, 19, 291
Aeroplane Experimental Unit *see* Martlesham Heath:
Air Accident Investigation: 9, 80, 118
Air Battalion, Royal Engineers: 33, 61
Air Board: 36, 49, 52-54, 62, 83, 266-267, 296
  aircraft/engine names: 39
  aircraft specifications WW.1: 50, 52, 62-64, 68-69, 73, 75, 77-78, 84
  Ball-bearing Committee: 36, 37
Air Ministry: 36, 80, 83-84, 86, 96, 117, 122, 269
  Joint Aero-engine Committee: 174
Alan Muntz Ltd: 251
Albion motor-cycle gearbox: 104, 105
Allchin, Tommy: 134, 156
Allen, Titch: 101, 145, 202
Allied Iron-Founders Ltd: 188-189
Aluminium alloys: 43, 48-49, 81
  cylinder heads: 49
  pistons: 32, 34, 100, 173, 178, 180
American production methods: 169, 295-296
Anderson free-piston toroidal engine: 252
Antoine motor-cycle: 4
Antoinette aero-engine: 13, 15
  monoplane: 9, 11, 13, 46
Anzani aero-engine: 20, 32, 34, 42, 46, 140, 151, 176
  British Anzani aero-engine: 89
  Astra car: 225
Armistice: 61, 67, 70, 84, 108
Armstrong motor-cycle gearbox: 103
Armstrong-Siddeley *see* Siddeley:
Armstrong Whitworth: 19-20, 23, 62-63, 65, 96
  ABC aero-engines: 19-23, 25, 35, 150
  Ara: 62-64
  Armadillo: 62, 63
  propellers: 150
  Siskin *see* Siddeley:

Army Aircraft Factory, Farnborough: 33, 34, 61
Army Balloon Factory: 7
Army Council: 270
Arnfield, J & E, Ltd: 228, 229
Arrol-Johnston Co: 18, 56
d'Ascanio, Corrandio: 127
Ashby, Count: 134-135, 156
Associated Automatic Machine Corp: 185, 188
Associated British Engineering Group: 151
Audi 5-cylinder engine: 148
Auster B3 target drone: 95
Austin Motor Co; 14, 30, 63-64, 209, 218, 221, 252, 302
   Greyhound: 63, 94
   Mini: 210,
   Seven: 173, 176, 213
   V-12 aero-engine: 30, 295
Austro-Daimler aero-engine: 34, 56, 108, 158
The Autocar: 20, 141, 145, 148, 150, 152, 154, 166, 171, 180, 182-183, 232, 247, 259, 263, 292, 298
The Automotor Journal: 15
Aumeteyer, Alfred: 206, 208, 257
Automobile Association: 124, 236
Auto-Rides Ltd: 186
Autoys Ltd: 211, 214, 230
Avery, W S Ltd: 187
Aviators, pioneer: 7-14
AVRO: 28, 64-65, 163, 170
   A V Roe & Co: 12, 14
     Hamble Works: 28, 64
   AVRO-Canada: 254
   500: 26, 27
   504: 14, 23, 27, 57, 64-65, 89, 295
   523 Pike: 29, 64
   529: 64
   531 Spider: 64
   532: 64
   533 Manchester: 29, 64
   548: 65
   Type D: 47
   Type F: 59
   Type G: 27, 35
   cycle-car: 109

B

Bailey, S L: 35, 99-101, 114
Baldwin, Stanley: 267, 269-270
Bamboo Cycle Company: 285, 287
Barnes, Freddie: 101, 103, 130, 131, 134
Barnwell, Frank: 59, 69, 83, 84
Bartelt, F L: 12, 91
   Ornithopter: 12, 91
Barter, F: 101
Barter, J: 114
Bartlett, Ken: 119, 136
Bath, W W: 95
BBC: 193-195, 256
Bean motor-car: 14, 169, 171, 174, 209
Bean Industries: 174
Beardmore Ltd: 42, 48, 55, 290

ABC engines: 55, 56
Beart-Bradshaw motor-cycle: 136
Beaumont, Monty: 145
   radial engined car: 145
Beck, Frederic: 205-207, 256
Bellamy, Leslie: 145
Belsize Motors Ltd: 56, 131, 174, 177, 183
   ABC engines: 56
   Belsize-Bradshaw car: 86, 171-181, 62
   receivership: 181
Bentley, Horace: 100
Bentley, W O: 34-35, 48, 57, 80, 83-84, 100, 149, 264, 294
Bentley Rotary BR1/2: 34, 45-48, 53-57, 62-65, 68, 71, 74, 83, 100, 149, 294
Berwick, F W Ltd: 55-56
   ABC engines: 55-56
BEW (Brooklands Engineering Works Ltd): 115, 175
Beyersdorf, Fr: 239
BHP see Siddeley
Birfields: 224
Black, Sir John: 25, 174, 219, 221, 252
Blackburn, Robert: 46
Blackburn Aircraft Co: 30, 109, 163, 210
   Beverley: 95
   Mersey monoplane: 47
   Perth: 94
   Sidecar: 42
   Mercury aero-engine: 46
   motor-car: 109
Blackburne motor-cycle engines: 134-135, 138
   Burney-Blackburne: 140
   Burney brothers: 140
   Tomtit aero-engine: 87
Blake Bluetit: 42
Blériot, Louis: 8, 98, 107, 140
   monoplane: 13, 35, 108
Bluebottle boat: 15, 291
BMC: 213, 244
BMW aero-engines: 82
   motorcycle: 109, 119-120, 146-147
Bond, Lawrie: 124, 210, 214, 231
Bond Aircraft & Engineering Co; 210
Bond Cars Ltd: 163-164, 175, 195, 213, 217, 224-225, 231, 236, 242, 245, 256
   875: 213
   Bond Minicar: 124, 207, 209-213, 215-216, 219, 302
   toroidal engine: 214-215, 230
   Bond Minitruck: 213
   Commercial: 212-213
   P1 scooter: 127, 163-164
   Power Ski: 256
   Ranger Boat: 256
   Unicycle power unit: 213
Boultbee, Harold: 93
Boulton-Paul: 68-69, 163
   P7 Bourges: 68-69
   P8 Atlantic: 69
   Phoenix: 89
Bournemouth: 8, 16, 291
Bradshaw, George: 1
   Bradshaw Railway guide: 1-2

Bradshaw, Ewart Gladstone: 2, 3, 5, 127, 164, 189, 209-211, 214, 217, 230-231, 235, 239-240, 245, 272-273
   Bradshaw's Motor House Ltd: 209
Bradshaw, Geoffrey: v, 2, 37, 137, 148, 173, 194, 217, 225, 246, 256, 257, 259, 273-275
   Advanced Auto Aids 'Gripswitch': 275
Bradshaw, Granville Eastwood: 291, 293
   addresses: 276
      Darby House, Sunbury: 29, 99, 169, 183
      Grey Cottage, Isle of Wight: 239, 245
      Jermyn Street offices: 86, 126, 130, 155, 177, 184
      Lowfield Park Manor (Hyde Croft): 183, 189, 246, 276
   ancestors: 1-2, 272
      Bradshaw & Son (optologist): 2, 3, 272
   apprenticeship: 3
      Edinburgh: 3-5,
   autobiography: vi, 183, 239, 257
   bankruptcy: 166, 188-189, 239, 253
      H M Treasury/tax-man: 266-267
   Braddles: 3
   Bradshavian: 147, 263
   ill-health: 240, 243, 245
   Methodism: 3
   OBE and knighthood: 85
   obituary: 3
   parents and siblings: 2, 259
      Pamela: 259, 273
      Peter: 189, 256, 257, 273
      Vivien: 273
   Muriel Mathieson: 198, 273:
   Violet Partridge: 273
Bradshaw designs and theories:
   advertising/amusement machines: 183-187
   aerial barrage/mines: 199
   aeroplane: 9, 18, 19
   air pump: 163, 204
   BBC Any Questions; Tonight: 256
   Bombs etc: 38, 39
   Bumblebee V-twin: 122, 127, 137-139, 196, 220
   captive bolt: 37, 167
   car heater: 225
   car lights: 152, 196
   clutch-stop: 178-179
   company promotions: 183
   constant velocity joint: 224
   consulting engineer: 107, 152, 169, 183, 199, 219, 241, 262
   Granville Bradshaw Ltd: 86, 183
   de-icer mat: 96
   direction indicator: 224
   engines:
      3-cyl radial engine: 145, 149, 202
      acoustic engine damping: 129
      fuel economy device: 215-218, 219, 222, 225
      gas cycle/gas generator: 226, 250
      glass engine block: 129
      nacelles: 82

oil cooled engine *see* oil cooled engine:
    pulsation motor: 154, 226-229, 230, 250
    steam generator engine: 153-154
general arrangement drawings: 55, 261
golf balls: 197
heart monitor: 186
hill-holder brake: 154
invalid carriage: 166
lectures: 120, 146-148
Little Stockbroker machine: 184-185
man powered flight: 97-98
motorcycle:
    500cc V-twin: 160
    £20 motorcycle: 142, 147-
    frames: 103-105, 141, 158
    post-war 'ABC': 142-143, 146
    patents: 3, 48, 82, 97, 103-105, 110, 111, 129, 132, 138, 142, 152, 154, 167, 178, 183-197, 203, 217, 219, 224-225, 229-230, 235, 239-241, 243-244, 246, 250, 253, 270, 277-284
    People's Car *see also* Shay, J E: 196, 214-218, 221-225, 237
    radial engined: 148, 149, 166
prophecies: 141-154
propellers: 150
    mounting: 40, 60
    driven cars: 150-152
    driven motor-cycle: 150
    variable pitch: 9, 97, 203
royalties: 137, 156, 168, 177, 222, 254, 261, 266-271
scenic railway: 185
shaft drive: 111
ship's stanchion: 163, 202
spanner, adjustable: 5, 6
stereo photography/TV: 192-195
suspension:
    leaf spring 103, 110-111, 120, 142, 148, 160
    torsion bar suspension: 142, 148, 301
thief-proof screws: 190-192
toroidal engine *see* toroidal engine:
torpedo propulsion: 202
torque equaliser: 97-98
waste disposal unit: 195, 196, 219, 220
writer:
    book on the slide rule: 259
    Jottings from an inventor's notebook: 257
    Jottings from a designer's notebook: 51, 209, 221, 256
    Mystery of the old manor clock: 183
    Technical problems discussed: 141, 143
    'Designer' nom de plume: 141, 199
    'Technius' nom de plume: 209, 211, 221
Bradshaw Patent Screw Co: 191
Bradshaw Prime Movers Ltd: 205
Brasier et Cie: 83
Brazil-Straker: *see* Cosmos
Bristol Aeroplane Co: 27, 83-84, 96, 119
    Cherub aero-engine: 87, 89-90, 173
    Hercules radial aero-engine: 203, 205
Bristol & Colonial Aircraft Co: 59, 69-70, 83-84
    Babe: 53, 42, 59
    Badger X: 69-70, 78
    Bullet: 84
    F.2B Brisfit: 58, 69, 75
    F.2C Badger: 63, 69
    Jupiter aero-engine *see* Cosmos:
    Scout: 83-84
    liquidation: 84, 115
British Aerial Transport Co: 42, 49, 61-62, 65-68, 267
    FK.22 Bat: 49-50, 65
    FK.22/2 Bantam: 50-51, 62, 66, 73
    FK.23 Bantam: 61, 66-68
    Sport: 67
    FK.24 Baboon: 66, 68
    FK.25 Basilisk: 66-68
    FK.26 Commercial: 68
    FK.28 Crow: 42, 68
British Aerospace: 116
British Aircraft Corporation:
    BAC 1-11: 96
    Concorde: 96
    TSR-2: 96
British Automatics Co: 184, 188
British Caudron Co: 71
British motor-cycle industry collapse: 143
British Motor Trading Corporation: 170, 182
British Photomaton Trading Co: 189
British Radial motor-cycle: 144
Briton motor-car: 10, 14
BRM racing car: 229, 237, 262
Brockhouse engine: 212
Bromega Ltd: 235-246, 253
    Omega engine *see* toroidal engine:
Brooklands: 8, 9, 12-14, 18, 23, 26, 28, 31, 34, 47, 73, 74-75, 89, 99, 101, 104, 107, 116, 150-151, 155, 166, 173, 259, 291
    Blue Bird café: 14, 28, 99, 173
    Gazette: v, 257
    propeller testing: 18
    motor racing: 13, 16, 99, 117
    RFC Air Acceptance Park: 28, 33, 104
    RFC Experimental: 37
    Shed No.11: 12, 14, 18, 25, 31
    test hill: 112, 150
    TT: 102
Brotherhood, Peter Ltd: 202
Brough, George: 101, 145
    Brough Superior: 145
Brown, R: 40, 122-123
Bruce Peebles Ltd: 4
Brush Motors Ltd: 90, 254
BSA Ltd: 62, 214, 249
    Wankel: 208
    Winged wheel: 122
BT-H: 252
Bulman, Maj. George: vi, 53, 80, 264
Bulman, Paul 'George' W S: vi, 67, 264
Burgess-Wright biplane: 27, 28, 107, 108

C

Camm, Sydney: 118
Campbell, Eric: 170, 200
Camper Nicholson Ltd: 42
Campion, J: 173
Campling, Gilbert: 40, 51, 86, 122, 123, 125-126, 130-132, 155, 168, 177, 183, 262
    Campling, Gilbert Ltd: 40, 123, 125-126
    liquidation: 126
Carbon: 127, 263
Caspar C.17: 88
Castrol oil: 44, 106
Cave-Brown-Cave, H M: 28
Celerity valve springs: 114
Central Aircraft Co: 68, 89
    Centaur: 89
    Coupe: 89
Charteris, Ronald: 13-14, 16-20, 26-28, 35, 51, 86, 113, 131, 150, 169, 174, 267-268, 291
    Lady Louisa (née Keppel): 16, 86
    Louisa Eileen (née Knox): 13, 86
Churchill, Winston: 36, 85, 148, 199, 269
    'leave room for criticism': 148
Cierva, Juan: 91, 97
Cissac, Henri: 4
Civilian Aircraft Company: 93
    CAC Coupe: 93
Civil Aerial Transport Committee: 108
Classic Motor Cycle: 263
Clerget aero-engine: 34, 44-45, 48, 53-54, 64, 75, 83, 294
Clyno Engineering Ltd: 56
    ABC aero-engines: 56, 74, 75
Cody, 'Colonel' Sam: 7-9, 18, 107, 257
Coin Operated Machines Ltd: 185, 188
Collier Bros (Matchless motor-cycle): 101
Comper Swift: 89
Compton & Herman body: 170, 173, 175
Continental Motors aero-engine: 94
Cooper, Fred: 203, 240
Corporation & General Securities: 188, 189
Cosmos Engineering Co: 39, 70, 84, 265
    Brazil-Straker: 52, 79, 83-84
    Mercury aero-engine: 39, 52, 54-55, 79, 81, 84
    Jupiter aero-engine: 39, 53-54, 57, 69-71, 78, 80-81, 83, 84, 117-119
    liquidation: 84
Count Zborowski: 117, 166
    Chitty racing car: 117, 166
Coventry-Climax engine: 219, 255
Coventry-Simplex engine: 182, 200
Craig, Joe: 120, 147
Crang, Dr Terence: 242
Crawford monoplane: 94
Crossley Motors Ltd: 34, 56, 109
    ABC aero-engines: 56, 77
    radial motor-car engine: 144
Curtiss, Glen: 57, 294
    Glenn Curtiss Aeroplane Co: 35, 38, 57, 59, 294-295, 297
    Curtiss OX-5 aero-engine: 52, 83, 295

Curtiss Wright company: 59, 91, 254
CW.1: 91
Wankel: 247, 249
Cycle-cars: 176
Cyclogyro: 98
Cycle-motors *see* Scooter:

D

The Daily Mail: 7-8, 14, 18, 86, 107, 237
   Harmsworth, Alfred (Lord Northcliffe): 7, 18, 69
   Circuit of Britain: 9, 14, 18, 295
   Gliding competition: 86
   Light aeroplane competition: 86-90
Daimler-Benz: 34, 247-249, 256, 294
Daimler Motors: 253
The Daily Sketch: 237
The Daily Telegraph: 185, 236-237
Darracq 4-cyl aero-engine: 91
David Brown Ltd: 163, 218, 290-291, 293
Davies, Sammy: 168
Dawson, Ben: 185, 187-188
Day, Joseph: 206
Day-Leeds motor-car: 291-292
deHavilland, Geoffrey: 33, 61-62, 91, 140
   motor-cycle engine: 140
deHavilland Ltd: 61-62, 95, 96
   Airco Aircraft Manufacturing Co: 38, 61-62
   AT: 38
   DH.4: 58
   DH.9: 14, 41, 62, 170
   DH.10: 61
   DH.11 Oxford: 61-62
   DH.12: 62
   DH.20: 62
   DH.53 Hummingbird: 87
   Gypsy aero-engine: 94
deHavilland Propellers Ltd: 95
Delaney, Laurence: 186
Dennis, Thomas A: 86, 91, 96, 169, 174-176, 267, 271
Dennis, Rene: 96
Depth-O-Graph Ltd: 192
Depurdussin monoplane: 13
   Depurdussin Syndicate: 65
   Flying School: 34
DFP Doriot, Flandrin et Parant motor-car: 100
DFR motor-cycle: 135
Dick, Kerr & Co: 3, 35, 86, 210
   Dick, W B: 3, 106
Diesel, Rudolph: vi, 261
Diesel engines: 137, 148
Disposal & Liquidation Board: 269, 270
Dodgem cars: 186
Dorman, Alfred: 287
Dorman, Thomas: 285
Dorman Engineering (DEC) Co: 286-290
   Whirlwind motor-cycle: 287-289
Dorman, W H & Co: 18-19, 24, 130-132, 137, 289, 291
   500cc motor-cycle engine: 136
   Adams V-8 aero-engine: 19, 23, 291

Bradshaw oil-cooled engine: 130, 132, 135
   Sports: 132-133
diesel engine: 137
Dorman-Bradshaw Hornet aero-engine: 130, 131
KNO twin-cam motor-car engine: 24-25, 170, 293
DOT-Bradshaw motor-cycle: 134-136
Douglas company: 32, 100-101, 114
   aero-engine: 87-88, 94
   motor-cycle: 99, 100
   Vespa scooter: 127, 147
Dover, Horace: 287
Dover Ltd: 287, 290
Downing, Spencer: 287-288, 290
Dowty hydraulics: 163
Dowty Marine Ltd: 240, 242
Driggs Ordnance motor-car: 299, 301-302
Dudley, Leslie: 193-194
Dunstall Park aerodrome: 11-12
Dutheil-Chambers 4-cyl aero-engine: 91
Dynamometer: 58, 112, 133, 165

E

Eastwood Automatic Co: 186
Eaton Manufacturing Corp: 254
Edmund motor-cycle: 101
Edwards, Courtney: 237, 274
Electric car: 126, 166
Electric Construction Corporation: 6, 8-9
Electric Railway and Tramcar Carriage Company: 3
Elliott, R B C: 88, 244
Emerson, Jack: 101-102, 113-114, 134, 224, 229
EMI: 194
Enfield Cycle Co; 126, 288
   electric car: 126
   motor-cycle: 161
Engines, aero:
   50-hour endurance test: 59, 78, 83, 265
   air brake: 228
   bearings: 36
   carburettor icing: 46, 73-74, 81
   cooling: 36, 48, 81, 82
   cylinders, machine turned: 4, 44, 48, 82
   copper jackets: 12, 15, 20, 48
   copper plating: 20, 48-49, 80-82, 128
   cylinder, heads alloy: 49, 81
   Poultice: 44
   machine-gun interrupter:
      Constantinescu: 24, 78, 123
      Kauper-Sopwith: 123
      Scarff-Dibovsky: 24
   nacelles: 82
   power to weight ratio: 54
   radial aero-engines: 43, 46-48
   rotary aero-engines: 43-46
   service life: 43, 45, 48, 55, 82, 83
   simplicity of manufacture: 53
   synchronous torsional vibration: 69, 80, 82-83, 264, 297
Engines, motor-car:

horsepower: 51, 54, 165, 209
   RAC rating: 165-166, 209
   taxation: 165
laminated leaf valve spring: 157, 158
obturator rings: 45
spiral induction port: 156
ENV aero-engine: 18, 26-28, 34, 107, 173
English Electric Company: 3, 27, 70, 86-87, 137, 174
   P.10: 70
   Wren: 70, 86-88
English Electric Manufacturing Co: 3,
Ernest Theodore White & Co: 125, 130
Esselbé, Bauart: 206, 256
Excess Profits Tax: 84, 115, 170, 268
   Finance No.2 Act, 1915: 115

F

Fairey Swordfish: 82, 163
Fairmile Marine: 163, 200-205, 230
Fairy Fée motor-cycle: 114
Falahee, Jack: 100, 168
Farman, Henri: 8
   biplane: 10, 38, 61, 90
   Moustique: 42
Faulkner, William: 29, 168
Fauvel, Charles: 91
Fedden, Roy: 79, 81, 83-84, 149, 265
Fenton, Graham: 119
Ferguson, Joe: 218
Ferguson J B Ltd: 56,
   ABC aero-engine: 56
Ferguson, Harry: 56, 216, 218-219, 237
   tractor: 218-219
   World Car: 218
Fiat Motors Ltd: 32, 266, 268
Fitchel ünd Sachs: 208, 247, 249
Flanders, Howard: 26
   B2 biplane: 26, 27, 35, 47
Fletcher, A A: 89
Fexitor suspension *see* Moulton:
Flight: 11-12, 19-20, 24, 39, 96, 98, 141, 183, 296
Folland, Harry Philip: 33, 37, 70, 71, 118
Folland Aircraft Ltd: 71
   British Marine Aircraft Co: 71
Ford, Henry: 117, 169, 209, 218
Ford Motor Co: 14, 30, 142, 149, 170, 182, 209, 215, 218, 221, 244, 249
   Rolls-Royce Merlin: 79
   Consul: 166, 215, 218, 219
   Prefect: 254
   V-8: 166, 182, 207, 242
   Zephyr: 218, 274
Franklin aero-engine: 94
Free-piston engines: 205-207, 234, 238, 250-251, 254
Free Piston Engine Co: 251

G

Galloway Engineering Ltd: 56, 62
    Adriatic aero-engine: 62
Gardner diesel engine: 137
Gas turbine engines: 207, 226, 230, 250, 252, 254
GB (Nottingham) Ltd: 145
GB radial engined motor-cycle: 144
General Amusement Co: 186
General Electric (USA): 252, 295
General Electric Co (UK): 174,
General Machinery Corp: 205
General Motors: 205, 209-210, 238, 250, 252, 295
    Vauxhall/Bedford: 209-210
Gialdini, John: 188, 189
Gibb, Matthew: 235, 239-240, 244-245
Gibson, Prof A H: 49, 58, 81, 128
Gillett Stephen & Co: 138-140
Gilmour, Graham: 11, 140
Glass-fibre: 164, 213, 219, 223
Gloucester Aircraft Co (Gloster): 71, 118
    Gladiator: 71, 82
    Grebe: 71, 72
    Mars Nightjar/Nighthawk see Nieuport:
Gnôme rotary aero-engine: 9, 18, 27, 34, 38, 43-45, 47, 50-51, 79, 83, 108, 122-123, 158, 294
    Monosoupape: 34, 45, 50, 66-67, 106
    Walthamstow factory: 61, 294
Gnôme-Castrol oil: 106
Gnôme et Rhône: 45, 66, 84, 119, 136, 268
    ABC motor-cycle: 119
Gordon England, Doris: 173
Gordon England, Eric: 173
Gosport Aviation Co: 42, 51
    Shrimp: 42
Gossamer Condor: 98
    Albatros: 98
Gow, David: 159, 161
Grahame-White, Claude: 11
Grahame-White Aeroplane Co: 65, 170
    Buckboard: 170-171
Granville, Cmmdr W: 183
Gray, Col Reg: 137, 164, 205, 210-211, 213, 216, 229, 235-237, 239, 242, 244-245, 252, 254, 256, 261, 272
Grey, Christopher (Baron Glenconner): 194
Green, Gustavus: 72
    Green aero-engine: 18, 27-28, 34-35, 48, 94
Green, Major Frank: 58, 72
Griggs motor-cycle: 126
GSD motor-cycle: 115
GT (Geoffrey Taylor): 175
Guest Keen Nettlefolds (GKN): 174, 191-192
    Bradshaw tamper-proof screw: 191
    Pozidrive: 192
Guy, Syndey: 14-15
Guy Motors: 14-15, 56, 84, 174, 251, 252
    ABC aero-engine: 56
Gwynne's Ltd: 34, 48, 57, 83, 100

H

Hackbridge Electrical Construction Co: 174
Hadfield Steel Ltd: 157, 174
Haddon, Walter: 132
Haffner, Raoul: 91
Halford, Frank: 23
Handley-Page Ltd: 72, 170
    V/1500: 81
    HP-23 Sayers: 87, 88
Harley-Davidson motor-cycle: 120
Harper-Bean, John: 15, 168, 171-172
Harper-Bean Ltd: 59, 86, 131, 168-171, 174, 262, 271
    Bean Industries: 174
    liquidation: 174
Harries, Betty: 199
Hart, Norman: 201
Hart Engine Co: 47, 49, 268
Hatry, Clarence: 3, 166, 187-189, 240, 262
Hawker, Harry: 9, 27, 38, 47, 53, 71, 75, 80, 107-108, 112, 116-118, 264
Hawker Aircraft Co: 118
Hawker Engineering Co: 14, 116-119
    Cygnet aero-engine: 88, 89
    motor-cycle: 117
Hawker-Siddeley Ltd: 96, 254
Heath Parasol: 89
Helicopter/Helithopter: 91
    Revoplane: 91
    Rotachute gyroglider: 91
    Rotabuggy gyroglider: 91
Henderson-Glenny Gadfly: 89
Hendon aerodrome: 33, 35, 61, 67-68, 95, 117-118, 170
Hendy Hobo: 89
Henriques, Sir Philip: 268, 270
Heron, Samuel D: 49, 58-59, 81
Hey, Ben: 156
Hillman Motor Co: 174, 213
Hispano-Suiza aero-engine: 34, 48-49, 55, 57-58, 75, 77, 79, 81, 83
Hobourn Aero Components Ltd: 241, 242
Hoffman bearings: 271
Holt-Thomas, George 61: 294
Hooker, Peter Ltd: 294
Houlding, Bert: 130-131, 133, 181
Hounsfield, Leslie: 101, 116
Howard-Wright biplane: 26, 107, 108
Hucks, Bentfield K: 30
    Hucks starter: 30
Huff-Daland HD-4: 52
Humber Ltd: 34, 48, 56, 83, 155
    ABC aero-engine: 56
Humphries, Ernest: 101
Hunslet Engine Co: 46
Hyde-Beadle, F P: 51-52
Hydroplane hull: 16

I

Imperia motor-cycle: 134
Inderwick, Harold: 220
Indian motor-cycle: 120
Inglis, R S, rocker gear: 114, 175

Institute of Automobile Engineers: 128
Institute of Motorcycle Engineers: 143
Isaacson, Rupert: 34, 46-47, 268
Isaacson aero-engine: 18, 34, 46-47
Issigonis, Alex: 178, 210, 221, 224, 236, 300
    Mini: 178, 224, 236, 300
Isle of Wight Marine Engine Co: 241
ITA/ITV: 194-195
Ives, Frederick: 192, 193
Ixion (Canon Basil Davies): 104, 109, 142, 146, 262-263

J

Jaguar Car Co: 218, 224, 229, 244. 252
JAP Ltd: 13, 140, 146, 214
    JAP-Avroplane Company: 13
    motor-cycle engine: 134, 210
    V-twin motor-cycle engine: 5, 9, 151
    V-aero engines: 13, 34
Jarvis of Wimbledon: 117, 175
Jones, George: 130-133
Junkers aero-engine: 163

K

Katzmeyer effect: 97, 98
Kenworthy, John: 63, 94
King, Francesca: 194, 211, 222, 224, 235, 239, 244-245, 256, 273
Kingswell monoplane: 87
Knox V-12: 296-298
Kodak: 193
Koolhoven, Frederick: 62, 65, 67-68, 267
Kremer, Henry: 98
    man-powered flight prize: 98

L

Lanchester, Frederick: 11, 162, 259
Latimer-Needham, C H: 90
Lawrence, Charles: 58
    radial aero-engine: 42, 58-59
Laxtonia Engineering Co: 297-299
    V-12 engine: 297-299
    V-24 engine: 298
Lea Francis cycle-car: 129, 178, 182
Leach, Frank: 156-157, 159, 164
Legh, Peter: 67-68, 80, 264
Lemery, Douglas James: 139
    chain-saw: 139, 235
Leyat, Marcel: 151
Leyat Hélica: 151
Leyland Motors: 84, 116, 174, 213, 252, 256, 301
    Leyland-Trojan: 116
Light Car & Cyclecar: 293
Lisle, Edward (snr): 10-12
    Edward (jnr): 10,
    Joe: 6, 9, 12, 14-15
Locke-King, Hugh: 13, 28
Longman, Frank: 135
Louis, Harry: 225-256
Low, Dr Archibald: 37-39, 187
    radio control aircraft: 37

Lowin, Rex: 240
Loxham Garages: 133, 209, 235
Lubbock Committee: 269
Lucas, Ralph: 163
    Valveless car: 163
Lucas, Joseph Ltd: 224, 256
Luton Minor: 89-90, 243
Lycoming aero-engine: 241
Lyon & Wrench Ltd: 30, 32, 266

M

McCook Field, USA: 41, 59, 75
Maachi 16: 42
de Marcay 'Passe-par-tout': 42
Macklin, Noel: 200, 204
    Silver Hawk Motors: 200
Macmillan, Norman: 80, 264
Maitland, Andrew: 151
Man-powered flight: 97-98
Manning Wardle & Co: 46
Manning W O: 86
Manufacturers' Union: 160, 181
Marconi airborne transmitter: 31
Marians, Bertram: 146, 155
Martin-Jap motor-cycle: 100
Martin, Glenn: 57-58
Martin, James V: 41-42
    Martin K.III Kitten: 41-42
Martin and Handasyde: 14, 89, 102, 118
    Martinsyde fighter: 50, 77
Martlesham Heath: 34, 41, 53, 61, 64, 66-75, 77-81, 86, 89
    Aeroplane Experimental Unit: 34, 61
    Boscombe Down: 61
Masefield, Peter: 85, 254
Matador motor-cycle: 130-131, 133
    Matador-Bradshaw: 133, 135-136
Matchless motor-cycle: 101
Mather & Platt Ltd: 56-57
    ABC aero-engine: 56-57
Maudsley Motors Ltd: 56
    ABC aero-engine: 56
Maxim, Hiram: 11, 96, 259
Mazda (Wankel) rotary: 248-249
MCC trials: 9, 113, 117, 168, 171
Meadows, Henry Ltd: 133, 182, 200, 210, 251-252
Mendip motor-car/engine: 24-25, 291-292
Michelin Cup air-race: 18, 26-28, 47, 108
Midlands Aero-Club: 11
Midget Motor Ride Ltd: 186
Mignet 'Flying Flea': 89
Miles, F G: 94
    Miles Aircraft Co: 94
Military aircraft trials: 17, 23, 26-27, 35, 47
    Alexander Prize: 17-18
Minerva motor-car/engine: 150, 286, 288, 293
Mineur motorcycle: 132
Ministry of Munitions: 36, 46, 49, 51, 66, 85, 266-267, 269
    Aircraft Production: 268

Ministry of Supply: 36, 210
Moore, Richard: 155, 160, 204
Moore, Walter: 114
Moore-Brabazon, Charles: 8, 28
Morris Motors Ltd: 14, 142
Mortimer, Joe: 157-158, 161-162
Moto-Guzzi motor-cycle: 120, 143, 160
The Motor: 3, 180, 298
Motor Union Insurance Co: 167, 170
MotorCycling: 120, 127, 139, 141-142, 147, 183, 236
The Motor Cycle: 32, 51, 104, 109, 113, 123, 141, 146, 221, 225, 262
Motor-cyles, feet-forward: 141, 147
Motor Sport: 130, 181
Moulton suspension: 300
    Flexitor: 212, 300
Moveo: 133
Moveo 6-cyl car: 133
Mundy, Harry: 232, 247
Munitions of War Act: 36, 108, 114-115
Murphy, Frank: 62
Musgrave, H P: 108

N

Nagler, Bruno: 91
Napier: 4, 163, 252
    Lion W-12 aero-engine: 64, 68-69, 79, 82, 148, 200
    Napier Culverine aero-engine: 163
    Nomad aero-engine: 252
    Napier-Railton racer: 148, 200
National Factory Scheme: 108
National Flying Services Ltd: 89
National Physical Laboratory: 43, 151
National Research Development Corporation: 235, 239
Navarro, J G: 90
    Navarro Aircraft Co: 90
        Burton Aircraft & Manuf. Co: 90
    Navarro Aviation Co: 90
    Navarro Safety Aircraft Co: 90-91
        Navarro Chief tri-motor: 89, 91
    Navarro-Wellesley Ltd: 90
        Wellesley-Brown Aircraft Co: 90
New Haven Evening Register: 286, 294
New Service Organisation for Motorists: 167
Nieuport & General Aircraft Co: 55, 70-72, 109
    Gloster Mars: 71
    Goshawk: 71, 117-118
    London: 72
    Nieuhawk: 70, 84, 117
    Nighthawk: 62, 71-72, 83
    Nightjar: 71
Nightingale, Florence: 198
Noel, Ernest: 172, 267, 271
Noel le Parmentier 'Wee Mite': 88-89
North British Locomotive Co Ltd: 56, 57
    ABC aero-engine: 46
    Ruston Proctor Ltd: 46, 57
Norton motorcycles: 101, 114, 120, 147, 214

Norton-Villiers-Triumph Wankel rotary: 208, 249
NSU motor-cycle/motor-car: 247-248

O

Official Secrets Act: 76, 117
Ogilvie, Alec: 27
Oil-cooled engines: 86, 88, 128-139
    aero-engine: 130-132
    single cylinder: 128-132
    horizontal twin: 129-130, 132, 177, 182
    motor-cycle users list: 133-134
    V-twin: 129, 176-180
Oil lubrication: 105
    cooling: 36, 128, 182, 203
    castor oil: 36, 105-106
OK Supreme motor-cycle: 101, 132, 135
Olympia shows: 8, 11-12, 17-18, 113, 158, 176, 180
Omega toroidal engine: *see* Toroidal engine:
Ord-Hume, Arthur: 90, 98, 241, 243, 245
Ornithopter: 12, 91, 97-98
Orr-Ewing, Hugh Eric: 200

P

Packard/Hall Scott engine: 79
    Liberty V-12 aero-engine: 41, 55, 58, 64, 297
    marine engines: 201-203, 205, 230
Parke, Lt Wilfred RN: 9, 35
Patents Office: 183, 259, 260-262
Pathé News: 236
Pemberton-Billing, Noel: 14, 90
Perkins engines: 137, 248, 252
dePescara, Raoul: 205, 250-251, 253-254
Perry-Bean motor-car: 169
Perry-Vale motor-cycle: 101, 134
Petters Ltd: 77, 244, 254, 255
Petrol: 102, 105
    Benzole: 102, 105
    Petroil: 106
    pinking: 102, 105, 216
    Pratt's: 102
Peugeot motor-cycle: 4, 5, 134
Peyret-Mauboussin PM-10: 91
Phelon, Joah: 155
Phelon & Moore: 55, 79, 119-120, 134-135, 146, 155-157, 163-164, 202, 204, 207, 230, 245,
    Bi-flex spring frame: 160-161
    swinging arm suspension: 161
    Bradshaw in-line twin: 142, 161-163
    Panther ohv: 146, 156, 263
    Panthette: 120, 141, 146, 157-160
    Panther-Villiers: 159
    Phelon & Rayner: 138
    P&M 3^1/$_2$hp: 155
    Princess scooter: 164
    Red Panther: 142, 159, 160
Phoenix Dynamo Co: 3, 35, 86
Phoenix Aircraft Co: 90, 241
Photobooths: 185, 187

Photomaton photobooth: 187-188
Piaggio S A & Cie: 127
    Vespa scooter: 127
Phipps, Pickering P: 285, 287
Picture Post: 195
Pixton, Howard: 23
Pobjoy radial engine: 89
Pomeroy, Laurence: 120, 143, 209
Porte, Lt John, RN: 34, 51
    Porte Baby aeroplane: 34
Power Specialities Ltd: 220-221
Pratt & Whitney Wasp aero-engine: 59
Precision Motor Company: 288, 293
    motor-cycle engine: 15, 286, 288-289
    motor-cycle: 286, 289
'Precision' motor-cycles: 290
Prestwich, John Alfred: 13, 187
Pride & Clarke: 160
Public Records Office: 266
PV.7 Eastchurch Kitten: 40-41, 59
PV.8 Grain Kitten: 40-41, 55, 59

Q, R

Radley-England: 9
Raleigh Safety Seven: 176, 213
Rank Organisation: 194
Ransome, Simms & Jeffries Ltd: 56-57
    ABC aero-engine: 56-57
Raynham, Fred: 23, 27-28, 89, 118
Read, Herbert: 126
Records:
    air: 28, 53, 71, 107-108, 118
    land speed: 148, 151, 200
    motor-cycle speed: 100-101, 113
Redifussion: 126
    Radio Furniture & Fittings Ltd: 126
Redrup, Charles: 43, 144-145
    aero-engine: 43
    motor-cycle engine: 144-145
Regent Carriage Co: 170
Reliant Motors Ltd: 141, 174-176, 213-215
Renault V-8 aero-engine: 14, 51, 65
Rentschler, Fred: 58-59
le Rhône aero-engine: 34, 44-45, 50, 54
Ricardo, Harry: 23
    Ricardo-Halford-Armstrong V-12: 21, 23
Richelieu Motor Co: 300-302
Ridley, Lord: 94
Robinson Redwing: 93-94
Rolls, Charles: 8, 11, 33, 257
Rolls-Royce Ltd: 8, 39, 59, 83, 163, 170, 243-244, 253, 267
    aero-engine: 34, 64
    Packard/Ford built: 79
    Eagle V-12: 39, 55, 68, 81
    Falcon: 39, 79
    Merlin: 79, 252
    motor-cars: 85, 166, 182-183, 209
Rolt, L T C: 154, 180, 261
Rolt, Major Tony: 218-219
Roots blower: 163, 204
Rover Car Co: 171, 174, 177
    8hp cycle-car: 171, 176

gas turbine: 207, 234, 250, 252, 256
Rovin-Bradshaw motorcycle: 135
Royal Aero-Club: 8, 9, 11, 13, 33, 89
    Aero-engine Manufacturer's sub-committee: 19
    Certificates: 8, 11, 14, 16, 107
    Eastchurch: 33, 40
    Leysdown: 33
Royal Aeronautical Society: 34, 39, 85, 105, 140
Royal Aircraft Establishment, Farnborough: 51, 54-55, 58, 61, 64-65, 67, 71, 74-75, 77-78, 80, 83-84, 86, 91, 243-244, 264
Royal Aircraft Factory, Farnborough: 26, 37, 49, 51, 53, 58, 61, 65, 72, 140
    RAF No.8 aero-engine: 52, 58, 72
    AT: 36-37
    BE.2 fighter: 23, 31, 33, 61, 63
    SE.4: 70
    SE.5/5a: 37, 61, 70, 83
Royal Air Force: 34, 51, 58, 61, 63-64, 83-84, 87, 95, 108, 116, 163, 170
    Museum: 12, 95, 170, 264
    Uxbridge: 78
Royal Automobile Club: 33, 155, 157, 184
Royal Commission on Awards to Inventors: 84, 201, 266-271
Royal Flying Corps: 8, 28, 33, 36-37, 56, 104, 108, 155, 163
    Brooklands: 31
    Central Flying School: 27, 61, 65, 108
    Naval Wing: 33-34
    Special Reserve: 28
    Upavon: 34, 38, 61, 108
Royal Navy: 163, 199, 201, 203
Royal Naval Air Service: 28, 34, 64, 71, 170, 291, 294
Royal Naval Air Stations: 33-34
    Calshot: 8, 9, 33
    Eastchurch: 32-33, 41
    Felixstowe: 33-35
    Grain: 33, 35, 40-41, 78
    Netheravon: 38
    Orfordness: 33, 41, 67, 77
    Port Victoria: 40
Royal Naval Volunteer Reserve: 19, 28, 200
RWD-1 sports aeroplane: 91

S

St. Thomas' Hospital, London: 198, 273
Salmet, Henri: 20, 21
Salmon, Peter: 88
    monoplane: 88
Salmson aero-engine: 44, 46, 69
Sanchez-Besa Multiplane: 91
Sangster, Jack: 171
Santos-Dumont Demoiselle: 42
Sarkis, Joe: 135, 157
Sartoris, George: 96
Sassoon, Ellis Victor: 18
    Birdling aeroplane: 18
Saunders, Samuel: 15, 16
Saunders, S E Ltd: 35, 42, 51, 89, 109, 126

Kittewake: 51, 52, 55
Saunders-Roe Ltd: 94, 109, 244
    London: 94
Sawyers, W H: 87
Schleicher, Rudolph: 120
Science Museum, London: 85, 119, 127, 264
Scooters: 121-127, 143, 164
    d'Ascanio, Corrandio: 127
    Autoglider/Auto-ped scooter: 126
    Gloster Unibus: 126, 127
    Vespa: 127, 164
    cycle-motors: 121-122
Scott, David: 241
Selsdon Aero & Engineering Ltd: 51, 84, 122, 125-126, 266
Selsdon Engineering Ltd: 40, 85, 125-126, 130, 266
    Brighton Road: 122
    Southern Counties Garage: 123
    Camco Works: 122, 125
    Hooper W-24 aero-engine: 123
Shadow factories: 55-57, 268
Sharp, Paul: 210
Sharps Commercials: 164, 209-210, 231, 239, 245
Shay, J E Ltd: 139, 196, 219-225
    Bradshaw Peoples' car: 196, 218, 221-225, 237
    3-wheeler: 223
    sports car: 223 224
    500cc in-line twin: 214-216, 218, 221-222, 224, 231
    Lansing-Bagnall Ltd: 196, 220, 225
    Rotoscythe: 139, 196, 220, 239
    Rotocultivator: 223
    Rotogardener: 220
    BUX engine: 220
Sheffield-Henderon-Bradshaw: 135
Sheffield Simplex Ltd: 55-57
    ABC aero-engine: 55-57, 74
Shoreham aerodrome: 47, 94
Short Brothers: 35
    Satellite: 89-90
    Scapa: 94
    Singapore: 94
    Sunderland: 95
Shuttleworth Collection: 87
Siddeley, John: 59, 109
    Armstrong-Siddeley: 59, 84, 118, 183
    BHP-Siddeley Puma: 42, 55, 58, 61-62, 64, 69, 72, 75, 264
    Genet aero-engine: 93
    Jaguar aero-engine: 52, 54, 57-59, 71, 72, 80, 83
    Lynx aero-engine: 39, 54, 58, 67, 78
    Siddeley Deasey: 52, 58, 63, 72-73, 109
    Sinaia: 79
    SR.2 Siskin: 53, 58, 63, 72
    Tiger: 79, 264
Siemens Bros : 3
Siemens-Schuckert: 37, 46
    Albatros/D.IV/D.V Scout: 46, 50, 53
    Siemens und Halske aero-engine: 46, 53
Sigrist, Fred: 27, 107, 108, 116

Simplex Automobile Co (USA): 57-58
Skootamota: 121-127, 211
Smiths Static radial aero-engine: 58
SNECMA: 119
Society of British Aircraft Constructors: 113, 267
Sopwith, T O M: 3, 28, 38, 86, 107-108, 114, 116, 118-119, 254
Sopwith Aviation Co: 23, 57, 73-77, 84, 107-116, 125, 252
    AT: 36-38
    Atlantic: 108
    B1: 35
    Bulldog: 53, 73, 75
    Camel: 68, 73, 116
    Cobham: 72, 75
    Dove: 108, 113
    Dragon: 74, 75
    Gnu: 108
    Pup: 73, 108
    Salamander: 116
    Schneider: 117
    8F.1 Snail: 50, 67, 73-74
    Snapper: 76, 117
    Snark: 53, 72, 76
    Snipe: 65, 70, 73-75, 108
    Sopwith-Burgess: 108
    Sopwith-Farman: 108
    Sparrow: 38, 40
    Strutter: 77
    Albemarle Street showrooms: 113, 125, 130
    Canbury Park, Kingston: 108, 115, 116
    Ham Common Works, Richmond: 108, 116
    South Molton Street showrooms: 113
    voluntary liquidation: 84, 116
Sopwith-ABC motorcycle: 86, 109-116, 136, 158, 175, 263
    Designed in 11 days: 109
Southern Martlet: 93-94
SPAD S.7: 41, 58
Standard Motor Co: 174, 219, 221, 252, 256
    Standard-Triumph Car Co: 174-175, 213, 215, 223
The Star: 237
Star companies: 6, 9, 10-15, 107
    4-cyl aero-engine: 11-12, 91
    monoplane: 10, 18
    Star motor-car: 10, 14, 18, 169, 177
    Starling motor-car: 10
    Shed No.11, Brooklands: 12, 14, 18, 25, 31
Stereo photography/3D TV: 192-195, 235
Stereocopic Processes Ltd: 192-194
Stereo Television Co: 194
Stewart, Major Oliver: 80, 264
Strikes: 114-116, 171, 181
    General strike, 1926: 188
    Moulders: 114-115, 181
Suez Crisis: 213
Suzuki oil-cooled motor-cycle: 137
Sulzer engines: 96, 251
Sunbeam Motor Co: 15

Arab V-8 aero-engine: 64, 69, 79, 82-83, 264
Superchargers: 23, 120, 142, 145, 147, 226
Supermarine aircraft: 14, 109
    PB.31E Nighthawk: 32
Swift Cars: 170-171
Szekely SR-3-0 radial: 91

T

Tait-Cox, Major J H: 71
Tamper-proof Screw Co: 191
Teignmouth Pier: 186
Telescopic sail: 197
Television *see* stereo photography:
Temple Press: 221
Tombs, Montague: 148
Toroidal Engines Ltd: 242, 245, 257
Toroidal rotary engine: 163, 204-205, 207, 214, 230-248
    Admiralty/P&M: 163, 204-205, 207, 240
    aero-engine: 98, 232, 237, 241, 245
    demonstration model: 230, 232-233
    diesel: 232, 234, 253, 255-256
    eccentric bush: 234, 241-242
    gas cycle/gas generator: 226, 250
    helicopter engine: 244
    Hummingbird: 207-208
    marine engine: 242-243
    motor-cycle engine: 245-246
    New Action: 207, 231
    Omega: 205, 230-240, 252
    on-going development: 208, 257
    pulsation motor: 243-244
    racing car engine: 237, 245
    silent aero-engine: 246
    twin chamber: 231, 240, 250-258
Toreador Engineering Co: 133
    ohc oil cooled engine: 133
Torrens: 263
Townend, H C H: 82, 259
    Townend ring: 82
Tracked motorcycle: 127
Traction Aérienne *see* Wind Waggon:
Tragatsch, Erwin: 262
Trenchard, Sir Hugh: 108
Triumph motor-cycle: 120, 132, 160-161, 174, 176, 239, 288
    500cc sv motor-cycle: 9, 99
Triumph motor-car: 174, 213
Trojan Ltd: 101, 163
    Mini-motor: 122
TT races: 101, 133
Turner, Edward: 120, 146, 160, 208, 218, 239
Turner Manufacturing Co: 163, 204
    air-pump: 145, 163, 204

U

United Electric Car Co: 3
Universal Aeroplane Co: 18
    Birdling aeroplane: 18
US Army/Army Air Force: 38, 41, 58-59, 67, 75

US Navy: 42, 57-59, 202
US Patents: 254, 256
Usborne, Vice-Admiral Cecil: 200-201, 203-204

V

V-1 Doodlebug flying bomb: 37, 226
le Vack, Bert: 134
Valentine, James: 13, 47
Vauxhall/Bedford/Chevrolet *see* General Motors:
Verdon Roe, Edwin Alliott: 13-14, 26-27, 109
Viale, Spirito: 46, 48, 59
    Viale aero-engine: 46
Vickers & Co: 5
Vickers Aircraft Co: 14, 27, 33, 47, 49, 56, 96, 109
    Brooklands: 14, 33, 96
    Crayford: 33, 96
    Erith: 33
    FB.12: 49, 96
    FB.16: 49, 96
Vickers-Armstrongs (Aircraft): 85, 96
    Vanguard: 96
    VC-10: 96
    Viscount: 96
Vickers-Armstrongs (Engineering): 96
    ABC Motors Ltd: 96
Villacoublay airfield, France: 13, 67
Villiers 2-stroke engine: 87, 159, 164, 211- 212, 215, 220
Volkswagen: 149
Vulcan Motor Manufacturing & Engineering Ltd: 56, 170-171, 182
    ABC aero-engine: 56
Vulcan Iron Works: 200

W

Wade, Geoffrey: 240
    GE Motors: 240, 255
    water-jet propelled boat: 240
Wakefield, Charles C: 106
Wall, A G: 117, 146
Wall, A W: 121
    Autowheel: 121
Wall Street Crash: 15, 141, 188-189, 210
Wallis, Barnes: 257, 265
Walmsley, James & Co: 130-133, 136, 177, 272
Walton Motors Ltd: 39, 40, 55, 57, 74, 79, 84-86, 123, 168, 175, 264, 266-271
    Royalties: 266-271
Wankel, Dr Felix: 247-248
Wankel rotary engine: 230, 241, 243, 247-249, 250, 253-257
War Break Clause: 268
War Office: 26, 32, 53, 104, 267, 271, 291, 295
    aeroplane, "no military value": 26, 35, 291
Waring, Samuel: 49, 65, 266-267, 269-271
Water propulsion: 240, 256
Weir 'Pixie' aero-engine: 94

Weir, J G: 53, 56
Weir, Sir William: 49, 53-56, 79
Wellworthy Piston Co: 224, 231, 243, 247
Weslake, Harry Ltd: 255
Westland Ltd: 77-78, 244
    Wagtail: 50, 67, 77-78
    Weasel: 50, 63, 77-78
    Widgeon: 88, 93
    Woodpigeon: 90
Whitcombe, Jack: 167-168, 172
White, H T: 184-185
White, J Samuel & Co: 125-126, 242-243, 265
    Somerton Works: 125-126
White & Thompson No.2: 51
    Norman Thompson: 51
Williams, Tom: 176, 213
Willans & Robinson Ltd: 3, 140
Wills, Harry: 241-242, 244, 256
Wills, H A & Co: 241, 243, 255
Wind-Waggon: 18, 25, 29, 59-60, 148, 150-152
    armoured car: 151
    Beacon motor-car: 151
    Hélica: 151
    sledges: 152
    Traction Aérienne: 151
Wolseley Motors Ltd: 16, 34, 56, 58
    ABC aero-engine: 56, 75
    Hispano Suiza aero-engine: 83
    Viper aero-engine: 83
Wong Tong Mei biplane: 20
Woodnutt, Mark MP: 241
Woolley, Brian: 263
Wright Brothers: 7, 38, 48, 57
    Dayton Wright Co: 38, 58
    Bug AT: 38
    Wright Aeronautical Corp: 58, 59, 299
    Wright-Martin Corporation: 54, 57, 59, 268
    Wright radial: 59
        Whirlwind radial: 57, 59
Wright, Howard T: 26, 107

X, Y

Yorke, Maurice: 86

Z

Zenith motor-cycle: 101, 103, 130, 134, 182
    Zenith Gradua: 103, 132, 135
Zeppelin airships: 32, 37, 41, 294